THE ARCHITECTURE OF CLOUDS

The Architecture of Clouds

School of Meteorology, University of Oklahoma

OXFORD
UNIVERSITY PRESS

Great Clarendon Street, Oxford, OX2 6DP,
United Kingdom

Oxford University Press is a department of the University of Oxford.
It furthers the University's objective of excellence in research, scholarship,
and education by publishing worldwide. Oxford is a registered trade mark of
Oxford University Press in the UK and in certain other countries

Published in the United States of America by Oxford University Press
198 Madison Avenue, New York, NY 10016, United States of America

British Library Cataloguing in Publication Data
Data available

Library of Congress Control Number: 2023931899

ISBN 9780198870548

DOI: 10.1093/oso/9780198870548.001.0001

Printed and bound by
CPI Group (UK) Ltd, Croydon, CR0 4YY

Cover image: Howard B. Bluestein

To my wife Kathleen, my rhetorissima, for all her love and support;
To my mother, father, and my grandparents, for bringing me into the world of clouds.
December 2022

Contents

Preface

How many of us have first encountered clouds when we were very young, lying on a field or a beach, gazing up and watching clouds drift by, and then imagining that their shapes represent peoples' faces or animals? We have all looked at the inside of clouds, like fog, which has no shape. When we have flown in an airplane and have looked out a window, we have been afforded a view of clouds from many different perspectives: from underneath, from above, from various viewing angles, and also from within, but with only tunnel vision. Most people seem to neglect noticing the grand show of clouds in the dome above us as the clouds move from one place to another, not noticing the sky, half of our field of view. Alas, for many this hemispheric view is substantially blocked by buildings and trees. At the beach, over the ocean, on the open prairie, from mountaintops above timberline, this dome is much more noticeable. When I moved from Boston to Oklahoma over four decades ago, I was first struck by the beauty of the sky suddenly made visible in all its grandeur.

Painters such as the 19th century German-American Albert Bierstadt, among many others, have portrayed the majesty of clouds in paintings; Bierstadt produced many striking paintings of storm clouds gently touching the Rocky Mountains. The great 20th century American photographer Ansel Adams captured towering cumulus clouds and convective storms over the San Francisco Bay area in black and white. In a different way, many artists have produced capriccio landscape paintings highlighted by mood-shaping clouds. The majesty and beauty of clouds is amplified by the stark contrast between areas of light and darkness. Clouds can both shape and reflect our mood.

My first interest in clouds probably began when I lived in South Miami Beach in the mid to late 1950s. As a young child, I was mesmerized by the tantalizing thunderstorms that formed over the sea-breeze front just inland from the edge of the Florida Peninsula to the west, but never made it to the east, where I lived. I recall fantasizing about thunderstorms while looking through photographs in the magazine *Arizona Highways,* arguably an unlikely magazine to look for photographs of convective storms, since I was not aware at the time of the Southwest Monsoon in desert regions of the US and did not know that desert areas could have thunderstorms. I vividly remember sitting at the beach one morning while watching the sun-outlined ("silver-lining") portion of the top of a tropical towering cumulus cloud approach from the east and then briefly dump heavy rain on me. My interest was further piqued by a small book, *Weather,* by Lehr, Burnett, and Zim (Golden Press, 1957), which included many color drawings of clouds. Though I had experimented with photography of the moon, planets, etc. using my telescope when in the 5th and 6th grade of elementary school, I did not take many cloud photographs, at least that I can recall or find.

After moving back to the Boston area at the beginning of the 1960s, my view of clouds was shut down in large part, owing to the obscuring urban landscape and air pollution, the latter being especially bad during the summer when most tall convective clouds occur. As an undergraduate at M. I. T. in the late 1960s, while majoring in electrical engineering, I became seriously interested in cloud photography for the first time, especially when visiting my father in South Florida and noticed (again) how spectacular the clouds looked there. As a senior I took a photography class in Doc Edgerton's laboratory, but not from him directly; Doc Edgerton was a much beloved professor who perfected stroboscopic photography. Perhaps you have seen one his famous photographs of a splattering milk drop frozen in time. Having taken many color photographs of clouds using an old rangefinder camera and having had the negatives developed and printed commercially, I began to experiment with a new single-lens reflex camera and began to take many black-and-white photographs. I developed the negatives in a darkroom both at M. I. T. and in my undergraduate dormitory, Baker House, and for the first time I gathered a set of cloud images, for a class project. I also took many photographs of political and cultural events going on around Boston and Cambridge during the turbulent years of 1969 and 1970; some were published by the M. I. T. student newspaper, *The Tech*. The most memorable photographs I took were not of clouds. Some of my favorites included US senator Ted Kennedy speaking at Kresge Auditorium, the Grateful Dead performing on the M. I. T. Student Center steps, peace marches, demonstrations, the linguist and political activist Noam Chomsky, giving a lecture, the oldie's group Sha-Na-Na at the Boston Tea Party, and my various friends. My interest in photography and digital signal processing, the latter discipline's being relatively new at the time, led to my B. S. thesis, in which I devised a technique for unblurring photographs blurred due to motion, using spatial Fourier transforms.

During my early years of graduate school, when I studied both electrical engineering and meteorology, I began to take 35 mm color slides, especially during summer visits at the National Hurricane Research Laboratory (NHRL) (now the Hurricane Research Division—HRD) in Coral Gables, Florida, where I became captivated by cumulonimbus clouds and waterspouts. It was there in the early 1970s that I met Wendell Mordy at the University of Miami, through his daughter, who showed me his superb collection of cloud photographs. I had much encouragement also from my friend and colleague Bob Burpee at NHRL, and avid cloud photographer Ron Holle, who was at the Experimental Meteorology Laboratory (EML). Joanne Simpson, an eminent and pioneering meteorologist then at EML took me on a flight inside and around big thunderstorms over central Florida during a randomized cloud-seeding field experiment, further piquing my interest in cumulonimbus clouds, from which I later acquired the middle nickname "Cb." Joe Golden took me up in a small plane looking for waterspouts in the Florida Keys. My interest in clouds intensified after I moved to Oklahoma in the mid- 1970s, and later for parts of the year in Colorado, where I saw for the first time, tornadoes, majestic thunderstorms, and enigmatic mountain-wave clouds in wide-open settings.

The main purpose of this book is to describe clouds in a visual, poetic, and *personal* way, with some elementary explanatory scientific material on why clouds form and the meteorological context in which the clouds appear. I will try to relate clouds to one's everyday life and the weather while attempting to interest the reader in becoming more observant of them. The clouds to be discussed will range from placid, ground fog to violent tornadoes and the inside of the eye of a powerful hurricane.

I took nearly all the photographs, so consequently this book will *not* be a compilation of the best photographs from a wide variety of sources, as in some other books. Therefore, although they will not represent an exhaustive list of all types of clouds as in a formal cloud atlas, a broad spectrum of cloud types will still be represented. Instead of trying to represent every possible type of cloud, I intend to show photographs I have taken which can be appreciated both on an esthetic level and which bear some important meteorological message. In this sense I hope to invoke the belief of the eminent late eighteenth- and early nineteenth-century German writer, poet, and scientist, Goethe, that art and science go together.

While most clouds were photographed in Oklahoma and Colorado, where I spend most of my time, I have also included photographs I have taken in many other parts of the world, including most regions of the US, and parts of Canada, Europe, Greenland, Iceland, China, Japan, Australia, New Zealand, Argentina, Africa, as well as the tropical ocean near and adjacent to West Africa, the North Atlantic, and the South Pacific. While most photographs were taken from the ground, some were taken aboard an airplane or helicopter, and some were taken from the summit of mountains, which I climbed, or on cross-country ski or snowshoe trips on established trails or into the backcountry, at high elevation, from which I feel I earned the right to take photographs while engaging in strenuous but joyful exercise.

Although a vast majority of the photographs I have taken were serendipitously obtained during various outdoor, travel, and field-program activities, during the final stages of preparation for this book I took special pains to target certain types of clouds and also be vigilant (never leave home without a camera in addition to an iPhone camera). An unexpected consequence of the recent and ongoing COVID-19 pandemic (2020–2022) has provided me with many more opportunities to take photographs, especially in the Boulder, Colorado area. Although I have often taken sabbatical leaves in Boulder and have spent summers and winter breaks in Boulder, owing to the pandemic, my cloud photography got a boost because I was not inside a temporary, windowless office at NCAR or out at a late-afternoon movie, or dinner. Even while working on my laptop inside my residence, I could easily run to a window or go outside to take cloud photographs, something that is impossible to do, for example when attending a seminar or meeting or commuting.

This book is in large part my attempt to update, in my own personal style, the pioneering book by the prominent cloud dynamicist Richard Scorer *Clouds of the World: A Complete Colour Encyclopedia*, with some added stimulation from my friend and colleague Kerry Emanuel's book on hurricanes, *The Divine Wind*. In addition, this book is an update on my 1999 Oxford University Press book, *Tornado Alley: Monster Storms of the Great Plains*, with a generalization to include not only severe-storm photographs, but also most other clouds in general. The reader will be served color radar images and some line drawings in addition to many color photographs. I have tried not to repeat photographs included in my 1999 book, but I found it necessary to include a few because in the past two decades or so I have not been able to surpass them in their beauty and representativeness. Since cost and space limitations make it impossible for me to include many of my favorite reproductions of paintings of clouds and some of my favorite, witty, cloud/weather-related cartoons from *The New Yorker* magazine, I am including some website addresses where readers may find them on the internet.

I invite the reader to enjoy this journey through the clouds on parade as the skyliners of our lives (in my opinion, most appreciated while listening to doo-wop music).

Doo-lang doo-lang doo-lang, doo-wop doo-wa, and dip da-dip da-dip[1]

Howie "Cb" Bluestein
Norman, Oklahoma, and Boulder, Colorado
June 2022

---

[1] With thanks to the Chiffons, the Five Satins, and other Doo Wop groups of the 1950s and early 1960s.

Acknowledgments

A lot of friends and colleagues played both direct and indirect roles in my working on this book. They are thanked especially for making it possible for me to take many of the photographs herein, for providing valuable advice, and for intellectually stimulating exchanges of scientific ideas. I particularly thank the National Science Foundation for providing over four decades of financial support for the research activities undertaken by my graduate students and me, some risky and brand new, especially those which have involved making visual, in situ, and Doppler-radar measurements in convective storms and tornadoes. The National Oceanic and Atmospheric Administration (NOAA) also provided support for my early days of research while I was learning how to storm chase and while learning to work with Doppler radar data. I am most indebted to Ed Kessler (deceased), the first director of NOAA's National Severe Storms Laboratory (NSSL), who enticed me to Norman, Bob Burpee (deceased), a great friend and supporter both at MIT when I was a graduate student and at the National Hurricane Research Laboratory and Hurricane Division in Coral Gables and Virginia Key, Florida, and who also facilitated my flights into hurricanes, to Bob Davies-Jones, retired from NSSL, who enabled my early storm chases, to Dave Jorgensen, retired from NSSL, and Roger Wakimoto, who enabled many of my flights near convective storms. Summers and sabbaticals with support from NCAR in Boulder, Colorado supplemented my support from the School of Meteorology at the University of Oklahoma at Norman and also provided me with an excuse to take photographs in and around Boulder. At the University of Oklahoma (OU), I am indebted to all the graduate students, undergraduate students, and others who participated in our nearly annual spring field programs. In particular, I recognize the relatively recent efforts of Jake Magraf, Trey Greenwood, Zach Wienhoff, Dylan Reif, Jeff Snyder, Jana Houser, Robin Tanamachi, Chris Weiss, Mike French, Herb Stein, Jim LaDue, Doug Speheger, Bill McCaul, Greg Byrd, and longer ago, Lou Wicker and Brian Jewett, who made many of our field programs possible, and many others unnamed, who worked with me in the field. Andy Pazmany and Ivan PopStefanija at ProsSensing in Amherst, Massachusetts, Bob Bluth at the Naval Postgraduate School in Monterey, California, Steve Frasier at the University of Massachusetts, Amherst and a number of his graduate students, Wes Unruh (deceased) at the Los Alamos National Laboratory provided various mobile Doppler radars and a mobile Doppler lidar. I thank all my department chairs at OU who have supported me, such as Rex Inman (deceased), Claude Duchon, and Bill Beasley, but especially Jeff Kimpel (deceased), who first made it possible for me to get involved in storm chasing. Fred Carr, I thank for his long friendship, companionship, stoking the flames of our mutual love of hiking and skiing, and years of support while he was our chair. I also thank David L. Boren, former OU president and Elbert (Joe) Friday, former director of the National Weather Service, for their vital support of my work at OU. Lance Bosart at the University of Albany I thank for his long-time friendship, kinship in our interest in the weather, and for providing me with a superb model of how a university professor should be conducting himself/herself while teaching and doing research. Dale Durran at the University of Washington, Seattle, graciously provided comments and advice on my section on clouds associated with mountain waves. Rich Rotunno at NCAR provided long-term friendship and taught me many things about various aspects of atmospheric dynamics in the most stimulating way and shared with me his love of Doo Wop music from the 1950s and 1960s. Morris Weisman at NCAR also provided long-term friendship, taught me many things about the dynamics of convective storms and weather forecasting, and also sharpened my verbal sparring abilities and love of Chinese buffet lunches, which were necessary to fuel photography expeditions successfully. Kerry Emanuel at MIT provided long-term friendship and many stimulating discussions about meteorology and life in general, and with whom I shared a belief in the mutual dependence of art and science on each other. John Brown at NOAA in Boulder has been my friend since graduate school and is someone who selflessly taught me many things about meteorology, beginning when he was my TA (teaching assistant) and continuing when I became a colleague and long-term friend and hiking companion. Joe Golden, retired from NOAA, enabled my flying around waterspouts in a small aircraft and in a helicopter, and shared his love of meteorology. He is one of the inspirations for scientific storm chasing, having led a pioneering team to photograph features in a tornadic supercell in central Oklahoma in the early 1970s while correlating the features with remote Doppler radar data from NSSL. Joanne Simpson (deceased) at NOAA graciously invited me aboard some research flights when I was a graduate student, while she was engaged in a cloud-seeding experiment over Florida and showed me what a scientist does. Pauline Austin (deceased) and Speed Geotis (deceased) taught me about weather radar when I was a graduate student. I am most indebted to Fred Sanders, my thesis advisor at MIT, who sanded my rough edges (at least a little bit) and stimulated and mentored me to have the wonderful career I have had. Roger Wakimoto at UCLA has provided an excellent example of how to conduct research using observations based on both photographs and radar data, and has carried on the work of his advisor and someone I have always admired, Ted Fujita. Peggy LeMone, a fellow cloud enthusiast and long-time colleague at NCAR, who has provided sound advice; Peggy was memorably a companion for viewing waterspouts with back when I was just a graduate student. I thank my first book editor, Joyce Berry, for her expert guidance through my first three books for Oxford University Press. Particular thanks go out to Sonke Adlung at Oxford University Press for his also-expert guidance during the beginning of this book project, and to Katie Lakina for moving this book to production. In addition, I am indebted to Giulia Lipparini, Senior Project Editor at OUP, John Smallman at OUP, Anya Hastwell, freelance copyeditor, and Ishwarya Mathavan, production manager at Integra Software Services. John Lewis at the Desert Research Institute in Reno, Nevada, provided some help with a few historical details. And last, but certainly not the least, I thank my brilliant and beautiful wife, Kathleen Welch, a most accomplished expert on classical rhetoric, among other things, and my mother, my father and my grandparents, my paternal grandfather especially, whom I particularly loved, and all who made it possible for me to pursue my dream of studying the weather and teaching meteorology, activities for which I can hardly believe I am still being paid to do.

1

Introduction

The Physical Theory and the Esthetics of Clouds

> *I am the daughter of Earth and Water,*
> *And the nursling of the Sky;*
> *I pass through the pores of the ocean and shores;*
> *I change, but I cannot die.*
> *For after the rain when with never a stain*
> *The pavilion of Heaven is bare,*
> *And the winds and sunbeams with their convex gleams*
> *Build up the blue dome of air,*
> *I silently laugh at my own cenotaph,*
> *And out of the caverns of rain,*
> *Like a child from the womb, like a ghost from the tomb,*
> *I arise and unbuild it again.*
>
> "The Cloud," Percy Bysshe Shelley (1820)

> *"Clouds always tell a true story, but one which is difficult to read."*
>
> Ralph Abercromby (1887)

> *Cloud-maidens that float on for ever,*
>   *Dew-sprinkled, fleet bodies, and fair,*
> *Let us rise from our Sire's loud river,*
>   *Great Ocean, and soar through the air*
> *To the peaks of the pine-covered mountains where the pines*
>   *hang as tresses of hair.*
>
> Oscar Wilde's first published poem, a translation of a
> chorus from Aristophanes' *Clouds*

1.1   What are clouds and why do they matter?

Clouds are opaque or partially translucent collections of tiny droplets of water or ice crystals suspended in the air, wafting across the sky with the wind or stubbornly remaining

The Architecture of Clouds. Howard B. Bluestein, Oxford University Press. © Howard B. Bluestein (2024).
DOI: 10.1093/oso/9780198870548.003.0001

fixed in the sky, even in *spite* of the wind. Visible to us during the daylight hours as sunlight is scattered back to our eyes by the outer edge of collections of cloud particles, they line the sky's horizon like skyscrapers in a large city or may command our attention as isolated monuments or sculptures floating above us, or may completely envelop the sky overhead in a "cloud" of mystery. Sometimes they hinder our view of anything but themselves, especially when we are inside them. Some are subjectively beautiful and majestic, while others are non-descript and complex, or even terrifying. I love the cartoon by the late cartoonist Gahan Wilson that appeared in *The New Yorker* magazine many years ago,[1] in which what looks like an older man, holding the hand of a young boy, who is probably his grandson and is pointing at a cloud within a field of clouds, and saying "That's a cloud, too. They're all clouds." I can't say precisely what I love about this weird cartoon, other than to comment that I find it funny without knowing exactly why. But the subject of the cartoon is the subject of this book, in which I would like to say to the reader exactly what Wilson is having the man say to the boy, as *I* point to the clouds.

Regarding clouds, Percy Shelley's famous poem about clouds says it all: The "daughter of Earth and Sky," "the nursling of the sky," "I change, but I cannot die," "I arise and unbuild it again." A vital part of the water cycle in the atmosphere, they undergrow vast transformations, appearing when water vapor, which has evaporated from bodies of water at the ground or from raindrops in the air, condenses or turns into ice crystals directly, a process known as "deposition." The cloud particles may evaporate or "sublimate" (turn directly from ice crystals into water vapor) back into the atmosphere and disappear from view, or they may conglomerate into masses heavy enough to fall to the ground, where the water is absorbed by the ground and eventually evaporated back into the air above: They don't die, but rather rise and build again. They are like timeless ghosts, enduring cycles of a type of reincarnation.

Oscar Wilde's translation of a choric ode from Aristophanes' play *Clouds* also expresses, centuries earlier, some of the sentiments put forth in Shelley's poem: "Cloud-maidens" floating on forever; "Dew-sprinkled, fleet bodies," rising from . . . "Great Ocean."

Ralph Abercromby, a nineteenth-century Scottish meteorologist, is quoted as noting the difficulty we have in understanding clouds and what they mean ("which is difficult to read"). Clouds provide rain needed to sustain our food supply and life, and significantly affects the Earth's radiation budget, which in turn affects how warm or cold it is. The nature of the cloud particles (liquid water droplets, ice crystals) affects how the cloud manages radiation. Clouds can foretell the weather ("tell a true story")

---

[1] https://books.google.com/books?id=Yro0AwAAQBAJ&pg=PA71&lpg=PA71&dq=%22That%27s+a+cloud,+too.+They%27re+all+clouds.%22+Gahan+Wilson&source=bl&ots=WRwxPGEfyj&sig=ACfU3U2yCJJPUs1ndgW1e8o1RntM-GkoUQ&hl=en&sa=X&ved=2ahUKEwiUmuy5u-PyAhWBlmoFHegkDY8Q6AF6BAgDEAM#v=onepage&q=%22That's%20a%20cloud%2C%20too.%20They're%20all%20clouds.%22%20Gahan%20Wilson&f=false; p. 71. If this website address or any other given in this book is no longer active, the reader is encouraged to use his/her favorite search engine to find it elsewhere.

on short time scales, warning us of potentially catastrophic natural hazards like tornadoes, flash floods, destructive straight-line winds, and heavy snow. Artists have rendered
clouds in paintings to invoke strong emotional responses, just as the real phenomena
associated with them sometimes do. The responses may involve not just the feeling
one gets in awe of nature's power, but also moods of light and darkness, calm and
foreboding.

In this book I will attempt to evoke emotional responses from the appearance of
clouds, while at the same time providing some understanding of the dynamics of air currents that produce them on scales larger than the clouds themselves, their cloud-scale
physics, thermodynamics, and "microphysics," the physics on the scale of cloud particles: why they exist and what type of weather they harbor. Johann Wolfgang von Goethe,
the German, late eighteenth- and early nineteenth-century poet and scientist, among
other things, promoted *both* science and art together and in fact it has been said that he
believed in the "marriage" of the two.[2] It is in this spirit that this book is intended. To
begin, I ask the reader to consider the different moods each of these very-different looking clouds induces and to ponder the vastly different physical mechanisms responsible
for each of them:

The object of much of my scientific research has been tornadoes, so it is fitting that
we begin with a tornado photograph. The top-left pane in Figure 1.1 shows a classic,
US Plains tornado. It was created in a supercell, when the air was very buoyant and the
wind speed and direction changed rapidly with height, especially in the lower portion
of the troposphere, and rain evaporated into unsaturated air underneath the anvil of the
supercell. Are you frightened? Do you get the feeling of a violently rotating column of
air about to suck up buildings and trees like a vacuum cleaner and destroy everything in
its path and heading toward you?

The top, right panel in Figure 1.1 shows early-morning, shallow, radiation, ground
fog. Does this photograph instill a feeling of stillness and tranquility as if you are awak-
ening from a deep sleep (at dawn), the opposite of how you feel when viewing the
approaching tornado in the top, left panel?

The left panel on the second row of images in Figure 1.1 shows two stationary altocumulus lenticularis (shaped like a lens) clouds, which look like flying saucers. They are
formed when currents of air that alternately rise up and sink down in a wave-like fashion, particularly after what happens sometimes when air has been forced up and over a
mountain. Do you feel a bit apprehensive about whether or not these are flying saucers
about to land, with aliens coming out of them to greet us, like in the famous early 1950s
sci-fi movie, *The Day the Earth Stood Still*,[3] rather than these "flying saucers" just being
harmless clouds?

Mamma pendant from a cumulonimbus anvil are seen in the right panel in the second
row of images in Figure 1.1. They are slowly sinking patches of cloud in between which

---

[2] A. Wulf, 2015, *The Invention of Nature: Alexander von Humboldt's New World*, Vintage Books, p. 41.
[3] https://en.wikipedia.org/wiki/The_Day_the_Earth_Stood_Still

Figure 1.1 *Some examples of photographs of clouds that may evoke distinctly different moods. (top, left) Classic appearance of a tornado (the cloud is the condensation funnel) inflicting damage in Selden,*

the air is rising. Do they look very unusual? Do they instill a sense of awe and foreboding? Is the world about to end as a dimpled sky seems to be inverted, turned upside down?

The left panel in the third row of images in Figure 1.1 shows the back side of a cumulonimbus cloud, with a "backsheared" anvil and penetrating top. Can you feel the *power* of this storm's highly buoyant updraft, like a fist punching its way up into the stratosphere?

The right panel in the third row of images in Figure 1.1 shows how colorful the underside of a cloud can be at sunset. Are you startled by the flaming colors? Are you dazzled by the complex, banded, yet irregular texture of the cloud? In some places there are lines (striations) of cloud material, in others there are irregular sections of cloud.

Finally, consider the bottom panels in Figure 1.1, a photograph of a stationary ("standing") altocumulus lenticularis, orographic wave cloud (left panel), in which there are many layers of smooth-looking, elliptically shaped plates of cloud. The plates look like a stack of separated pancakes. Does this image spark your interest as to how such a weird-looking cloud can form and why it doesn't move, as a solid object suspended in

---

Figure 1.1 (Continued) *Kansas, on May 24, 2021. Some fallout of the debris from the debris cloud is visible on both the left-hand and right-hand sides of the condensation funnel. Helical striations are visible in the lower portion of the condensation funnel as a result of the swirling, upward motion of a segment of a condensation funnel. (top, right) Morning radiation ground fog in Norman, Oklahoma, on Oct. 24, 2011. (second row, left) Isolated, flying-saucer-like lenticular clouds hovering just south of Blue Lake, in the Indian Peaks Wilderness, west of Ward, Colorado, on June 22, 2021. (second row, right) Cumulonimbus mamma over Boulder, Colorado, late in the day on July 10, 2011. (third row, left) A large cumulonimbus cloud, viewed to the southeast, from Norman, Oklahoma, on May 21, 2011. The University of Oklahoma—NOAA—State of Oklahoma National Weather Center (NWC) is seen at the lower left. Convective towers are penetrating above the anvil level set by previous convective towers that have since come to their equilibrium level (level of non-buoyancy or neutral buoyancy). This cloud looks like an atomic bomb explosion. In fact, there was a scientific paper in the late 1950s in which it was actually suggested without irony that one could study cumulus convection by filming clouds produced by atomic explosions![a] Note how the antenna on top of the NWC pierces the sky above it, just as the penetrating top in the cumulonimbus pierces the stratosphere. (third row, right) The colorful, complex pattern of orographic wave clouds in the lee of the Rocky Mountains, over Boulder, Colorado, on Jan. 3, 2021. The view is to the north-northwest at sunset. These clouds are associated with vertically-propagating gravity waves, since they lean backward against the flow with height (leaning to the left with height). See Chapter 2 for more details on what this means. (bottom row, left) A pile-of-plates lenticular cloud over Boulder, Colorado, on Jan. 21, 2021. These clouds are probably associated with vertically-trapped gravity waves. See Chapter 2 for more details on what this means. (bottom row, right) A multi-layered orographic wave cloud over the snow-covered Foothills of Boulder, Colorado at sunrise on Dec. 23, 2022.*

Photos courtesy of author.
[a] Levine, 1959.

the air would, but instead the wind blows right through it, leaving cloud particles behind? The photograph in the right panel shows the startling effect of lighting at sunrise on the both a wave cloud and the snow-covered mountains below.

Now that we have a preview of some clouds that provoke a wide range of feelings, we'll disregard esthetics for a bit and take a brief look at how clouds form, before returning to esthetics.

1.2 Saturating the air: Atmospheric thermodynamics and cloud initiation

Air pressure (the force per unit area exerted by a collection of air molecules in motion owing to their temperature) in general decreases with height because the weight of the atmosphere decreases with height[4]: there is less mass of air on top of you the higher up you go. It is more difficult to breathe on a high-elevation mountain than at sea level for most of us, unless we are physically conditioned for higher elevations. "Hydrostatic" pressure, which is most of what constitutes pressure, is just the downward force exerted by the mass of air above due to gravity, per unit area, in the absence of any force imbalances that would instigate air motion. Think of the mythological Greek God Atlas carrying on his shoulders not the entire Earth, including its atmosphere, but just a column of its atmosphere, extending from the ground up to its "top," where the density of air decreases to the point that pressure has no meaning, because air molecules are so far apart that the air is no longer a gas. Air near the bottom of the pile is compressed more relative to the air higher up, where the downward force of gravity is less, so the *rate of decrease* of pressure with height itself decreases with height. This situation was graphically illustrated in an uncaptioned, untitled cartoon by cartoonist Bob Mankoff that was published in *The New Yorker* magazine many years ago,[5] in which a column of people is stacked into a pile, with the person at the bottom compressed in depth and expanded in width the most, and the person at the top expanded in depth and compressed in width the least. Perhaps the cartoon means that society is stratified so that some people (at the bottom) bear most of the weight of the world, while others (at the top) have a much easier life.

There is also a small contribution to pressure that is associated with air motion (the wind). Imagine the pressure you feel on your hand when you stick it out the window of a car when you are speeding ahead down the road. This pressure is *non-hydrostatic* (i.e., a result of causes that are not hydrostatic), and is also called the *dynamic* part because it depends on air motion. In the reference frame of your hand, unless you have a lot of holes

---

[4] In Chapter 4 we will see how vertical shear also affects pressure, in supercells in particular. Also, in just a bit we will see how pressure is affected by buoyant clouds. The main contributor to pressure, however, is just the weight of the column of air above.

[5] I was unable to find this cartoon freely on the internet. It appeared on p. 73 of the Jan. 7, 1980 issue of the *New Yorker*, which can be accessed online from *The New Yorker* archive online if you have a subscription.

in it, air approaches it and must come to a screeching halt, because it cannot penetrate through it. What is the dynamic contribution to pressure? Since air must decelerate (to zero) just before it reaches your hand, there must be a pressure-related force acting on it to slow it down. This force is called the pressure-gradient force. It is one of the most important forces in determining how the winds blow in the atmosphere.

To understand this pressure-related force, imagine that one person is trying to force his/her way into a room by pushing against a door (to the left), while another person is on the other side of the door trying keep the person *from* pushing the door in and in fact might be trying to push the door open outward (to the right). If both people push with same force per unit area onto the door, then the door will not budge; in this case there is no change in pressure (force per unit area) exerted across the door: there is no gradient in pressure across the door and it stays put. However, suppose that the person outside the room pushes more forcefully inward (to the left) than the person inside pushes outward (to the right). Then the door will be forced inward. In this case there is a gradient in pressure, such that the force inward (from the right) exceeds the pressure outward (from the left). The door is therefore accelerated from right to left in the direction *opposite* to that of the pressure gradient (which is directed from lower pressure to the left of the door to higher pressure to the right of the door). Now, replace the door with a volume of air and you understand the *pressure-gradient force*.

The main thing to remember is that *the pressure-gradient force* (PGF) (and the acceleration it induces) always acts in the direction *opposite* to that of the pressure gradient, *i.e., from higher to lower pressure* and represents the difference in pressure acting across a volume of air.

Now we return to the problem of how air comes to halt as it approaches your hand: There must be relatively high pressure on it, so that there is a PGF acting in the direction opposite to the wind on your hand created by the motion of your car in order to slow the air down. In a cloud, regardless of whether the contribution to pressure is hydrostatic or dynamic, the gradient of pressure (difference in pressure across a certain distance) induces a force, part of which is always vertical, and a much smaller part which is horizontal. Because the total pressure (the sum of the hydrostatic part and the dynamic part) varies much more rapidly in the vertical than in the horizontal, it is the vertical hydrostatic pressure gradient that contributes most to the PGF. For example, the pressure varies by about 1, 000 hPa (hectopascals[6]) in the lowest 10 km of the

---

[6] A hectopascal (hPa) is also known as a millibar (mb), which is 1/1, 000 of a bar, but a bar is not the official S. I. unit. The pressure in S. I. (in French, Système International) units, i.e., in terms of meters, kilograms, and seconds (M-K-S), is 1 kg m s^{-2} m^{-2} (pressure = force/unit area = mass X acceleration/unit area) = 1 Pa (Pascal). One bar = 10^5 Pa, so although it is not an S. I. unit, it is proportional to an S. I. unit. Pressure is also given in terms of inches of mercury, where one inch of mercury = 33.864 hPa, the depth of mercury that supports a pressure of 33.864 hPa (i.e., it is not actually a pressure unit) at a temperature of 0° C and a standard acceleration of gravity. At sea level the pressure is around 30 inches of mercury or 1, 000 hPa or mb. Most meteorologists use hPa or mb as the pressure unit, the former being the accepted S. I. unit. Weather maps are typically given at various "standard/mandatory" pressure levels, such as 925, 850, 700, 500, 400, 300, 250, 200, 150, and 100 hPa/mb, etc. Altimeters usually give pressure in terms of inches of mercury, for historical reasons—barometers used to be given mostly by mercury instruments.

atmosphere (roughly the distance between the surface of the Earth near sea level and the average height of the tropopause, which separates the troposphere below from the stratosphere above); in the horizontal, however, pressure varies about only 1–10 hPa over 1, 000 km in the horizontal (unless you are in strong tornado). We'll see later that in highly buoyant clouds and clouds embedded in an environment in which the wind speed and/or direction changes abruptly with height, the vertical dynamic pressure gradient can be more important than vertical hydrostatic part. The dynamic part of the pressure field in fact plays a dominant role in the behavior of supercell convective storms.

If air is lifted upward to lower pressure, then work is expended as the air "parcel" (a small chunk of air, a fixed collection of air molecules within a small volume) expands and pushes outward on the air surrounding it. The work expended as air is pushed aside is manifest as a loss of heat energy, which is accompanied by a drop in temperature. The amount of water vapor that air can hold increases rapidly with increasing temperature: Hot air can hold a lot of water vapor; cold air can hold much less water vapor. The excess water substance is condensed into liquid or solid form as water droplets or ice crystals.

Air that is warmer than its environment is also less dense than the air in its surroundings at the same pressure, and is therefore buoyant and forced upward, while air that is denser than its surroundings is negatively buoyant and is forced downward. An analysis by the third-century BCE Greek mathematician Archimedes is commonly used in physics and meteorology courses/textbooks to explain how buoyancy works. In essence, objects of lesser density than that of the fluid in which they are immersed experience a greater upward-directed PGF than the downward force of gravity. I am reminded of the popular segment on the old David Letterman late-night talk-show television program in the US called "Will it float?" in which various objects are tossed into a tank of water and one is asked to guess, in essence, if they will be buoyant and remain at the top of the tank, or if they will be negatively buoyant and sink to the bottom of the tank. One might argue that this "contest" is somewhat similar to the weather-forecasting problem of "might there be a thunderstorm later today or not?"

If air is lifted enough so that it is cooled to saturation (the point beyond which no additional water vapor can exist in vapor form, the dew-point temperature) and if the air becomes cooler than that of its environment (the region outside of which air is being lifted), then the parcel of air is "stable," since its density is greater than that of its environment, according to the "ideal gas law." The ideal gas law states that density is proportional to pressure and inversely proportional to temperature; so, if the pressure inside the air parcel (a small collection of air molecules) is the same as it is in the environment, just outside the air parcel, cold air is denser than warm air.

If lifting suddenly were turned off, the air parcel would accelerate downward, back toward where it had been originally, because its buoyancy would be negative, acting in the downward direction. In this case the forced displacement of the air (upward) and the displacement of the air from the force acting on it (downward) would be in the opposite directions. Since the air is stable, small-scale, turbulent motions are unlikely owing to the resistance of the air to vertical motion; in this case the cloud will likely have a smooth, "laminar" appearance. Most of the clouds described in Chapter 2, which considers mainly non-buoyant clouds, especially the ones in the lower half of the troposphere, will therefore be relatively laminar in appearance. Think of "smooth," melodic jazz versus dissonant, improvised jazz: the former is restrained, subject to rules, and rather regular, while the latter is much more all over the place, without much restraint, following the whims of the musician(s). In the case of clouds in the atmosphere, the restraint is from stability. If the air is stable, but air is given a slight initial push upward, then as air is driven back toward its equilibrium level (where the net vertical force is zero), it will have enough kinetic energy to overshoot its equilibrium level and become negatively buoyant (cooler than its surroundings), resulting in an oscillation, a bobbing up and down of the air.

On the other hand, if air is lifted enough so that it is cooled to saturation (i.e., to the dew-point temperature), but the air is now *warmer, not cooler* than that of its surroundings, then the "parcel" of air becomes "unstable," because its density is less than that of its environment: If lifting were suddenly turned off, the air parcel would accelerate upward, in the direction it was originally displaced, because its buoyancy would be positive, upward. In this case the forced displacement of the air (upward) and the displacement of the air from the force acting on it (upward) would be in the same direction. Turbulent motions will ensue as the air accelerates and there is a strong gradient of wind (rapid changes in wind over very short distances, wind "shear") at the cloud's edge, so the outer edge of the cloud will *not* have a smooth appearance, but instead will have many small-scale wrinkles, as in a head of cauliflower (see, e.g., Fig. 3.26), like the top and sides of a cumulus cloud. Buoyant clouds are the subject of Chapters 3 and 4 and it is there that we will consider in more detail how the vertical profile of moisture, in addition to the vertical profile of temperature, affects convective (buoyant) clouds. The reader is cautioned that the real world of clouds is not so easily divided into those that are buoyant and those that are not, since some clouds (such as cumulonimbus clouds) may have both buoyant and non-buoyant regions.

1.3 Nucleation and the growth of liquid water droplets and ice crystals: Cloud microphysics

Forming a cloud is more complicated than just cooling the air to saturation. At relatively warm temperatures, for example in the lower region of the troposphere, water vapor requires aerosols known as "condensation nuclei" onto which the water vapor condenses. Cloud condensation nuclei (CCN) facilitate cloud formation at warm (generally above freezing) temperatures because water droplets have surface tension, which requires more energy than that available when water vapor just condenses into liquid water with no consideration for anything else. Salt particles are common condensation nuclei, but there are many possible others, including dust and pollen. Below freezing, water vapor may still condense onto nuclei but not freeze and therefore become "supercooled" liquid water cloud droplets. In reality, the air must be "supersaturated" for liquid water in order to condense from water vapor, but the presence of CNN reduces that amount of supersaturation needed to only a small amount.

At very cold temperatures, below around −40° C (which happens also to be −40° F), ice crystals may appear without any particle nuclei at all. This process is known as "homogeneous nucleation." At warmer, but still sub-0° C temperatures, ice-crystal formation is facilitated by ice-nuclei particles. So, the aerosol content of the air and the temperature are very important factors in producing clouds. Air cannot be too "clean" or it would be more difficult for clouds to form. On a macroscale, when water vapor turns into the ice phase on a cold surface such as a tree (or, dangerously, on an aircraft wing) solid, "rime" ice may form.

Rimed ice particles that form in a cloud on snowflakes and fall out are known as "graupel" (Fig. 1.2, bottom, right panel) or "soft hail." Hoar frost is a deposit of feathery ice crystals on vegetation when the air is near saturation and below freezing (Fig. 1.2, all panels but the bottom right). Hoar frost, when deposited on entire trees in a forest, gives the forest an enchanted, fairytale-like look.

Cloud particles may consist only of liquid water droplets, at temperatures greater than or less than 0° C, and at temperatures below 0° C of ice crystals only, or of a mixture of liquid water droplets and ice crystals. When cloud particles consist of liquid water droplets only, there may be a spectrum of sizes, ranging from the very smallest (~5 μm[7]) to the largest (~50 μm). When cloud particles consist of ice crystals, there can be a fascinating, wide variety of shapes and sizes to the crystals. The reader is referred to specialized books on snow crystals and snowflakes, which detail the many types of ice crystals found in nature. Ice crystals forming on a solid surface when air is cooled to saturation at night and it is below 0°C is called frost (Fig. 1.3) and may therefore be considered analogous to ice deposits on solid objects below freezing.

Sometimes there is a mixture of water droplets and ice crystals. When this happens, ice crystals grow at the expense of the water droplets because the saturation vapor pressure is lower over the ice crystals: liquid water droplets then evaporate because the air is subsaturated with respect to water and grow onto the ice crystals, on which the air is supersaturated.

Rain can form as cloud droplets grow in size and become heavy enough to fall out, snow can form as ice crystals grow and become heavy enough to fall out, or rain can form as ice crystals form, fall out, and then melt on the way down to become raindrops. The process by which rain forms in the absence of ice crystals is called the "warm-rain" process, while the process that forms rain when there is a mixture of liquid water and ice crystals is called the "cold-rain" process. The cold-rain process used to be called the Bergeron–Findeisen process, after Tor Bergeron and Walter Findeisen, but now is

---

[7] 1 μm (micron) $= 10^{-6}$ m $= 10^{-3}$ mm)

Figure 1.2 *Ice crystals on vegetation and clumps of icy spheres. (top, left) Trees with ice crystals deposited on their branches (hoar frost) on a mountain near Santa Fe, New Mexico, on March 13, 2006, shining brightly in the sunlight as low clouds and fog are breaking up. (top, right) Hoar frost in Boulder, Colorado, on Jan. 2, 2023. (top, right) Close-up view of needle-like, feathery structures, hoar frost, on vegetation in Boulder, Colorado, on Jan. 12, 2015. (bottom, left) Hoar frost blanketing a pine tree in Boulder, Colorado on March 7, 2023 (bottom, right) Graupel at around 10, 000 feet elevation on the Tesuque Creek Trail, near Santa Fe, New Mexico, on Oct. 27, 2022.*
Photos courtesy of author.

known as the Wegener–Bergeron–Findeisen process, with Alfred Wegener also being recognized. The warm-rain process occurs when there is a spectrum of rain droplet sizes, as larger raindrops fall more rapidly than smaller raindrops and the smaller raindrops stick to the larger raindrops through a process known as "coalescence." On the other

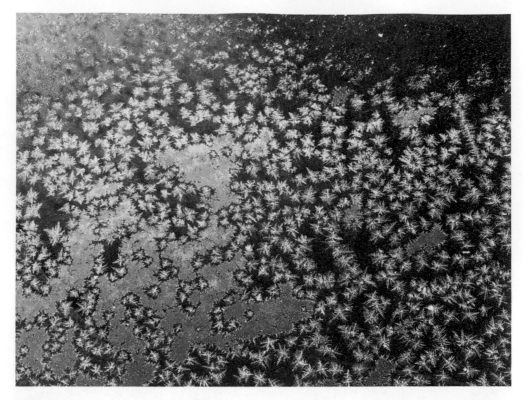

Figure 1.3 *Frost on the windshield of my car in Norman, Oklahoma, early in the morning of Feb. 21, 2020.*
Photo courtesy of author.

hand, collisions among raindrops can also break up drops into smaller drops, essentially the opposite of coalescence.

When rain falls through a column of air that is mainly warmer than 0°C and hits the ground and the ground is below 0°C, the rain freezes on contact and is termed freezing rain (solid ice, not ice crystals), potentially producing a dangerous, ice storm (Fig. 1.4).

When rain falls through a column of air that is colder than 0°C it may freeze on the way down and become ice pellets, or as known colloquially in the US as sleet. In convective storms water droplets may be lofted to levels well below 0°C by strong updrafts and freeze into ice particles, which then may accrete more liquid water drops during both vertical and horizontal excursions, forming hailstones (Fig. 1.5), which fall out when they become large enough to fall more rapidly than the speed of any updraft in which they're caught. Since the "terminal fall speeds" (as a hailstone falls it plows through air and there is a drag force on it acting in the upward direction, opposite

Figure 1.4 *Aftermath of an ice storm. (top) An icy playground in Norman, Oklahoma, on Dec. 27, 1987, after an ice storm. (bottom) Ice-storm encrusted tree branch crushed over onto the ground on the University of Oklahoma, Norman, campus on Dec. 26, 1987.*
Photos courtesy of author.

Figure 1.5 *Large hailstones. (left) An ellipsoidally shaped, smooth, hailstone, which fell from a cumulonimbus in south-central Kansas on May 31, 1999, and which was over 3.5 inches (almost 9 cm) in diameter; (right) hailstone with spikes and other irregular, bulbous protrusions, that fell from a cumulonimbus in southwestern Oklahoma on May 29, 2013. This hailstone measured almost 3 inches (7 cm) along its major axis.*
Photos courtesy of author.

the direction the hailstone is falling; when the upward drag force becomes equal to the downward-directed force of gravity, the falling hailstone slows down to its "terminal" fall velocity) of large hailstones may be as high as 35 m s^{-1}, it takes very powerful updrafts to keep large hailstones suspended aloft long enough to accrete enough liquid water for large hail to form. Obviously large hailstones can inflict serious damage and injuries. The largest documented hailstones to fall in the US are around 6 inches in diameter[8] (9.7 cm). By this standard, those seen in Figure 1.5 are small, yet among the largest that most storm chasers typically find, which are typically no greater than about 3–4.5 inches in diameter)

If the air is well below freezing, then the dominant makeup of the cloud is ice crystals, which can grow into snowflakes. The snowflakes may fall to the ground as snow (Fig. 1.6), or may melt to become raindrops. The efficiency with which ice crystals form and aggregate depends largely on the temperature and the amount of moisture in the air. When the air temperature is near 0°C, giant snowflakes are sometimes observed.

[8] One could argue that the circumference of the hailstone or its weight is a more representative measure of size, since hailstones are not always spherically shaped.

Figure 1.6 *Heavy snow. (top) Heavy snow falling in Boulder, Colorado, on March 14, 2021. (bottom) Cross-country skiers and their dog on the sidewalk in Boulder, Colorado, on March 14, 2021.*
Photos courtesy of author.

1.4   The beauty and morphology of clouds: What does a cloud look like?

How a cloud appears to the eye depends on how much light is scattered by the cloud particles along its surface edges back to the observer and also what colors are preferentially scattered. Electromagnetic radiation at visible wavelengths is absorbed and re-radiated, interacting with all the neighboring cloud particles. That is, some of the radiation scattered off one particular cloud droplet/ice crystal is then absorbed and re-radiated by others before it gets back to the observer. This process of "multiple scattering" is not simple: The concentration of the cloud particles (how many there are in a fixed volume of space and how far apart they are separated from their closest neighbors), the nature of the cloud particles (water or ice or water-coated ice), the size distribution of the cloud particles (relative number of large ones to small ones: how uniform or non-uniform the sizes are), and the sizes of the cloud particles with respect to the wavelength of sunlight (when relatively small, the scattering is associated with "Rayleigh" scattering, while large it is associated with "Mie" scattering, each of which causes the degree of scattering to vary with the wavelength of the light in different ways) all play a role in the scattering process. In addition, the shapes of the cloud particles (spherical, ellipsoidal, in a regular crystal pattern of various types, highly irregularly shaped), the angles at which sunlight hits the cloud and the angles at which it is scattered back to the observer, relative to the sun and the observer, the differences in phase of the scattered light by each group of particles, and the polarization of the light, can also affect how the cloud looks: Is it bright? dull? of a certain color or ranges of color? There are many books written on the details of how light is scattered multiple times by particles, but I refer the reader especially to those by or co-authored by Craig Bohren at Penn State University to learn more about the process of scattering and how complicated it is. In addition, how certain regions look in comparison to neighboring regions (bright, dark, etc.), may also contribute to our esthetic sense.

In addition to brightness and color, the texture of the edges and surfaces of clouds vary. Some have sharp edges, the boundaries between the cloud material and the adjacent clear, unsaturated air. Others have fuzzy edges. The relative dryness of the air outside the cloud and the nature of the cloud particles may affect how sharp the edges are, since drier air stimulates evaporation more readily than does humid air. Sharp edges may be associated with small cloud droplets because they all can evaporate quickly at the cloud edges, while fuzzy edges might be associated with clouds having a broader spectrum of particle sizes since the evaporation of the smaller droplets occurs first, leaving larger gaps between the larger, remaining droplets. The edges may be smooth if the air motions are laminar, but ragged looking if there is turbulence and/or changes in water vapor outside the cloud that are relatively abrupt.

While a cloud is a cloud, regardless of the time of day or what the quality of light is or what the surrounding terrain is, a particular cloud can look vastly different depending on the quality (angle of incidence, intensity, etc.) of light on it. The French painter Claude Monet famously painted the Rouen cathedral in France many times under different lighting conditions. Elaine Sciolino, in her 2020 book *The Seine*, noted that Proust had speculated as to whether or not a writer could do in writing what Monet had done with paintings. He later suggested that it is the ideal that we seek and that "after experiencing the beauty of physical sites through Monet's eye, the reality of visiting them may disappoint."[9] It has been noted that Charles Dickens once said that "It would be well, if Americans loved the Real less and the Ideal more."[10] In this book I have in some instances revisited certain types of clouds and that while I have tried to represent the Platonic ideal of some, a different type of beauty, perhaps each of its own Platonic ideal, can be seen for different lighting conditions—front lit, back lit, silhouetted, at night, etc. For many of the clouds, in addition to attempting to convey different feelings for different lighting conditions, the nature of the surroundings . . . mountains, oceans, buildings, etc., may also influence how one feels. In many instances I show more than one photograph of each type of cloud to suggest on one hand not only the generality of each, but also how they can vary in appearance, and that there are few, if any, *sui generis* types of cloud. We'll consider cloud classification itself shortly.

1.4.1   Clouds as depicted by artists

The visual properties of clouds as snapshots, frozen in time, usually with little regard for the motion of the air or the microphysical properties of the clouds, have long fascinated artists. An introduction to some of the best paintings that depict clouds may be found in great museums such as The National Gallery of Art and the Smithsonian American Art Museum in Washington D. C., the Museum of Fine Arts in Boston, the Metropolitan Museum of Art in New York, the National Gallery and the Tate Gallery in London, the Art Institute in Chicago, the de Young Museum in San Francisco, and the Rijksmuseum in Amsterdam, to name just a small (non-exhaustive) number that I have visited, some as often as I can when I visit the cities in which they reside. Many others undoubtedly are found elsewhere, but I cannot personally attest to them.[11]

Early (pre-nineteenth-century) paintings often showed clouds as relatively indistinct features in the background, or as puffy, floating cushions inhabited by cherubs. In the French Baroque painter Nicolas Poussin's "The Dance to the Music of Time" (1640), a horse and human figures defy gravity by resting on top of a cloud.[12] The Dutch painted many beautiful representations of clouds in the seventeenth century. Perhaps one of the most famous early cloud devotees and sky flaneur was John Constable, the great British artist, who painted his "Cloud Studies"[13] in 1821–22 during his so-called skying period

[9] Sciolino, E., 2020: *The Seine*. W. W. Norton & Co., 370 pp. (p. 264)

[10] From a "Comment" piece "Blunt Trauma," in the March 30, 1998 issue of *The New Yorker Magazine*, in the context of how "America has always been a place devoted to the literal, the unambiguous, the explicit."

[11] For more on clouds as depicted by artists, I recommend *The Soul of All Scenery* by Stanley Gedzelman (2014, Google Books) and related articles in *Weatherwise*, listed in the reference list.

[12] https://en.wikipedia.org/wiki/A_Dance_to_the_Music_of_Time_(painting)#/media/File:The_dance_to_the_music_of_time_c._1640.jpg

[13] https://www.tate.org.uk/art/artworks/constable-cloud-study-n06065

at Hampstead (Thornes 1999). The cloud studies were accompanied by a complete description of the weather conditions, the day, and the time.

Most artists did not specialize in clouds; for most the clouds they painted were just part of what happened to be on the canvas. JMW Turner, another great nineteenth-century English artist, painted pictures that evoked "descriptions of atmospheric conditions." His paintings often showed the "vanity of human effort in the face of the awesome power of nature."[14] The "interplay of light and dark" seen in paintings was seen as a battle between good and evil. His pictures were often a visual poem or a historical or biblical event. Edmund Burke, the British statesman and philosopher, in the mid-nineteenth century proposed a theory of the "Sublime," in which "things that are huge, obscure, and terrible arouse feelings that invigorate and elevate the mind." In addition, there are also the "Beautiful," things that are smooth, unthreatening, and pleasurable." The mysterious quality of the Sublime keeps some things partially hidden and obscure. The Sublime was described by some of the poetry of John Milton. Turner liked to represent an "elemental vortex," in snowstorms and deluges. On the other hand, in 1891 Oscar Wilde exclaimed that "Art is our spirited protest, our gallant attempt to teach Nature her proper place." So, does art tame nature by revealing its true "nature," or does nature simply leave us in awe?

The nineteenth-century British artists John Martin and Francis Danby also painted cataclysmic events. Other earlier, notable European artists who included clouds in fantastic landscapes (or seascapes), sometimes with vivid contrasts in lighting, include Claude Gellee (more commonly known as Le Lorrain or Claude), Ludolf Backhuysen, Francisque Millet, Jacob van Ruisdael, and Aelbert Cuyp in the seventeenth century, Gustave Courbet and Vincent Van Gogh, in the nineteenth century. The eighteenth century does not appear to me to have stimulated as many artists to do as much with clouds as did the preceding and following centuries. During the eighteenth century, Joseph Wright and Fragonard, however, appear to be some of the few who did.

Nineteenth-century American landscape painters such as Thomas Cole, Albert Bierstadt, Thomas Moran, and Frederick Edwin Church also sometimes showed clouds and fantastic lighting, particularly represented in the American West or the Hudson River valley, often with towering mountains mirroring towering clouds. They adapted the European painting techniques for the North American landscape. These pictures captured the spirit of the times, in which the rawness of nature, the western frontier, and religion all played roles. The "pristine beauty" of the landscapes was depicted as the appearance of the land was changing in response to "modern improvement" from railroads and the construction of industrial buildings.[15] Albert Pinkham Ryder in the late nineteenth century showed clouds glowing at night as the setting for psychologically unsettling topics. Winslow Homer, at the turn of the century, painted a

---

[14] https://www.tate.org.uk/art/artworks/turner-snow-storm-hannibal-and-his-army-crossing-the-alps-n00490

[15] From a descriptive panel entitled "Manufacturing and the Wilderness" at the Museum of Fine Arts, Boson, Massachusetts, May 28, 2022.

vivid image of a waterspout threatening a lone man on dismasted boat, which was surrounded by sharks ("Oh, the humanity!").[16] Thomas Hart Benton, an American artist in the first half of the twentieth century, painted images of the American landscape, in which Robert Hughes, the art critic, wrote in his book *American Visions*,[17] that "…the very clouds in his landscapes flex their biceps" (as in Fig. 1.1, left panel, third row). The "regionalist" John Steuart Curry, also painted similar clouds about the same time as Benton, including some frightening tornado and severe-storm clouds, while wonderfully capturing the feeling of the landscape and people experiencing the weather, not to mention including some allegorical meaning from time to time. Renee Magritte, the surrealist, painted a trompe l'oeil of a cloud in the early twentieth century. Georgia O'Keefe, in the mid-1960s, painted semi-abstract cloud patterns as viewed from above.[18] In all, during the seventeenth, eighteenth, nineteenth, and early twentieth centuries, many artists included clouds in landscape paintings with mountains, oceans (and shipwrecks), volcanoes erupting, biblical catastrophes, and, for contrast, pastoral fields. Very few concentrated on clouds as their main subject. Beauty, color, form, and mood played roles of varying degrees.

More contemporary artists such as Doug West, Dale Terbush, Jim Alford, John Rogers Cox, Tim Cox, David Holland, Bruce Cascia, Dave Kennedy, and Wilson Hurley, just to name some that I admire, have also represented clouds in striking and illuminating ways, despite the recent century's preoccupation with abstract art. West, Alford, and both Coxes in particular have depicted the beauty of clouds in the clear, dry skies of the western and southwestern parts of the United States. West focused on clouds in many of his serigraphs. In one he describes an image as follows:

To West, some clouds announce tension to be released in some dramatic happening. Tension in clouds may be likened to the tension that appears in many fiction-literary works. Ter Bush's breathtaking paintings appear to be an enhanced version of Cole's and Bierstadt's. Washington, DC-based artist Amy Marx and others have painted some beautiful representations of tornadoes and severe storms based on photographs of tornadoes and severe storms she collected from storm chasers. I know there are many other paintings like hers that I have seen at arts festivals, but cannot recall them exactly for this book.

Like some of the nineteenth-century painters, some of the early (nineteenth-century) photographers were also fascinated by clouds. Perhaps the first clouds photographs, which were daguerreotypes, were those taken in the 1840s in the United States by Bostonians Albert Sands Southworth and Josiah Hawes. The Frenchman Gustave LeGray, in Europe in the mid-nineteenth century, took his "Seascape with Cloud Study."[19] Another

---

[16] A memorable excerpt from a quote from a radio reporter when the Hindenburg zeppelin exploded in flames in 1937 over New Jersey. I use it here parenthetically, in a semi-ironic way, to describe the overly emphasized dangers depicted in the painting.

[17] Hughes, R., 1997: *American Visions: The Epic History of Art in America*. The Harvill Press, 635 pp.

[18] For example, "Above the Clouds I" (1962/1963), Georgia O'Keefe Museum, Santa Fe, New Mexico, https://prints.okeeffemuseum.org/detail/458772/okeeffe-above-the-clouds-i-1962-1963

[19] https://www.getty.edu/art/collection/objects/61914/gustave-le-gray-mediterranean-seascape-with-cloud-study-french-1857/

Frenchman, Charles Manville, also produced some photographic cloud studies, as did Ralph Abercromby. The German–American Alfred Stieglitz produced his "Songs of the Sky"[20] and then "Equivalents"[21] in the US. In the latter there is a 1922 set entitled "Music: A Sequence of Ten Cloud Photographs." Karl Struss took a photograph he called "White Cloud,"[22] also in 1922. Ansel Adams's "Moonrise, Hernandez, New Mexico"[23] in 1941 shows a vivid mountain-wave cloud (a phenomenon to be discussed in detail later in Chapter 2). For years I wanted to see a cloud like the one Adams captured, almost as much as I had wanted to see an actual tornado. Adams also has taken some beautiful, stark-looking towering clouds over San Francisco Bay and other places. All of the aforementioned photographs were taken in black and white. More recently, Eliot Porter, in 1958, took an untitled photograph showing tiny clouds. Even more recently, black-and-white cloud photography has made a comeback with the extraordinary images in Florida taken by Woody Walters. Nicholas Trofimuk has recently exhibited some beautiful black-and-white photographs of clouds in New Mexico. Other commercial artists such as John Fielder in Colorado have produced more conventional, but brilliant color images in the context of their mountainous backgrounds.

Amateur storm chasers and sky enthusiasts, in the latter half of the twentieth century and beginning of the twentieth-first century, have taken some striking color photographs of clouds, many of which have been reproduced in books, magazines, and weather calendars. In 2004 the US Postal Service, in collaboration with The Weather Channel, the American Meteorological Society, and the National Weather Service, issued a collection of 15 postage stamps entitled *Cloudscapes*, which featured spectacular cloud photographs. Among the best and most prolific color-cloud photographers, in this author's opinion, in part because either he has seen a lot of their work and/or knows or used to know them, are Dave Hoadley, Al Moller (deceased), Tim Marshall, Jim Reed, Roger Hill, Warren Faidley, Dave Blanchard, Bill McCaul, Ron Holle, David Holland, David Mayhew, Eric Nguyen (deceased), Greg Thompson, Martin Lisius, Jeff Snyder, and Trey Greenwood.[24] This list is *not* anywhere close to being exhaustive and I have seen many incredible cloud photographs posted on social media very frequently from many other photographers. Mitch Dobrowner, unlike most, has exhibited and published relatively recent, spectacular, black-and-white storm photographs, in a style not currently embraced by most photographers, who favor color images. Other famous and spectacular black-and-white photographs, some from as far back as the late nineteenth century, have been taken of tornadoes and waterspouts (see NOAA's NWS photo library).[25]

---

[20] https://www.metmuseum.org/art/collection/search/267455

[21] https://archive.artic.edu/stieglitz/equivalents/

[22] https://collections.mfa.org/objects/268575/white-cloud;jsessionid=125CF8DDC9FA5909EB8E43A700 70EE71?ctx=43807b51-96a7-45db-9a11-881e78fd68a7&idx=5

[23] https://www.moma.org/collection/works/53904

[24] I apologize to the many other fine photographers among the community of storm chasers that I am *not* recognizing here.

[25] https://photolib.noaa.gov/Collections/National-Weather-Service/Meteorological- Monsters/Tornadoes, under "Meteorological Monsters" and "Tornadoes" and "Waterspouts."

Dave Hoadley[26] was one of the first amateur storm chasers, to have taken spectacular storm photographs as early as the late 1950s. He was my original inspiration for
taking severe-storm photographs, with some additional inspiration earlier for cloud photographs in general from Wendell Mordy; he was, at the time, a retired professor of
atmospheric sciences at UCLA, whom I met in Coral Gables, Florida during one of my
summers at NOAA's National Hurricane Research Laboratory, when I was a graduate
student. Most of the amateur photographers are probably not well-known much beyond
the community of storm chaser enthusiasts, since for most of them their primary occu-
pation is not cloud photographer. Their contributions to what we know about the cloud
structure in severe thunderstorms of the Plains, however, have been substantial. Their
work has appeared frequently in weather calendars and in some books and may also
be found in social media. For every photograph of mine in this book, one can most
likely find others of each weather phenomenon or cloud type illustrated here that are
better. My intent is not to produce a set of the "best" cloud photographs, but rather a
set representing my personal vision and experiences.

There is one important difference between paintings and photographs: The imagination is let loose more on the former than the latter. Edward O. Wilson, the prominent
scientist and writer who is famously known for his studies of ants, in his 2017 book *The
Origins of Creativity*[27] pointed out that "Where scientific observation addresses all phenomena existing in the real world, and scientific theory addresses all conceivable real
worlds, the humanities encompass all three[28] of these levels and one more, the infinity
of all fantasy worlds." It is perhaps the fantasy world that many photographers really
pursue, the cloud with the not-yet-seen perfect lighting and perfect form. Perhaps it is
the ideal that is the fantasy, but not existing in reality. Many of the paintings in which
spectacular clouds have been represented are capriccios, fantastical representations of
fictional objects in fictional places or composites of real objects from many places. My
early interest in clouds can probably be traced to the Golden Nature Guide *Weather* by
Paul Lehr, R. Burnett, and Herbert Zim, originally printed in 1957,[29] in which there are
drawings of cloud capriccios.

The photographer Jeff Wall has noted that "Most photographs cannot get looked at
very often. They get exhausted. Great photographers have done it on the fly."[30] I suspect that his qualified statement excludes spectacular cloud photographs. Arthur Lubow,
a writer for the *New York Times*, has said that "the photographer is understood to be
waiting for the right convergence of subject, lighting and frame before clicking the shutter." He noted that the photographer Henri Cartier-Bresson called this the "decisive

[26] Dave Hoadley for many years, especially in the 1970s and 1980s edited a paper newsletter, *STORM
TRACK*, in which he published his own professional-grade sketches of tornadoes and severe convective storms.
I met him at an American Meteorological Society conference on severe local storms in Omaha, Nebraska, in
1977. He is not only a superb photographer, but also a gifted sketch artist.
[27] Wilson, E. O., 2017: *The Origins of Creativity*. Liveright Publications, p. 187.
[28] I somehow, however, note that only two are mentioned.
[29] Lehr, P., R. Burnett, and H. Zim, 1957 (1975 ed.): *Weather*. Golden Nature Guide, 160 pp.
[30] "The Luminist," Arthur Lubow, *The New York Times Magazine*, Feb. 25, 2007, pp. 56–63, 138, 140,
152–153.

moment." Oscar Wilde, in describing "modern Romantic poets," also noted its dealing with the "momentary aspects of nature." So, photography, unlike painting, is much more spontaneous and you take what you can get. This is certainly the case for cloud photography.

Peter Thuesen, in his 2020 book *Tornado God: American Religion and Violent Weather*,[31] pointed out from a section of Psalm 104:3 in the Bible, that referring to God, "you make the clouds your chariot, you ride on the wings of the wind," as evidence that God controls the weather, and by inference, the clouds. Years ago (in 2000) I saw an intriguing painting *"Hope is: Wanting to Pull Clouds,"* by a German artist, Sigmar Polke (1941–2010), in Washington, DC,[32] which turned out to be a twentieth-century painting (1992) possibly inspired by this psalm, which to me shows the relationship between the wind, clouds, and the imagination. When I first saw this painting, I completely missed the possible allusion to scripture and mistakenly thought it was a Renaissance painting, simply depicting how we have tried to harness the wind, since it shows a person in what I thought were Renaissance-style clothes trying to rein in two clouds, with the obvious siting of a sailboat in between the two clouds. I think that this painting is especially relevant to this book in suggesting that we can understand clouds as both art and a manifestation of the laws of physics. On the other hand, Arthur Hertzberg, in his book *Jews in America*, 1989,[33] noted that in 1893 a moralist, European rabbi, Israel Meir Ha-Kohen, in his book *Hafetz Haim* (The Seeker of Life) claimed that there were destructive tornadoes, floods, etc. beyond the sea and that they represented the wrath of the Lord, thus completely ignoring the secular laws of physics and thermodynamics.

Beyond paintings and photographs, artists have created representations of time-dependent small clouds in three dimensions, accessible near the ground, in closed spaces. In a recent biography of Oscar Wilde by Matthew Sturgis,[34] I learned that "nebulators" were used to make clouds on stage for theatrical performances (ostensibly only for background effect, not for their performances themselves). The Dutch artist Berndnaut Smilde, for example, has recently created clouds using smoke and mist machines in museums and other spaces around the world.[35] In fact, some of his creations have been photographed, so that his works may actually be seen represented in more than one medium. The artist David Rockwell produced an LED cloud on the ceiling of the FAO Schwarz building in New York City.[36] The American conceptual artist, born in Spain, Inigo Manglano-Ovalle, produced a titanium-coated mobile sculpture in the shape of a

[31] Thuesen, P., 2020: *Tornado God: American Religion and Violent Weather*. Oxford University Press., p. 28

[32] After an extensive search, I recently located this painting, no longer on display, at the National Gallery of Art in Washington, DC, thanks to Katherine Crowe, a reference archivist at the National Museum of Natural History of the Smithsonian. Since the painting is not owned by the National Gallery of Art, I attempted to contact the artist's estate or representatives to seek permission to reproduce it, but was unable to track them down. https://www.nga.gov/audio-video/audio/collection-highlights-east-building-english/hope-is-wanting-to-pull-clouds-polke.html

[33] Hertzberg, A., 1989: *Jews in America*. Columbia University Press, p. 145.

[34] Matthew Sturgis, *Oscar Wilde: A Life*, A. Knopf, 2021.

[35] D. Stone in *National Geographic*, March, 2019, pp. 8–14.

[36] D. Solomon, *New York Times Sunday Magazine*, Nov. 28, 2004, https://www.nytimes.com/2004/11/28/magazine/toystore-story.html

cumulonimbus.[37] In addition, I came across an intriguing sculpture of a tornado by an artist/sculptor from Broomfield, Colorado, Heide Murray, which showed a funnel cloud with a house hanging from its side, all perched atop the head of a child.[38] The sculpture, which, like the photographs of Smilde's clouds, is a representation of a cloud, in this case, the condensation funnel of a tornado. To paraphrase the Belgian surrealist artist René Magritte's title for a painting of a pipe, "Ceci n'est pas une pipe," all the cloud photographs shown herein perhaps should be titled "Ceci n'est pas un nuage." These examples are only those that I have serendipitously come across and undoubtedly there are many others.

Finally, I note that some cloud modelers have simulated on a supercomputer the remarkable representations of virtual cumulonimbus clouds and tornadoes, which resemble actual clouds with an almost frightening realism. Specifically, I am amazed at the depictions of tornadic supercells done by Leigh Orf[39] at the University of Wisconsin, Madison, which look very much like the ones we see in the real world when we storm chase. I have seen other impressive virtual representations of storm clouds produced more recently, especially by my colleague Ming Xue's research group at the Center for the Analysis and Prediction of Storms (CAPS) in the School of Meteorology at the University of Oklahoma (OU).

1.4.2   How large are clouds?

There is also the element of spatial scale to consider: How large are clouds? To an observer in outer space, the atmosphere is composed of areas of clouds, some of which are truly vast. During the Northern Hemisphere summer, there is an extensive area of low clouds over the ocean west of California. If we zoom closer to the cloud area, we might find that the area is composed of smaller clumps of clouds with the structure of bands or a regular grid. If we zoom in even closer, we may find that each cloud contains protuberances and other types of sub-structure. If we zoom in even closer yet, we can see individual cloud droplets.[40] Cloud droplets are to clouds what air molecules are to dry blobs of air. We don't see each individual cloud droplet; we see collections of cloud droplets as continuous masses of cloud material. That clouds as a whole are not continuous matter allows us to penetrate right through them. Think of a painting made by the Impressionist Claude Monet: Viewed from a distance one sees beautiful, colorful forms; viewed from close up, one sees individual blotches of color.

Meteorologists understand air motions and model the atmosphere for the purpose of predicting the weather by assuming that the air is a fluid. To understand and predict

---

[37] *Sunday New York Times*, Arts and Leisure Section, March 27, 2005, p. 6.

[38] Heide Murrays' sculpture "Tornado," was made from "wool, slik, suri, plastic, composition, wool, and wire with found objects. Needle and wet felted." It was exhibited at the Dairy Arts Center in Boulder, Colorado and viewed by the author on Dec. 13, 2014.

[39] L. Orf, et al., 2017: Evolution of a long-track violent tornado within a simulated supercell. *Bull. Amer. Meteor. Soc.*, **98**, 45–68.

[40] This hypothetical exercise reminds me of the 1968 version of the documentary film *A Rough Sketch for a Proposed Film Dealing with the Powers of Ten and the Relative Size of Things in the Universe*, by Charles and Ray Eames, I saw when I was an undergraduate in college. I recall it now only as the *Powers of Ten* movie.

clouds, meteorologists need to look at air as if it were a fluid, which is continuous, *and also* composed of individual cloud droplets and/or ice crystals. A knowledge of different scales is essential. The study of the latter is called "cloud microphysics." The microphysical makeup of a cloud is like looking at the molecules or atoms in solid objects. We also need to know something about thermodynamics, the exchange of energy that determines the temperature of the atmosphere and hence whether it's cold enough for condensation or the creation of ice crystals.

Since computer models that forecast the weather over the globe or over specific regions aren't able to follow each individual cloud droplet/ice crystal around (not as of the date of this writing anyway), the effects of all the cloud droplets/ice crystals must be taken into account from measurements of *collections* of cloud droplets/ice crystals. The process of estimating the *bulk* effect of cloud droplets based on measurements we *can* make is called "parameterization." Thus, we can say that if the temperature, humidity, and pressure are within a certain range, there is likely to be a certain type of cloud droplet/ice crystal, it will grow a certain rate, there will be a certain number of them, and a few will be large and many will be small; we don't look at the growth of each one in detail, at least not for the purposes of making a weather forecast. If we did so, we would "lose the forest through the trees" (or "lose the clouds through the cloud particles") and not detect important processes that are occurring on much larger scales. We analyze how a ball thrown up in the air will move without considering how all the atoms and sub-atomic particles respond to nuclear forces and consideration of quantum mechanics. The physicist Richard Feynman once pointed out[41] how "Things on a very small scale behave like nothing that you have any direct experience about. They do not behave like waves, they do not behave like particles, they do not behave like clouds, or billiard balls, or weights on springs, or like anything that you have ever seen." In this case, we may think of clouds as being on a very small scale, behaving in a certain way, while air motions are on the larger scale and behave in another way.

It is interesting that no matter what scale one looks at a cloud or at the motion of air, one sees similar structures. For example, a cloud viewed from outer space may look like long swaths of cloud circling the entire width of the Pacific Ocean from west to east. Within or near that swath of cloud, one may see a smaller, counterclockwise swirl (or comma) covering an area equivalent to that of a number of states (see Figs. 5.3 and 5.4). A closer look might find a thunderstorm, within which there is a mass of cloud circulating in a counterclockwise direction (see Fig. 4.121). Within that circulating cloud mass there may be a tornado (see, for example, Figs. 4.123, 4.128, and 5.9). Within the tornado there may be sub-tornado scale swirls (see Fig. 5.39), and so on. Since it doesn't matter on what scale we make our cloud observation, we always seem to find swirling structures, it is said that clouds behave like "fractals."

An intriguing characteristic of a two-dimensional fractal (a shape on a plane having an area and a perimeter) is that its perimeter increases as the scale on which it's being observed decreases. The old problem of finding the length of the coastline of Britain

---

[41] *Richard Feynman Lectures on Physics*, Vol. 3, Chapter 1, Quantum Behavior: https://www.feynmanlectures. caltech.edu/info/stories/gustavo_duarte_stroy.html

comes to mind: If one includes every tiny bend in the ocean shoreline the length gets bigger and bigger as the size of each bend decreases. Think of the surface area of a bubbly cumulus cloud: The surface area increases if one takes into account every tiny protuberance on the cloud surface; the smaller the protuberances, the larger the overall surface area of the cloud. In a sense, the finer scale structure there is in a cloud, the larger its surface area is.

1.4.3   How long do clouds last?

We now turn our attention to the time scale of a cloud. Most of the time we view clouds as if they were unchanging objects moving across the sky. In other words, clouds look like trees or rocks. The physicist Richard Feynman is reported to have said about clouds,[42] "As one watches them, they don't seem to change, but if you look back a minute later, it is all very different." With the exception of clouds viewed when the wind is extremely strong, only in time-lapse movies and videos do we see that clouds change their form as they are being sculpted by the wind. Typically, frames recorded every two to five seconds or so are needed to visualize the evolution of most clouds with sufficient continuity to see what is happening. This short time period contrasts sharply with the days necessary to see the evolution of foliage on a tree or the millions of years to visualize the change in a rock's appearance. To see the evolution of individual clouds and to track their movement, focus on a cloud that is adjacent to a nearby object such as a treetop or roof; then shift your focus to the nearby object and usually you will be able to detect changes in the shape of the cloud as it moves by. Many of us may recall lying on our back in a field or on the beach when we were young (or perhaps not so young) and watching clouds change shape as they passed by. In a sense, clouds rushing across the sky look free. But are they really free? Do they not obey the laws of physics and thermodynamics?

The time scale of a cloud can be measured by the time spent by the collection of cloud particles in the cloud, from the time the air enters the cloud and the cloud particles form, until the time the particles evaporate or sublimate and the air has exited the cloud. Or, the time scale of a cloud can be measured by the total lifetime of the visible cloud mass. Obviously, the latter is typically much longer than the former. A collection of cloud particles may only reside in the cloud for a matter of seconds or minutes, while the cloud may be visible for minutes, tens of minutes, hours, or even days.

Time scale can wreak havoc on one's suspension of disbelief when viewing movies or television programs that take place outside when the sky is visible. I constantly look at the sky in movies when it is visible. I may, for example, see a clear blue sky in the background when one character is speaking, and a field of clouds whenever another character, seen from another viewpoint, is speaking, and an even different type of cloud when yet another one is speaking or the same one is speaking, supposedly just several seconds later. My movie experience is thus partially ruined, knowing that the characters

---

[42] I first heard my colleague Rob Fovell at the University of Albany reference Feynman's comment, but he attributes it to James Gleick's book about Feynman "Genius," 1992, 1993 (paperback), p. 437, Vintage Books, 531 pp.

were *really* filmed on different days or times of the same day. Clouds change dramatically or disappear on time scales of much less than one day. Perhaps for the limited activity of watching a movie (or television program) it's too bad I look at the sky so much.

When we view a cloud, we use our eyes. However, we usually can't see inside clouds, which are opaque at worst, or only partially translucent at best. To get an X-ray view of a cloud and what lies in inside it, we use radars. The radars can detect that very tiny fraction of the microwave radiation aimed by the radar antenna at the cloud that is then "backscattered" to the radar. The radar makes measurements of the bulk properties of the cloud material. In a sense, a radar sort of sees the shadows of cloud (and larger) particles, like the view of the world of the people living in a cave described in Plato's allegory "The Cave."

Radars that concentrate their radiation in waves that are predominantly alternately or simultaneously, but separately, horizontally and vertically oriented, can be used to distinguish among particle types. These are called *polarimetric* radars (and will be described in more detail later). On the other hand, oddly, sometimes we won't "see" a cloud or precipitation because there are raindrops, which can be large, and/or hailstones, that are spaced rather far from each other. In this case, the radar will see what's there, even though the sky looks mostly translucent rather than opaque. We'll look in more detail at radar observations of clouds later.

1.5   The life of a cloud as experienced by a mobile battalion of air molecules

Consider now the life of a cloud as experienced by a parcel of air. Although days ago, the "parcel" of air, a tiny conglomeration of air molecules, may have been over the western portion of North America, and weeks ago over eastern Asia, and months and years ago almost anywhere in the Earth's troposphere, the lowest layer of the atmosphere, let's begin its journey, prior to its becoming part of a cloud, miles off the southeastern coast of the US, over the Gulf of Mexico. There it tumbles over and over in a turbulent fashion as it gets drenched by spray from breaking ocean waves just above the ocean surface. It warms up as it acquires heat from the ocean surface and gets more humid from the water vapor that has been evaporated from the sea surface. It heads north-northwestward, making landfall along the Texas coast. Its motion accelerates to almost 40 mph in response to falling pressure at the ground east of the Rocky Mountains, while westerly airflow aloft is warmed up as it descends along the lee slopes of the mountains. As the air parcel heads toward the Plains region of the central US, it rises very slowly as does the terrain, up to several thousand feet or more above sea level in western Oklahoma. While it cools off slightly, as it expands as a result of the lower air pressure at several thousand feet above sea level, its cooling is overwhelmed by heating as it comes into contact with hot ground during the daytime.

When it reaches far western Oklahoma, late in the afternoon, it suddenly encounters an invisible vertical wall, several thousand feet deep, behind which (to the west) the air is much drier and coming right toward it from the southwest at speeds of 40–50 mph. It is rudely deflected, like a football running back encountering a tackler. Now let's imagine that you are a massless passenger on the air parcel; as a rider of the air parcel, you fasten your massless seatbelts for the greatest natural amusement park ride on (or just above) Earth. You have warmed up, owing to your contact with the hot ground, to perhaps as high as the upper 80s or low 90s F, much warmer than you were when you tickled the 75°F-surface of the Gulf of Mexico. You shoot upwards and expand, like a bloated balloon, pumped up with air, and within minutes become saturated and consequently small water droplets form on tiny aerosols suspended in the air, some composed of salt from the ocean, the fog of atmospheric war, as the air parcel loses energy and cools as it does work against the surrounding air as it expands, but gains energy as water vapor condenses.

Pretty soon you break out into a region of clear, dry air, which cools off very rapidly with height. So rapidly does the air cool with height, that even though you continue to expand and cool, you become and remain warmer than the air around you. As a result of your newly acquired buoyancy, you accelerate upwards. More air condenses around you and forms a protective shell against the drier air, colder air which envelopes you like a blanket. You reach speeds as high as 100 mph straight upward, as you become 15–20° F warmer than the air outside. Some of your heat, which is lost as you expand, is replenished as more and more water vapor condenses. Liquid water drops appear, even though the temperature drops well below freezing.

Alas, as more and more water drops appear, you get heavier and heavier. You begin to lose your buoyancy: Water is a thousand times as dense as air in the troposphere, where the weather we experience happens. You slow down. Soon you are so high up that it is cold enough for ice crystals to form directly from the air around you, and you find yourself embedded in a world of both "supercooled" water droplets and ice crystals. Some of the water droplets begin to disappear, while the ice crystals begin to grow; it seems as if some of the water substance from the water droplets reappears on the ice crystals. Soon you are so loaded down with excess ice-crystal baggage, that you lose your buoyancy altogether. You are now at the apex of your journey, and like a package that has been shot into space by a rocket, like Icarus who has flown too close to the sun, you begin to fall back to Earth.

Your composition now changes, because your ice crystals don't fall out as quickly as your water droplets do. The ice crystals remain, high up in the atmosphere, 25, 000–50, 000 ft up (~7.5–15 km). As you fall, you warm as you encounter higher pressure and are compressed; your "balloon" shrinks. But your heavy weight still keeps you negatively buoyant. Matters are about to get worse. You find yourself falling into unsaturated air, so some of your water droplets begin to evaporate. As the water drops evaporate, you cool off, just as you do when you step out of a hot shower or out of a swimming pool. The cooling makes you even more negatively buoyant, so you accelerate downward, reaching speeds in excess of 60 mph. As you are about to hit the ground, you note that all your small drops have disappeared, leaving only the largest drops. You slam into the ground like a meteorite hitting a solid planet, but spread out like a pancake as you are deflected by the Earth's surface. Splat, but no crater and no end to dinosaurs . . . We end our hypothetical journey here, though we note that some of the water flows off in streams that ultimately ends up back in the Gulf of Mexico, while some evaporates into the air and is carried off elsewhere. It is noted that while clouds might seem free to roam as they please, in fact they are controlled by air motions, which are anything but free: They are constrained to appear according to the laws of physics and thermodynamics, to be discussed subsequently.

1.6   The life of a cloud as viewed by a stationary, ground-based observer

The journey of a cloud element just described was given from the point of view of the parcel of air. Such a reference frame, in which we follow a "parcel" of air along, is called a *Lagrangian* reference frame, so named after the Italian–French eighteenth and nineteenth century mathematician Joseph-Louis Lagrange. Let's now make believe instead that we're anchored to the ground in western Oklahoma, watching the sky. Such a reference frame, which is fixed to the ground, is called an *Eulerian* reference frame, so named in honor of the eighteenth-century Swiss mathematician Leonhard Euler. What might we see? The following description is a composite of many of the author's observations. First, we might see bands of low-level, flat-topped clouds streaming in from the southeast on the hazy horizon, but it would be clear directly overhead. Occasionally, large groups of birds, possibly making loud honking sounds, would fly by from south to north.

Around 5 p.m. local time, a north-south line of cumulus clouds might appear in the clear air, just to the west. Some of the clouds would build upward and lean to the northeast with height, but their skinny towers would become fuzzy looking and cut off from their bases. Such a sight would occur periodically every five minutes or so. Eventually, perhaps within 20–30 minutes, one of the towers, which is not easily identified in advance, does not get cut off from its base, but rather reaches great heights and puts out a veil of gray, hard-edged opaque or nearly opaque clouds consisting of ice crystals, an anvil, which rapidly spreads out to the east and northeast, like a veil covering the highest part of the sky ahead of where the clouds below are heading. Occasional flashes of lighting inside the anvil might be visible.

Soon, the region below the anvil and to the right of the cloud base might become less translucent, a sign that precipitation is reaching the ground. The first cloud-to-ground lightning flash is seen just before the precipitation becomes visible. As the precipitation shaft becomes opaque and dark, you can see that it has spread out at the ground, and assumes a flared-out shape. Just to the right of the flared-out precipitation shaft, and underneath cloud base, you might see a few cloud fragments, "scud," appear and disappear. Eventually more appear than disappear, and they join together and connect to the cloud base above, effectively lowering cloud base. A narrow band of clouds, like a tail, extends itself off to the right into the region of the precipitation.

The lowered cloud base then starts to rotate slowly in a counterclockwise direction. On the far eastern side of this lowered cloud base dust may be seen blowing from right to left, but eventually, whirls of dust are seen near the ground. Cloud elements in the foreground, however, are moving from left to right. An area of clearing is seen intruding into the cloud from its back side, so the cloud base assumes a horseshoe shape. Above one of the whirls, a funnel-shaped cloud appears to descend to the ground and heralds the formation of a tornado; first a whirl of dust is seen underneath the funnel cloud and then the funnel cloud connects to the center of the dust whirl. The tornado persists and about 20 minutes later leans to the right with height, becomes skinnier and ropelike, and then disappears. Precipitation falls from the entire cloud base and the cloud base disappears, leaving a region of rain, and possibly some hail, and anvil, which continues to move on the east and northeast.

The two perspectives of cloud formation just described in this and the preceding section are radically different. In the former, we move along with the air. In the latter, we remain fixed on the ground and watch the air move by. From the former perspective, we don't get to see much of what's happening around us; instead, we remain fixated on our own local universe and just "go with the flow," like a rider on a roller coaster. In the latter perspective, we get to see the big picture, the entire life history of a thunderstorm that produces a tornado, and see tiny portions of the movements of many pieces of air; we're cloud voyeurs, who get a more holistic view of the cloud system, looking at it from a distance.

The make-believe scenario we just described serves as an introduction to the subject of this book, clouds. Clouds, the bane of stargazers and eclipse watchers, tell an interesting story, no matter which perspective we have. The first scenario is useful for understanding how a cloud forms or decays. The second scenario is useful for describing the evolution of a cloud and the airflow into, within, and out from a cloud. Both scenarios illustrate some very important properties of clouds.

The first scenario (described in the Lagrangian framework) demonstrates for us that clouds are not like solid objects embedded within the air, simply being blown along by and moving with air currents. Clouds are in a sense *part* of the air. They are collections of tiny drops/droplets of water or ice crystals, or both, suspended in air and moving along with it. When the water droplets and ice crystals get large enough to fall out of the cloud, or at least from the location within the cloud where they are amassing, the droplets and ice crystals become precipitation particles and the air and the water substance separate from each other.

It is amazing to think that a cloud, while looking like a solid object from afar, is part of the atmosphere, which is a fluid. As such, the same collection of air molecules making up a part of a cloud can change shape. Not only can the collection of air molecules change shape, but it can also expand or contract as the pressure exerted on it changes. You can put your hand through a cloud and not feel it, while your hand might even disappear if the cloud is thick enough. Clouds can be strange and ghostlike.

The second scenario (described in the Eulerian framework) demonstrates for us that clouds can appear and grow as water vapor condenses into water droplets or is converted directly into ice crystals, or clouds can shrink and disappear as cloud material evaporates or sublimates. A cloud may appear to move, while what is actually happening is that cloud material is forming on one side of an existing portion of the cloud, and evaporating (or sublimating) on the other side. The cloud thus appears to be moving, but it really is not. Such apparent motion is called "propagation" and is like the apparent motion in an ocean wave. For example, the water surface at a beach may be just bobbing up and down, while the water underneath is just converging into a volume or diverging from a volume, depending on whether or not the water surface is rising or falling, but to an observer it appears that waves on the ocean surface are actually moving onshore. In most instances, however, there are both actual motion *and* propagation, just as in the ocean there are both water currents and wave propagation. *In nature, a cloud's movement is determined both by the air currents that move water droplets and ice crystals around and also by propagation, as some water droplets or ice crystals are forming and others are evaporating or sublimating.*

When a cloud forms more or less continuously on one side and dissipates more or less continuously on another side, the cloud may appear to be a coherent structure transported by the wind. There is therefore the illusion that these clouds are giant pieces of architecture in the sky (or airborne, "mobile" homes) moving along with the wind. Suppose that we moved along with the visible cloud mass, not with the air and not fixed to the ground. (We would need some aircraft to keep us moving along with the cloud.) In this case, our reference frame would be referred to as "quasi-Lagrangian," which is intermediate between a Lagrangian frame and an Eulerian frame. When cloud particles don't evaporate or sublimate, they just move along with the wind.

That some clouds have complicated but coherent structures, is an interesting observation. When clouds have an overall coherent structure, one can identify portions of the clouds and give them names. Thus, some clouds can be viewed as buildings with specific rooms that always are found in the same location with respect to the structure of the building. The kitchen is downstairs along with the living room and the bedrooms are upstairs. Even the different types of buildings have been given names. We'll later see that this is especially the case for clouds bearing convective storms. For example, there are "ordinary-cell" cumulonimbus, "supercell" cumulonimbus, "squall-line" cumulonimbus, "hurricane-eyewall" cumulonimbus, etc.

1.7 Passing through a cloud vertically, or through a cloud, darkly: Looking at clouds from both sides (now)

There are other perspectives on viewing clouds.[43] Clouds can be viewed from far aloft, from satellites looking down on them, or from aloft at closer range, at an angle, from aircraft. Clouds can be viewed from mountains, which may pierce through some clouds and also provide a view from above, or from a tall building, with the same effect. The different views of clouds from mountains are often spectacular and may be experienced by anyone in decent physical shape. Imagine that you are going for a hike in the Colorado Rockies during the summer. You might begin at sunrise so you can reach the summit of your mountaintop destination before noon, when nearly daily, afternoon

---

[43] The title of this section refers to Joni Mitchell's lyrics in 1966 and Judy Collins' 1967 hit record.

thunderstorms begin and scare the hell out of you, obviously because you could get struck by lightning. At sunrise, it is probably calm, and there could be some fog or a low cloud base overhead. You hike through the fog, where visibility may be restricted to just 50–100 feet. Eventually you sense more light and you suddenly break out of the cloud deck to view a deep-blue sky overhead. You soon reach timberline, above which there are no trees and you can get an incredible, unobstructed view. The top layer of the cloud deck glistens in the morning sun and streamers of cloud fragments rise upward out of the cloud deck and disappear. Richard Hamblyn, in his *The Invention of Clouds*, quotes from the diary of John Evelyn, an early mountaineer, a similar account of an ascent through clouds during a hike in the Alps.[44] The cloud deck below was aptly described as a sea of clouds, in which waves of clouds churn about. The top of the cloud deck viewed from above is called an *undercast*, the opposite of the bottom of a cloud, which when viewed from below is called an *overcast*. The entire cloud deck then gets higher and higher. Soon it breaks up, leaving a little haze below. Some wisps of cirrus clouds float by far above mountaintop, while a few patches of altocumulus clouds, also above mountaintop level, the remains of thundershower activity the day before, slowly disappear.

By midmorning cumulus clouds begin to form just above mountaintop level. They try to build upward, but fail, as narrow cloud elements become fuzzy looking and disappear. In an hour or so, some begin to look crisper and reach higher and higher into the sky. By late morning, some of the tops of the cloud towers take on a fuzzy appearance and leave behind a patch of anvil cirrus (to be discussed later). Soon, as you near your destination, the top of the mountain, you hear distant thunder. If you're lucky, the first storm of the day is not on *your* mountain peak. You look in the direction from which the thunder appears to be originating and see some precipitation falling. Occasionally, flashes of cloud-to-ground lightning are seen in the vicinity of the precipitation shafts.

In the best interests of your safety, you head on down, retracing your steps up to the top of the mountain. An anvil spreads out over your head, like a veil. Tiny chunks of soft hail (graupel) begin to bounce off your rain parka. Soon the ground becomes covered with the white graupel and the landscape takes on a winterlike aura. It cools off substantially, and the wind picks up. Patches of cloud form below and blow up the side of the mountain, enveloping everything in their path. As the clouds pass by, it seems as if someone has turned off the lights, and then turned them on again as sky clears. The sky, which had become totally overcast with anvil material, now begins to break up, and patches of blue appear. A rainbow might appear as the sun breaks through. The underside of the patches of anvil material might display mamma, pouchlike, downward protruding cloud elements that look like bubbly cumulus clouds turned upside down. If you're lucky, at sunset the mamma may glow bright red or orange, and you think you are in a painting by Albert Bierstadt, the great nineteenth-century German–American master.

---

[44] Hamblyn, R., 2001, pp. 18–119.

Our hypothetical day hike, modeled after many that I (and fellow hikers) have actually taken, shows how clouds can behave as a ceiling (overcast) when they have flat bases and are overhead, or like a water surface, when they intersect terrain, and when viewed from above are an undercast, shielding from our view all that goes on in the world below. Maybe you're not in good enough physical condition to climb a mountain, or are, but just don't want to exert the effort. I recall a breathtaking drive I once had on the island of Tenerife in the Canary Islands. My traveling companions and I began at sea level, where the vegetation was lush. We climbed up to the base of stratocumulus clouds above, where we continued through dense fog and a forest of evergreen trees. When we emerged from the cloud deck below, above the "Trade inversion," the sun was bright and the air was warm and dry. Continuing past a volcano, from which tiny clouds were steaming out of holes in the slate-gray ground, we eventually reached tall, desert-like sand dunes on the other side of the island.

The view of clouds from an aircraft is like that from a mountain, except that unfortunately the clouds pass by very quickly. Two main impressions one gets when flying by vast regions of the Earth's surface is that clouds are often layered and that similar types of clouds are seen bunched up in certain areas or appear in bands. That clouds are often layered is not obvious. Why should clouds be limited only to certain "stories" in the buildings in the sky, as "skyliners"? For example, there may be clouds on the ground floor (fog), a mid-level overcast (altostratus) in the upstairs bedroom, and a high-level overcast (cirrostratus) in the attic or penthouse. A plane taking off or landing may pass in and out of all the layers. One may pass over vast areas of land or ocean and see a checkerboard of the same type of cumulus clouds and then see a checkerboard of cumulus clouds that look different (have different width and depth, for example). While these observations are now commonly seen by even the most casual airplane passenger, observations similar to the aforementioned were probably first made by balloonists in the late eighteenth century, and must have made quite an impression on them. That clouds of similar visual appearance are often bunched up seems obvious, in that their visual appearance must be linked to similar atmospheric conditions in regions that are larger in area than that of individual clouds.

It is clear that our understanding and appreciation of clouds can depend on our perspective. Are we moving with the air, fixed to the ground, or moving with the cloud? To understand cloud dynamics, one has to envision "parcels" of air flowing into, inside of, and out from a cloud, at many different places in the cloud. To understand the cloud as a system, we need to look at the totality of different air trajectories. To relate the dynamics of clouds to what we see, we need to re-think what we know in a reference frame of a stationary observer on the ground. Like the Japanese filmmaker Kurosawa's movie *Rashomon*, in which a story is told from very different human perspectives, the cloud story can also be told from very different physical-location perspectives.

Even the visual aspects of a cloud depend on our perspective. An isolated cloud viewed from below at sunset may have a spectacular golden or red–orange tint, while the same cloud viewed from above in an airplane or in a satellite in space may look relatively drab. Some clouds look dull when viewed a low angle, but bright when viewed overhead.

1.8 Cloud classification

"The sky will next demand our attention. The soul of all scenery, in it are the fountains of light, and shade, and color. Whatever expression the sky takes, the features of the landscape are affected in unison, whether it be the serenity of the summer's blue, or the dark tumult of the storm. It is the sky that makes the earth so lovely at sunrise, and so splendid at sunset. In the one it breathes over the earth the crystal-like ether, in the other the liquid gold . . .

Look at the heavens when the thunder shower has passed, and the sun stoops behind the western mountains – there the low purple clouds hang in festoons around the steps—in the higher heaven are crimson bands interwoven with feathers of gold, fit for the wings of angels—and still above is spread the interminable field of ether, whose color is too beautiful to have a name."

Thomas Cole, "Essay on American Scenery" (1835)

Clouds are like art sculptures that appear in the sky museum or sky gallery. There is no admission price and the museum is always open, though the exhibit continuously changes. In this respect, clouds are like a slowly evolving play and the sky is the theater or stage in which or on which the play is performed. Before sunrise in a moonless sky, before clouds can be seen, the sky is like an undeveloped photograph,[45] and there is the anticipation of what will be revealed upon sunrise. I have gotten up many times in Colorado, before sunrise, eager to find what cloud tapestry will be found over the mountains, when the sun finally hits the sky to the west and perhaps reveals some colorful and spectacular orographic wave cloud.

Or, we may view the clouds the way a microbiologist views living organisms with a microscope or an astronomer views the night sky with a telescope, with the aim of finding order and discovering new features. As Darwin viewed animals, we may view the sky with the aim of classifying all the cloud animals that inhabit the sky zoo or natural habitat and try to make sense of how they are related to each other. It then seems logical to ask the following questions: How do we classify the types of clouds? Do we classify them on the basis of their appearance on the macroscale? Do we classify them on the basis of their composition (microphysics), i.e., are they composed of tiny water droplets around the same size, of different sizes, of ice crystals, etc.? Do we classify them on the basis of their height above the ground? Do we classify them on the basis of their environment? Are there cold clouds, warm clouds, mountain clouds, marine clouds, desert clouds, tropical clouds, arctic clouds, etc.? Do we classify them on the basis of the physical mechanisms responsible for their formation? In other words, are there highly buoyant clouds, clouds with just a little buoyancy, and non-buoyant clouds?

Clouds are like fingerprints (though fingerprints that change quickly with time); think of the fingerprint of a cumulus cloud, for example, as all its striations and protuberances. Each cloud form appears to be unique when viewed on very small scales, but when viewed on larger scales, clouds can be easily grouped into certain categories based on

---

[45] I realize that this simile is now outdated, since digital photographs require no time to be "developed."

visual appearance. It seems as if there probably must be a relation between their visual appearance and their composition, environment, and underlying dynamics.

The first recorded study of clouds may have been done in ancient Greece by Thales of Miletus (625–545(8) BCE), before Socrates lived. Thales is thought of as the first scientist within the context of Western civilization. The Greeks Democritus and Aristotle and the Romans Seneca, Pliny, and Lucretius all addressed the problem of cloud formation. After the fall of Greece and Rome, inquiries into the nature of clouds and science in general were shunted aside by religion. Descartes, however, in the seventeenth century, revived the consideration of how clouds are formed. Oliver Goldsmith and the German poet, scientist, and writer, among other things, Johann von Goethe, in the late eighteenth century both produced, from the standpoint of our current science at least, theories of cloud formation that must be considered bizarre. More popular and equally bizarre as Goldsmith's and Goethe's theories was the so-called vesicular or bubble theory, which involved hollow volumes filled with buoyant elements that rose, developed into clouds, until they burst, and rain fell out. It will later be shown that this weird theory contains at least a tiny element of truth to it.

Perhaps inspired by the Swedish taxonomist Linnaeus, who devised a classification scheme for plants, Jean Baptiste de Monet Lamarck, the evolutionary biologist, in France, and Luke Howard (Pedgley 2003), in England, both independently devised cloud classification schemes based on the visual appearance of clouds. Lamarck's classification scheme was not well recognized probably because it was published in an obscure journal in which the effects of the moon and the planets were used to forecast the weather (unsuccessfully) and because he fell into disfavor with Napoleon. Howard, however, a "chemist" (i.e., in the US, a pharmacist) and amateur meteorologist, wrote an essay "On the modifications of clouds" and presented it orally to the Askesian Society in London in December 1802. The Askesian Society was a small science club in London that operated like a university seminar course in that the speaker was challenged in the manner of a debate. In 1803 Howard's essay was printed in the *Philosophical Magazine*, which was at the time the best-known and most widely read magazine by both professionals and amateurs. Howard's essay was printed again in 1804 as a pamphlet, in 1811 in the *Journal of Natural Philosophy*, in 1812 in a best-selling meteorology book by Thomas Foster, and again in 1818 at the beginning of Howard's long book, in which his long record of weather observations was detailed. It thus appears that Howard's success in promulgating his classification scheme depended to a large extent on his ability, in an age without radio, television, and the internet, to disseminate his work widely through a legendary public lecture to an influential audience and through an essay published in an influential scientific journal and in an influential book.

Howard's taxonomy scheme was as follows: There are three main types of clouds, the (1) cirrus, (2) cumulus, and (3) stratus. The name of each type is expressed in Latin, as was accepted by convention at the time, and in order to convey information about the appearance of the clouds. The cirrus is so named because it looked like threads; the cumulus was so named because it looked like heaps of cloud material; the stratus was so named because it looked flat, like sheets. Four other names that combined the three basic forms were introduced as "cirro-cumulus," "cirro-stratus," "cumulo-stratus," and "cumulo-cirro-stratus." The latter was also known as "nimbus." The nimbus was the rain cloud and a combination of all three types of clouds. Some tried unsuccessfully to change the Latin names to English names. Since Lamarck had described clouds in French, which was not the accepted international scientific language, his scheme also lacked global appeal. The Latin names have endured two centuries of time. It will help your understanding of cloud names substantially if you have studied some Latin; the names may seem exotic if you have not: *Caveat spectator nubilorum, qui linguam Latinam nescit!*[46]

Howard's essay on clouds was translated into French and German. Incidentally, Goethe, in 1820, published poems in which Howard's cloud classification scheme was noted. A German clerk, Johann Christian Huttner, who assisted with the publication of Goethe's poems, it is thought, contributed his own interpretation of the poems: Stratus clouds were linked with rising clouds, cumulus with overturning, cirrus with scattering, and nimbus with falling. One might interpret this interpretation to be the life cycle of a rain shower. Such further publicity certainly must have helped promulgate Howard's classification scheme. The captain of a whaling ship, William Scoresby, Jr., during a trip to Greenland in 1810, may have been the first to use Howard's classification scheme.

Howard also tried to explain how clouds formed using the then-known laws of physics. His explanations were flawed, however, because he incorrectly thought that water vapor moved independently of the air; he also speculated about the role atmospheric electricity might play in cloud formation. To this day there are people who believe, contrary to evidence, that thunderstorm electricity is responsible for tornadoes (though in some instances changes in lighting activity may coincide with the formation of some tornadoes).

While Lamarck had first suggested classifying clouds according to their height, it was again Howard whose similar suggestions are remembered today. In a lecture in 1817, first published in 1837, he associated each type of cloud with its height. Cirrus clouds are the highest, cirrocumulus and cirrostratus are "intermediate" in height, and cumulus, cumulo-stratus and nimbus, and stratus are the lowest. It was not noted, however, that height is a relative measurement that depends on where the observer is; for example, an observer in a balloon or airplane or on a mountaintop might see fog or stratus, while an observer on the ground might look up and a cumulus cloud: Fog is simply any cloud that touches the ground, or where the observer is.

Following Howard's pioneering and most influential writings and lectures, the generally accepted cloud classification scheme slowly evolved into a set of ten types of clouds. In 1840, the German meteorologist Ludwig Kaemtz replaced Howard's cumulo-stratus with "strato-cumulus." The designation "fracto" was added by Poey in France in 1865. In 1879 the Reverend Clement Ley in England coined the terms "lenticularis" and "castellatus,"[47] and was also the author of a book entitled *Cloudland*. The types "cumulo-nimbus" and "strato-cirrus" were added in 1887 by the famous Victorian meteorologist Ralph Abercromby and the Swedish meteorologist H. Hildebrand

---

[46] "Beware of the cloud observer who speaks the Latin language!" (H. Bluestein, the author of this book).
[47] Also written now as "castellanus."

Hildebrandsson. According to Richard Scorer[48] and Richard Hamblyn,[49] the type "cumulonimbus" can be traced to Weilbach in 1880. In 1887 Hildebransson, Abercromby, and others assembled cloud photographs and a directory of 16 photographs was published in 1890. The basic cloud types as of 1887 were as follows: cirrus, cirro-stratus, cirro-cumulus, strato-cirrus, cumulo-cirrus, strato-cumulus, cumulus, cumulo-nimbus, nimbus, and stratus.

After a "Cloud Committee" was formed in Munich in 1891, a revised set of ten clouds was officially adopted and appeared in the first international cloud atlas, which was published in 1896, the so-designated International Year of the Clouds. Efforts were undertaken to see, based on journeys around the world, of how universal (or global) the existing cloud classification was. "Cumulo-cirrus" was replaced by "alto-cumulus." The prefix "alto-" means high in Latin. "Strato-cirrus" became "cirro-stratus." While the Frenchman Emilien Renou added the types "altocumulus" and "altostratus" in the mid-nineteenth century, these were not officially adopted until the publication of the 1896 cloud atlas. The ten basic cloud types in the 1896 atlas were as follows: cirrus, cirro-stratus, cirro-cumulus, alto-cumulus, alto-stratus, strato-cumulus, nimbus, cumulus, cumulo-nimbus, and stratus. The 1896 atlas was reprinted in 1910.

The International Commission for the Study of Clouds, which was formed in 1921, published a new edition of the international cloud atlas in 1932, which was designated as the first volume of a set. In 1932 the rain cloud type "nimbus" was renamed "nimbostratus."[50] Yet another new edition of this atlas, containing 101 cloud photographs taken from the ground and 22 from aircraft, was published in 1939, and followed an abridged version in 1930; a two-volume edition was published in 1956 by the World Meteorological Organization (WMO). The first volume consisted of text only and the second volume included 123 black-and-white photographs and 101 color photographs of clouds and related phenomena. It was the abridged version of this atlas that captured my imagination as a student and during the early part of my career in the 1970s. A revised version of Volume I was published in 1975.

A less formally structured, yet more scientifically detailed, cloud book entitled *Clouds of the World* was written by the prominent English meteorologist and fluid dynamicist Richard Scorer in 1972; this work also captured my imagination. In 1981 the WMO began to prepare a new edition of Volume II of the 1956 edition. During this time period I was honored to have some of my color-cloud photographs selected for inclusion in the new edition, including the cover-jacket photograph, which were published in 1987. My photographs were selected through the intervention of a member of the committee dealing with the new edition, Ron Holle, a scientist at what was then known as the Experimental Meteorological Laboratory in Coral Gables, Florida. Ron was working at the lab when I was in residence at an adjacent laboratory as a visiting graduate student. He knew of my interest in cloud photography. A few of my photographs were also published in

---

[48] Scorer, 1972.
[49] Hamblyn, 2001.
[50] I have never heard any of my colleagues use this name to describe a cloud. Typically, one would say "Look at that rain cloud," but not "Look at that nimbostratus." Maybe the Latinized version sounds too pretentious.

1987 in a marine cloud album, also under the auspices of the WMO. The most recent edition of the WMO cloud codes, which is separate from the cloud atlas, was published in 1995.

Since 1951, the following basic ten types or characteristic forms (*genera*) of clouds seen worldwide have been officially designated by the WMO: *cirrus, cirrocumulus, cirrostratus, altocumulus, altostratus, nimbostratus, stratocumulus, stratus, cumulus, and cumulonimbus*. The first three, containing -cirrus as suffixes, are considered high clouds; by high we refer to bases of 3–8 km (10, 000–25, 000 ft) in Polar regions, 5–13 km (16, 500–45, 000 ft) in midlatitudes, and 6–18 km (20, 000–60, 000 ft) in the Tropics. It is necessary to distinguish what is meant by high, middle, and low by latitude, since the depth and characteristics of the atmosphere vary according to latitude. The depth of the troposphere is deepest in the warm Tropics and shallowest in the cold Polar regions, as a result of varying solar and terrestrial radiation and of dynamical processes.

The next two, containing alto- as prefixes are considered middle clouds; by middle we refer to bases of 2–4 km (6, 500–13, 000 ft) in Polar regions, 2–7 km in midlatitudes (6, 500–23, 000 ft) and 2–8 km (6, 500–25, 000 ft) in the Tropics. Altostratus clouds, while mainly mid-level clouds, sometimes extend to higher altitudes. Nimbostratus clouds are characteristic of mid-levels, but can extend to the other levels also.

The seventh and eighth ("strato-" clouds) are considered low clouds; by low we refer to bases below 2 km (6, 500 ft), regardless of latitude. The penultimate, the cumulus, begin in the lowest level, but may extend up to the middle level or the high level; the last, the cumulonimbus, extend from low to middle levels, low to high levels, or middle to high levels.

The cloud genera are mutually exclusive; a cloud must belong to one and only one genus. Each genus, however, may be subdivided into *species*. Species are also mutually exclusive, though the same species can be applied to more than one genus. For example, a certain shape may describe more than one type of cloud. There are 14 types of species meaning such things as ragged-looking, lens-shaped, etc.

Clouds are also classified according to their *variety*, which "are related to the different arrangements of the cloud elements." Like species, the same variety may pertain to more than one genus. In addition, a cloud may exhibit the characteristics of more than one variety. There are nine varieties describing such attributes as exhibiting undulations (undulatus) and having elements arranged like human ribs or a fish skeleton (vertebrates).

In addition to genera, species, and variety, some clouds contain *supplementary features* or *accessory clouds*. These cloud features may be attached or are nearby, but separate from the main cloud body. A cloud may contain more than one supplementary feature or accessory cloud. There are six supplementary features and three accessory clouds. Examples of supplementary features include cumulonimbus anvils (incus) and funnel clouds (tuba).

Finally, it is recognized that a part of a cloud may grow vertically or laterally so that a new cloud, whose genus is different from that of the original *mother cloud* (not the mother *of* all clouds), emerges from the primordial cloud "soup." They are named by genus of the type of cloud forming, followed by the genus of the parent cloud with suffix "genitus" appended. There are seven mother clouds. Among them is the stratocumulus cumulogenitus, a cumulus cloud that flattens out into a stratocumulus. A summary listing all the official genera, species, variety, supplementary features or accessory clouds, and mother clouds, may be found in Table 2 of the 2017 online version of the WMO's *International Cloud Atlas* at http://cloudatlas.wmo.int, reproduced below as Table 1.1.

In addition to the standard cloud types and their sub-varieties listed in the table, there are also some special clouds we will also look at later. Some of these are man/woman-made, from fires, or induced by airflow over mountains.

Since the 1956 official WMO classification, storm chasers have identified many interesting and unique cloud features not described much, if any, in the international cloud atlases. It is not surprising, since prior to the last official cloud classification scheme, only relatively few serendipitous photographs of severe-storm features in the Plains of the US had been taken. In addition, routine reconnaissance aircraft on nearly annual research flights into tropical cyclones (hurricanes in the North Atlantic and Eastern Pacific and typhoons in the Western Pacific, for example) have allowed scientists to view some breathtaking cloud types seen only in the eyes of intense tropical cyclones.

We have talked about how clouds have been classified based solely on their appearance. Clouds may be classified also on the basis of certain aspects of their dynamics, that is, the forces that control them. We would expect that there is some relation between cloud dynamics and their appearance. Clouds that appear bubbly, like "heaps" of cloud material, such as the cumulus and the cumulonimbus, are "convective" in nature. Convective clouds are buoyant and transport heat vertically. Convective clouds are also referred to as "cumuliform clouds." The vertical motions in convective clouds range from 1 m s^{-1} in small clouds to as much as 50 m s^{-1} in severe-storm cumulonimbus clouds.

On the other hand, clouds that appear relatively flat and are developed more extensively in the horizontal are "stratiform" in nature. The stratiform clouds tend to be forced, that is they are not buoyant, but move because air on a larger scale is being moved horizontally, and slowly, vertically (upward). Buoyancy, on the other hand, tends to be local and the air is moved upward much more quickly. The convective clouds form in atmosphere that is "unstable," while the stratiform clouds form in an atmosphere that is stable. We'll discuss more precisely what we mean by stable and unstable in the next section. The vertical (upward) motion in stratiform clouds is usually less than 1 m s^{-1}.

We can also take a much wider view of clouds and consider the large-scale weather system in which they're embedded. For example, cirrus that accompanies the fast-moving current of air near the top of the troposphere, especially in midlatitudes from late fall to spring, are sometimes called "jet-stream cirrus." Cirrus may form in advance of mobile, upper-level disturbances, or at the top of convective storms. Stratocumulus may form over the ocean in the wake of the passage of a cold front, or as moist air is heated up over land in a stable atmosphere. Altocumulus may form as air is forced up and over

Table 1.1 Cloud Classification (*courtesy of the World Meteorological Organization*).

Genera	Species	Varieties	Supplementary features	Accessory clouds	Mother clouds and special clouds (most commonly occurring mother clouds are listed in the same as genera)	
					Genitus	*Mutatus*
		(listed by frequency of observation)				
Cirrus	*fibratus* *uncinus* *spissatus* *castellanus* *floccus*	*intortus* *radiatus* *vertebratus* *duplicatus*	mamma fluctus		Cirrocumulus Altocumulus Cumulonimbus Homo	Cirrostratus Homo
Cirrocumulus	*stratiformis* *lenticularis* *castellanus* *floccus*	*undulatus* *lacunosus*	virga mamma cavum		–	Cirrus Cirrostratus Altocumulus Homo
Cirrostratus	*fibratus* *nebulosus*	*duplicatus* *undulatus*	–		Cirrocumulus Cumulonimbus	Cirrus Cirrocumulus Altostratus Homo
Altocumulus	*stratiformis* *lenticularis* *castellanus* *floccus* *volutus*	*translucidus* *perlucidus* *opacus* *duplicatus* *undulatus* *radiatus* *lacunosus*	virga mamma cavum fluctus asperitas		Cumulus Cumulonimbus	Cirrocumulus Altostratus Nimbostratus Stratocumulus

Continued

Table 1.1 *Continued*

Genera	Species	Varieties	Supplementary features	Accessory clouds	Mother clouds and special clouds (most commonly occurring mother clouds are listed in the same as genera)	
					Genitus	*Mutatus*
		(listed by frequency of observation)				
Altostratus	-	*translucidus* *opacus* *duplicatus* *undulates* *radiatus*	virga praecipitatio mamma	pannus	Altocumulus Cumulonimbus	Cirrostratus Nimbostratus
Nimbostratus	-	-	praecipitatio virga	pannus	Cumulus Cumulonimbus	Altocumulus Altostratus Stratocumulus
Stratocumulus	*stratiformis* *lenticularis* *castellanus* *floccus* *volutus*	*translucidus* *perlucidus* *opacus* *duplicatus* *undulatus* *radiatus* *lacunosus*	vigra mamma praecipitatio fluctus asperitas cavum		Altostratus Nimbostratus Cumulus Cumulonimbus	Altocumulus Nimbostratus Stratus

Stratus	*nebulosus* *fractus*	*opacus* *translucidus* *undulatus*	praecipitatio fluctus		Nimbostratus Cumulus Cumulonimbus Homo Silva Cataracta	Stratocumulus
Cumulus	*humilis* *mediocris* *congestus* *fractus*	*radiatus*	virga praecipitatio arcus fluctus tuba	pileus velum pannus	Altocumulus Stratocumulus Flamma Homo Cataracta	Stratocumulus Stratus
Cumulonimbus	*calvus* *capillatus*	–	praecipitatio virga incus mamma arcus murus cauda tuba	pannus pileus velum flumen	Altocumulus Altostratus Nimbostratus Stratocumulus Cumulus Flamma Homo	Cumulus

mountains. Elongated bands of cirrostratus and other types of clouds may form at the rear of fronts. Cirrus form as anvils at the top of thunderstorms, while funnel clouds form underneath cloud base in the more severe storms. On a more local level, cumulus clouds sometimes form over fires and cirrus sometimes appear in the wake of jet aircraft. In subsequent chapters, we'll note the conditions under which each type of cloud forms.

In this book, we'll spend some time trying to understand more about how clouds form, and we'll focus on the beauty of some *selected* types of clouds and ignore or showcase very briefly most, but not all, of the not-so-beautiful ones, unless they evoke strongly a specific mood or are unusual in some way that they bear mention. A majority of the clouds described herein could arguably be described as the "greatest clouds on Earth," though some, like the gloomy stratus viewed from below are arguably not. To keep the scope of this book from becoming too wide and getting more out of hand, we'll limit ourselves mostly to ground-based (including from mountain tops) and some airborne photographs. Satellite cloud imagery is an entire subject in itself and would require a volume of its own. Instead, we will look at just a relatively small number of satellite photographs of selected types of clouds. A few noteworthy clouds we will not consider include noctilucent clouds in the stratosphere or Polar stratospheric clouds, mainly because I have never seen any, at least up to the time of this writing. In addition, while I will talk about banner clouds, I will not show any photographs of them also because I have never seen them yet definitively, by the time of this writing.

We'll first look at stratiform, non-buoyant clouds in Chapter 2 and then non-precipitating convective clouds in Chapter 3, and precipitating convective clouds in Chapter 4. Chapter 5 will contain a potpourri of topics related to rotating structures in phenomena of many different space and time scales, including tornadoes and tropical cyclones. In Chapter 6 we'll look at the current and future technology of observing clouds and measuring their properties, which is expected to further our understanding of clouds. There will be some overlap among the chapters: some clouds might be mentioned in more than one chapter. Although we will look at many different types of clouds, we will *not* try to describe exhaustively every type of cloud found in nature on our planet, but instead refer the reader to the *International Cloud Atlas* or other guidebooks for more information. In most instances I will show clouds of various types, with a number of different views of each type in order to show, when possible, how each looks different under different lighting conditions and also to show that the cloud type is viewable at different locations and times of the year. Enjoy your journey through the clouds. As we do, keep the following thoughts of the nineteenth-century American essayist Emerson in mind:

"The sky is the daily bread of the eyes. What sculpture in these hard clouds; what expression of immense amplitude in this dotted and rippled rack, here firm and continental, there vanishing into plumes and auroral gleams. No crowding; boundless, cheerful, and strong."

Ralph Waldo Emerson, Journal (1843)

References[51]

American Meteorological Society, 2002: Cloud watching with John Day. *Bull. Amer. Meteor. Soc.*, 83, 847–848.

Bentley, W. A., and W. J. Humphrey, 1962: *Snow Crystals*. Dover, 224 pp.

Bentley, W. A., 2000: *Snowflakes in Photographs*. Dover, 72 pp. (originally 1931, McGraw – Hill, as *Snow Crystals*)

Bohren, C. F., and D. R. Huffman, 1998: *Absorption and Scattering of Light by Small Particles*. Wiley, 530 pp.

Collier, C. G., 2003: On the formation of stratiform and convective cloud. *Weather*, 58, 62–69.

Galvin, J. F. P., 2003: Observing the sky – how do we recognise clouds? *Weatherwise*, 58, 55–62.

Gedzelman, S., 1998: Sky paintings. *Weatherwise*, 51, 64–69.

Gedzelman, S., 2002: Colors of the sky. *Weatherwise*, 55, 20–29.

Gedzelman, S., 2003: A cloud, by any other name... *Weatherwise*, 56, 24–28.

Gedzelman, S., 2014: *The Soul of all Scenery: A History of the Sky I Art*. Google Books, 311 pp.

Levine, J., 1959: Spherical vortex theory of bubble-like motion in cumulus clouds. *J. Meteor.*, 16, 653–662.

Novak, B., 1980: V. Meteorological Vision: Clouds. *Nature and Culture*. Oxford University Press, 78–100.

Pedgley, D. E., 2003: Luke Howard and his clouds. *Weatherwise*, 58, 51–55.

Pouncy, F. J., 2003: A history of cloud codes and symbols. *Weatherwise*, 58, 69–80.

Thornes, J. E., 1999: *John Constable's Skies: A Fusion of Art and Science*. University of Birmingham Press, Birmingham, U. K., 288 pp.

World Meteorological Organization, current online: *International Cloud Atlas: Manual on the Observation of Clouds and Other Meteors*, WM)-NO, 407. http://cloudatlas.wmo.int

---

[51]The lists of references in this and subsequent chapters not only refer to material quoted in the book, but also are meant to serve as a starting point for readers interested in learning, in more depth, about clouds, their characteristics, dynamics, and the meteorological conditions under which they form, including the weather features with which they are associated, etc. and are not meant to be entirely comprehensive or exhaustive. I have included what I consider to be representative references, in which other important references may also be found. I have tried to include some older, seminal journal articles, not always referenced in more modern publications, in order to, in some instances, restore their historical due. In other instances, however, I have not included the original reference because it may be too difficult to access online; in these instances, the original source may be referenced and an original figure reproduced in the later reference.

2

Non-buoyant Clouds in a Stable Atmosphere

In the first chapter of this book, I divided up the categories of clouds we see into buoyant and non-buoyant types. Completely buoyant clouds or clouds with only buoyant sections tend to have some bubbly, cauliflower-like surfaces, facing upward in the case of positively buoyant clouds and facing downward in the case of negatively buoyant clouds. Non-buoyant clouds, however, tend to be smoother looking and flatter, especially in lower portions of the atmosphere, where they are composed of tiny liquid water droplets. In what follows, more than one example of each type of cloud is often shown to impress upon the reader that there is a firm foundation for identifying distinct types of clouds and that most of the clouds are not just *sui generis*. I must admit, though, that many years ago, the prominent severe-storm expert and mesoscale meteorologist researcher Ted Fujita named some cloud features in tornadic convective storms in general based on a comprehensive set of photographs from just one case. Remarkably, the names he chose have mostly survived the test of time and persist, having been validated over and over again in many other storms; more on this in Chapter 4. I don't, however, have as much faith in *my* naming cloud features based on just one example.

2.1 Stratiform clouds

Stratiform clouds are rather flat, do not having much vertical development mainly owing to the stability of the atmosphere, which is responsible for its resistance to vertical motions. Sometimes these clouds appear in layers; sometimes they are expansive, covering much of the sky, while at other times they are localized.

2.1.1 Stratus and fog

Low-level stratiform clouds are called *stratus*. They tend to have low bases, and occasionally drizzle falls from them. They often appear to have a dull, grayish color, evocative of

The Architecture of Clouds. Howard B. Bluestein, Oxford University Press. © Howard B. Bluestein (2024).
DOI: 10.1093/oso/9780198870548.003.0002

the feelings expressed, coincidentally both in 1966, in records by Simon and Garfunkel "Cloudy" and the Mamas and the Papas "California Dreamin'."

When the bases of these clouds are actually at ground level, they are called *fog*. However, fog can technically occur whenever *any* cloud, stratiform or not, touches the ground; so even if you're inside a convective cloud that is, for example, touching a mountain, the "cloud" is still referred to as fog. Driving from below cloud base up a mountain road and into fog and then later above it can be a really exhilarating experience, as one experiences three radically different types of the appearance of the sky over a relatively short distance. In Chapter 1 I recalled a wonderful trip to the Canary Islands I took many years ago from well below cloud base, at sea level, up the slope of a volcano, watching the vegetation change in character as we entered a layer of fog, and then exited the fog above where it was sunny, bright, crystal clear, and much drier. We'll return to fog later after a brief diversion into a look at stratus from both above (Figs. 2.1 and 2.2) and below (Fig. 2.3).

While the undersides of stratiform clouds having low bases are usually flat, on rare occasions they sometimes exhibit pronounced undulations. Only as relatively recently as 2017, clouds with this feature have been given the special "supplementary"-feature designator *asperitas* (Fig. 2.4) (Harrison et al. 2017). The wavelike features look may look chaotic and are likely a result of gravity-wave motion in vertically trapped waves (see Section 2.2.3). However, most of the photographs of *asperitas* clouds that I have seen are not near any mountains, so if they were associated with gravity waves, they must not have been triggered by orography, but possibly as relatively warm, moist air has been forced up and over a shallow cold dome of air behind a front. Figure 2.5 shows what looks like *asperitas* clouds breaking up. Clouds with yarn-like undulations similar to those in *asperitas* clouds, but without a solid base, are seen in Figure 2.6.

Stratus and ground fog can form under a wide variety of atmospheric conditions. First, let's consider ground fog (Fig. 2.7). Fog adds a real softening quality to whatever's in the background owing to the scattering of light by small cloud droplets, but only when the fog is not so thick that the visibility actually goes to zero and there *is* no background at all to soften. Blowing snow and smoke can look like fog (Fig. 2.8). Fog, when viewed from just below the cloud base can look striking when sunlight illuminates the near ground and/or the far ground (Fig. 2.9), The most stunning images of fog, in my opinion, occur when there is a sharp contrast between the fog and clear, sunny, blue skies adjacent to the fog (Figs. 2.10 and 2.11) or when one can see both the top and bottom of the fog layer at once, when viewed from the side (Fig. 2.12), so that it is like seeing the floor of a house with a side wall having been cut out. When both the top and bottom of a fog layer can be seen in a cross section in the vertical, it is reminiscent of Judy Collins' 1967 song, written and also sung by Joni Mitchell a year earlier in 1966, about seeing clouds from both sides (now). Sometimes wisps of fog are sucked upward by air currents above (Fig. 2.13).

Radiation fog occurs when the winds are weak or non-existent (so that there is no turbulent, vertical mixing of cold, dense air at the ground with warmer, lighter air above the ground) and the ground cools to the dew-point (at constant pressure); the air becomes

Figure 2.1 *Patches of stratus. (top) Airborne view of "broken-sky" coverage of stratus, San Francisco Bay area, July 12, 2019. (bottom) Airborne view of fuzzy-looking stratus (and probably stratocumulus or altocumulus above) over southeastern Japan on March 4, 2016. Compare this panel with the top panel.*

Photos courtesy of author.

Figure 2.2 *The edge of stratus as viewed from above. (top) Airborne view of stratus to the north and its southern edge, over southern Louisiana, on Jan. 6, 2000. Its brightness when viewed from above is contrasted with the dull, gray which would be seen by an observer below. It looks as if a large sheet of white plastic has been cut out from a larger, parent cloud sheet. (bottom) Airborne view of a sheet of stratus very near the ground or fog, whose edge is illuminated by the sun near sunset, over the snow-covered surface of eastern Nebraska or Iowa, on Jan. 26, 2023.*

Photos courtesy of author.

Figure 2.3 *Stratus with undulations on the underside, in Norman, Oklahoma, on Jan. 27, 2014.* Photos courtesy of author.

saturated, but not just at the ground (if only the ground reached the dew-point temperature, then there would just be dew or frost on the ground and no fog). Radiation fog is more apt to form in a valley or other low-lying area (Figs. 2.14 and 2.15), a depression where cold, dense air tends to drain at night off the mountain slopes and where the air is protected from strong winds higher up which would tend to dilute the effects of cooling (as warmer air would be mixed with it). Radiation fog occurs when the sky is clear, or else clouds would absorb and re-radiate longwave radiation back downward, keeping the air just above the ground from cooling enough.

To me radiation fog is most impressive when treetops or structures such as utility poles, water towers, and buildings protrude upward above the top of the thin fog layer (Fig. 2.16) or at an extreme, when shallow ground fog can be so thin that one can stand in a field and have fog envelop one's feet, while having a view above the fog or the cloud base (the "ceiling") is so low that one can stick one's head into the cloud and look down and have trouble seeing one's feet (Fig. 2.17). This setup is beautifully depicted by photographs/drawings by Hugh Kretschmer, a photographer from California, that I saw in an advertisement in a magazine, which captured my imagination.[1] On many occasions I have run to fields early in the morning to catch shallow ground fog before the sun gets high enough to warm the air and dissipate the fog. At my late father's home in the

[1] I found one entitled "Heaven and Earth" online, which is similar to the one I saw in the advertisement, at https://www.hughkretschmer.net/STOCK/72/caption. This image, however, has the subject of the photograph with his head completely in the cloud.

Figure 2.4 Asperitas *clouds. (top) The bottom of a stratus deck with* asperitas, *Norman, Oklahoma, on March 13, 1982, well before such a cloud was recognized as being of a special type. The parallel wavy lines in the cloud base appear unworldly (or rather, unearthly). (bottom)* Asperitas *clouds over west Texas, north of Lubbock, on May 17, 2021.*

Photos courtesy of author.

Figure 2.5 *The undulating underside of a layer stratus breaking up, over Norman, Oklahoma, on Nov. 7, 2018. These clouds bear some similarity to the* asperitas *clouds seen in Figure 2.5, but with breaks showing clear sky.*
Photo courtesy of author.

western portion of Ft. Lauderdale, Florida, decades ago before the area became heavily developed, it was common to find shallow ground fog over grassy surfaces just before sunrise. As buildings were constructed and fields gave way to pavement, it became much more difficult to find early morning ground fog. Conduction in addition to radiation may play a role in losing heat at the ground. On very rare occasions the base of the fog was almost at the ground (Fig. 2.18), so the obscuration was just above ground level.

Sometimes fog will creep up a mountain as the slopes of the mountain are heated by the sun; in this case, what begins as fog will evolve into an ordinary, overcast cloud deck. Someone higher up, initially above the fog, will experience sunny skies and an undercast, followed by fog, and then followed either by the breakup of the fog or the migration of the fog to a higher elevation, and an overcast sky. I have often tried to climb above a fog layer in Boulder, Colorado, to take photographs, only to be thwarted by the upslope creep of the fog (Fig. 2.19) to my rear.

Figure 2.6 *Cloud elements with wavelike structures, but without a solid base. (top, left) Tangled-yarn-like stratus clouds with undulations, on Oct. 13, 2013, in Norman, Oklahoma. Some cirrus clouds are visible above on the left of the image. These yarn-like clouds bear some similarity to the* asperitas *clouds seen in Figure 2.4, but without the stratus overcast. (top, right) Patterns of concentrated smoke from a forest fire, viewed from Boulder, Colorado, on July 1, 2007. The filaments of smoke, while wavy, are more irregular than the waves seen in the* asperitas *in Figure 2.4. It is not certain, however, that these clouds are at low altitude because more recently, in 2021. I observed clouds similar to these, on another day, in Boulder, that clearly were accompanied by some scattered clouds at much lower altitude. (bottom, left) Wavelike filaments of smoke from a forest fire, viewed at sunset, from Boulder, Colorado, on Oct. 16, 2020. (bottom, right) Orographic wave clouds that look like tangled yarn, over Santa Fe ski area, New Mexico, on Dec. 19, 1996. Note the similarity of this cloud to the smoke patterns seen in the preceding two panels. It is likely that this cloud is glaciated because ice crystals caught in downdrafts persist longer than water droplets would. Smoke in the bottom, left panel has a similar appearance because smoke also persists in downdrafts.*
Photos courtesy of author.

What draws the fog up the side of the mountain? Am I the Pied Piper of fog? As the ground is heated, the air becomes warmer than that of its environment and positively buoyant. Imagine a paddlewheel whose axis of rotation is parallel to the surface of the mountain slope, aligned along a surface of constant elevation: Air will be forced upward along the ground, but not at all, some distance away from the mountain at the same

Figure 2.7 *Dense ground fog. (top, left) A roamin' in the gloamin' in fog in Norman, Oklahoma, on Dec. 2, 2013. This thick fog makes one feel the gloom of being totally immersed in the thicket of clouds, save for the penetrating headlights of an oncoming, unidentifiable vehicle, and a tree. (top, right) Thick fog just north of an outflow boundary/warm front in western Kansas, on May 24, 2021. (bottom, left) Sun breaking through early morning ground fog in Norman, Oklahoma, on Oct. 16, 2007. (bottom, right) Cable car to nowhere. In fog, Hua Mountains, near X'ian, China, on July 19, 2019. This cable car is about to disappear as it descends into the fog.*
Photos courtesy of author.

altitude. A result of this difference in upward acceleration of the air is that the paddle-wheel will rotate about a horizontal axis as in Figure 2.20. The net result of the sum of all the horizontal roll circulations is that air will be forced upslope along the ground and in the downslope direction away from the mountain. So, air will creep upslope and drag the fog up the mountain. Another example of fog that is forced up in elevation (but not illustrated here by a sequence of photographs) is seen in Figure 2.21.

Advection fog occurs when moist air is cooled to saturation as it flows (the properties of the moist air are carried along or "advected") over a cold surface. Perhaps the best-known example of advection fog in the US is that which occurs along portions of the

Figure 2.8 *Blowing snow. (top) A doppelgänger of fog in the form of blowing snow: Blizzard conditions I experienced while cross-country skiing around Long Lake, in the Indian Peaks Wilderness, west of Ward, Colorado, on Dec. 23, 2005. (middle) As in the top panel, but for ground fog in the incarnation of a ground blizzard. In this case, the visibility is lowered locally, only right near the ground. (bottom) Blowing snow looking like steam emanating from the top of a snow drift near Brainard Lake, in the Indian Peaks Wilderness, west of Ward, Colorado, on March 18, 2022.*

Photos courtesy of author.

Figure 2.9 *Bottom edge of fog/low cloud deck. (top, left) Fog over the Indian Peaks, Colorado, on Aug. 16, 2005 casting a gloom over the tundra, as viewed, by contrast, from a location bathed in sunshine. This view was toward the west, from the relatively flat terrain below Pawnee Pass. (top, right) Stratus hitting the Foothills of Boulder, Colorado, during the early morning of July 21, 2020, giving the impression of a low ceiling. (bottom) Low ceiling of stratus seen from Niwot Ridge in the Indian Peaks Wilderness, on June 29, 2021. In this photograph, the underside of the cloud is illuminated, not in shadows.*

Photos courtesy of author.

Northern California coast during the summer (Figs. 2.22–2.24), when relatively warm, moist Pacific air is cooled as it is advected over coastal waters, where cooler ocean water from below has been sucked up to the surface or "upwelled." Bridges across the San Francisco Bay area are great foils for the fog, as the tops stick out above or penetrate into and disappear in the fog layer (Fig. 2.25, top panel). On occasion, the fog may be so shallow that it can be seen flowing underneath a bridge (Fig. 2.25, bottom panel). By definition, advection fog requires some wind at the surface, while radiation fog is inhibited by wind, which promotes mixing of the saturated cloud air with drier air from above the fog layer. Warmer, moist air, when advected over a snow surface or hail covering can also be cooled to saturation, resulting in fog (Figs. 2.26 and 2.27). Advection fog is also common off the coast of New England during the summer when warm, moist air from the Gulf Stream to the south is advected over the relatively cool ocean offshore, and up near Labrador. Cape Cod (Fig. 2.28) and the islands (for example, Nantucket

Figure 2.10 *Fog, Hua Mountains, near X'ian, China, July 19, 2019. A seemingly endless stream of tourists is seen on foot ascending from and descending into the surrealistic, ghostly cloud layer. In this case, small patches of clear blue sky and the sun illuminating the fog deck (upper right) are seen in contrast to the gloom below (center). This fog probably represents the normal cloud deck that would be present even if there were no mountains, but the cloud is being forced to flow up the slope of the mountains on the right.*

Photo courtesy of author.

Figure 2.11 *A wall of fog looking like a giant tsunami of cool, moist air, gently sliding upslope, about to consume Boulder, Colorado, during the morning of July 17, 2004. Note the stark contrast between the darkness below and light from above. View is to the east or southeast.*

and Martha's Vineyard), are particularly pretty and comfortable during the fog episodes while hot and muggy conditions prevail over the mainland.

When I was in graduate school, I recall one occasion during the evening in winter, when based on a series of surface weather maps it appeared as if a warm front would soon pass by. My weather-enthusiast student comrades and I on the 16th floor of our building on the MIT campus in Cambridge, the Green Building, took the elevator down to the ground floor. Outside it was foggy and the temperature was around 20°F. As the warm front approached, we headed for the roof of our building, which is more than 18 stories tall. The air was much warmer than it was at the ground level, having risen into the 50s F. Since the warm front was sloped with height toward the colder air to the north, the top of the fog layer, which represented the frontal boundary, became lower, eventually revealing the tops of the buildings across the Charles River, in Boston, but not at the surface. The level of the top of the fog slowly descended and eventually the fog completely dissipated below as the warmer air engulfed us even down to the ground. What we did in effect was a vertical, "fog and reverse-fog chase," without ever having to travel horizontally away from our location.

Cloud fragments (scud) may also occur when precipitation falls from a cloud into an unsaturated layer of air below, evaporating and cooling it and possibly saturating it.

Figure 2.12 *Fog/shallow cloud deck hitting mountains. (top, left) Wall of clouds hitting the Continental Divide, obscuring the summit of Mt. Neva (12, 821 feet, 3, 908 m MSL, above mean sea level) near Nederland, CO, as fog, on July 1, 2004. Lake Dorothy, which is still partially frozen, is seen below on*

More often than not, just patches of "scud"-like cloud fragments appear rather than fog (Fig. 2.29), especially when a cool, moist air mass mixes with a warmer, dry air mass.

Steam fog can occur when two air masses, each of which is unsaturated, mix with each other (at constant pressure) and become saturated, to form fog.[2] Why this occurs is a consequence of the how the saturation properties of air vary with temperature (the saturation vapor pressure increases nonlinearly with height, exhibiting a convex shape, whereas turbulent mixing results in a linear variation of moisture content) (Fig. 2.30). Steam fog typically occurs when very cold air is "advected" (carried) over a much warmer body of water and water evaporates from the water into the air above it. The formation of steam fog differs from that of advection fog: for the latter, warm air flows (is advected) over a cold surface, while for the former the reverse is true, as cold air is advected over a warm surface. Steam fog is sometimes called *arctic sea smoke*, in recognition of the situation when very cold arctic air flows out over the much warmer Atlantic Ocean during the winter. When I was a graduate student, I had a standing invitation at my advisor's (Prof. Fred Sanders') home in Marblehead Neck, Massachusetts, early on any bitterly cold (temperature below 0°F) morning, to photograph arctic sea smoke and be treated to breakfast. Unfortunately, as a young graduate student who worked very late into the night, it was too difficult to get roused up early enough to make the trek out to the coast in time, even with the added incentive of a free breakfast of hot groats! Steam fog is also often observed over Lake Michigan, east of Chicago, during extreme cold waves. I have even photographed it in Norman, Oklahoma over a lake during a cold wave, when the air temperature was just below 0°F and the lake temperature

---

Figure 2.12 (Continued) *the left. Both cloud base and the top of the cloud deck are visible. It is summer, but ice and snow still abound at this elevation. (top, right) Fog from low-level upslope behind a cold front about to retreat from striking up against the Flatirons in Chautauqua Park, Boulder, Colorado, on March 2, 2003. The Flatirons are blocking the fog from going any farther to the west. (second row, left) Stratus clouds hiding the tops of the Foothills in Boulder, Colorado on March 30, 2021, as air is forced upslope after the retreat of a storm. The Mesa Lab of the National Center for Atmospheric Research (NCAR) is seen at the bottom, center. (second row, right) On Feb. 25, 2021, when the cloud deck has passed by to the west and is illuminated by the sun from the east, while the top of the Foothills are hidden by the cloud deck. (third row, left) Morning fog has lifted, but low clouds still obscure the Foothills in Boulder, Colorado, on June 1, 2016. The jogger seems to be heading toward the clouds. (third row, right) More than one layer of stratus against the Foothills in Boulder, Colorado, on Jan. 9, 2022. (bottom row). Low clouds/fog in Boulder, Colorado, on July 7, 2019, silhouetted against the Foothills. Both the top and bottom of the cloud are visible on the edge.*

Photos courtesy of author.

---

[2] For example, if at 1000 hPa (near the ground at sea level) we mix a parcel of very cold air at around −5°F (−20.5° C) which is unsaturated, having a mixing ratio (grams of water vapor/grams of dry air) of just 0.4 g kg^{-1}, equally with a parcel of much warmer air at around 60°F (15.5°C) which is also unsaturated and having been advected over a warm water surface and having a mixing ratio of around 9 g kg^{-1}, we get a parcel of air having a temperature of about −2.5° C and a mixing ratio of about 4.7 g kg^{-1}. This mixing ratio is greater than the saturation mixing ratio at 1, 000 hPa and −2.5°C of 3.2 g kg^{-1}.

Figure 2.13 *Strands of cloud extending upward from stratus/fog. (top, left) Stratus obscuring the tops of the Foothills, Boulder, Colorado, on Feb. 18, 2021. It appears as if a portion of the stratus cloud (center, top) is being sucked upward as a narrow column or tube of cloud material. (top, right) A series of upward bulges streaming upward in low clouds hitting the Foothills of Boulder, Colorado, on Jan. 9, 2022. (bottom) Low clouds obscuring the Foothills of Boulder, Colorado, on Jan. 9, 2022. NCAR is visible on the lower left. Two strands of clouds appear to be sucked upward and then curling in the direction of the wind.*
Photos courtesy of author.

was probably at least 55–60°F (Fig. 2.31): I have found empirically[3] that in Oklahoma and in Massachusetts, a difference in temperature between the water surface and the air of at least 40–50° F (> 20°C) is required for the formation of arctic sea smoke. However, it has been reported in the scientific literature that the threshold can be lower or higher, depending on the salinity of the water and the windspeed. Because the temperature difference between the air just above the water surface and the air aloft is rather large, the air is unstable and rapid turbulent mixing of the air occurs, making the fog edge change shape rapidly. Because very cold air with wind is required for steam fog, my attempts to photograph it have been physically challenging. First, as I previously noted, one must get up early to experience the minimum temperature before the rising sun has begun to heat the surface and second, it is cold and windy. Trying to avoid

---

[3] But with much too small a sample size.

Figure 2.14 *Ground fog. (top, left) Early morning radiation ground fog nestled in the valley below, in St. Adèle, Quebec, Canada, on Oct. 6, 2017, as viewed from the top of a nearby (closed) ski area, with some colorful, fall foliage appearing below. (top, right) Airborne view of radiation ground fog nestled in a lush, green valley in Germany, on Sept. 16, 2017. (bottom) Airborne view of radiation ground fog hitting higher terrain over the mountain west, somewhere between Seattle, Washington, and Denver, Colorado, on Jan. 27, 2011.*
Photos courtesy of author.

frostbite must therefore be a serious consideration and wearing gloves is essential . . . but gloves make it difficult to operate a camera. Owing to the lack of stability, steam fog perhaps does not really belong in this chapter, which is concerned with stratiform, stable-looking clouds. However, I felt that it is more logical to include it with all the other types of fog. As noted in Chapter 1, it is extremely difficult to divide sections of this book up just by the mechanism of cloud formation alone or just by appearance alone.

"Steaming fog," not to be confused with steam fog, can form on hard, wet surfaces when the air is relatively humid, and the sun suddenly comes out and heats the surface, such that some of the water is evaporated into the air; in doing so, the vapor pressure of

Figure 2.15 *Early morning radiation fog over the land and "steam" fog (see Fig. 2.38) over a lake, which lies at a local minimum in elevation, during a colorful autumn, at Val Morin, Quebec, Canada, on Oct. 13, 2013. In this case the trees are mostly above the lake fog, which is very shallow at the edge of the lake, but not above the land fog.*
Photo courtesy of author.

the air may increase enough that the air becomes saturated. I have seen and either photographed or taken videos of steaming fog on the top of fences at our house in Norman when the sun first comes out after a rain event and warms it, off the roof of our residence in Boulder after a heavy rain and the sun has just come out, and also on clear, wet roads at high elevation in Colorado, when the sun hits the road and warms it. In the latter case, I can attest to what a serious driving hazard it is, especially when the visibility suddenly goes to zero when you unexpectedly drive through it. I have found it very difficult to photograph steaming fog because the contrast between it and its background is usually weak.

Regardless of the physical mechanism responsible for fog, its appearance near mountains is often breathtaking. The sudden appearance of mountain tops or hills as one rises above the fog below can be stunning (Figs. 2.32–2.35 and 2.37). I have many times

Figure 2.16 *Poking through top of ground fog. (top, left) Two lonely water towers just barely sticking up above shallow ground fog, like a periscope in a submarine peering above the water level, in Norman, Oklahoma, on the morning of Jan. 29, 1992, viewed to the south from the Sarkeys Energy Center building on the University of Oklahoma campus in Norman. (top, right) Wind turbines and an instrumented tower poking out above the upslope fog, south of Boulder, Colorado, viewed from the Shanahan Trails in the Foothills of Boulder, Colorado, on Jan. 13, 2015. (bottom) Aerial view of a smoke stack and a few towers sticking out above ground fog somewhere in Germany, on Sept. 16, 2017.* Photos courtesy of author.

driven up through the cloud layer associated with a shallow cold front in Colorado to encounter a clear, bright and sunny sky and windy conditions above, especially when there is brilliant snow cover. The legendary mists and undercasts swirling around some of the steep mountains in China, which have been beautifully represented in Chinese paintings, or in the Alps (Fig. 2.35, bottom panel), are sights to behold.

Just as fog banks can be seen along coastlines (Figs. 2.22 and 2.23), they can also be seen inland, creeping up mountains when shallow cold air masses behind cold fronts are forced up to higher elevation (Fig. 2.36), driven by the difference in density between the dense, cold air below and less-dense, warmer air above (to be discussed in a later chapter). This mechanism for driving clouds up mountainous terrain is different from the mechanism by which clouds ascend mountainous terrain due to the heating of the sloped terrain (Figs. 2.19 and 2.20),

Figure 2.17 *Very shallow ground fog. (top, left) Very shallow morning, radiation ground fog in Norman, OK, on Oct. 7, 2014, in which the top of the fog layer is diffuse. (top, right) Early morning radiation ground fog in Norman, Oklahoma, on Oct. 11, 2011. (bottom) Morning radiation ground fog in Norman, Oklahoma, on Oct. 7, 2014, with the sun affecting sharp contrast at the top of the fog layer.* Photos courtesy of author.

2.1.2 Sheets of stratiform clouds at high altitudes: Altostratus and cirrostratus, and anvils from convective clouds

Truly stratiform, stable clouds are stratus clouds at midlevels, *altostratus*, and can be said with tongue in cheek, to consist of "fifty shades of gray," which is the title of a recent romance novel and movie. The distinction between altostratus and cirrostratus can sometimes be "nebulous." Both look like sheets of gray: the distinction is in the height of the cloud, which is not always apparent from visual observations from the ground, and at high levels, *cirrostratus*. Altostratus and cirrostratus may form when air is lifted gently over a broad area, typically by synoptic-scale disturbances, that is, those thousands of kilometers across and persisting for days, and with vertical velocities on the order of cm s^{-1}. Cirrostratus (Fig. 2.38), which are mostly at high enough altitudes that it is cold enough for them to have ice crystals, are sometimes shed from lower-level clouds that originated in regions of rising air and grew into precipitating clouds

Figure 2.18 *A very low cloud base (or "ceiling"): The bottom of a morning, radiation fog layer, Jan. 2, 1982, Ft. Lauderdale, Florida, that is slowly rising.*
Photo courtesy of author.

(nimbostratus) at higher altitudes or in regions of evaporating clouds at lower altitudes, leaving only high clouds behind. At sunrise and sunset, they can be very colorful. Some cirrostratus may be more localized as bands, some even having upper portions that may exhibit evidence of upward buoyancy (Fig. 2.39) No pictures of nimbostratus are shown in this book because they are usually at least partially obscured by the precipitation and are certainly not among the "greatest clouds on Earth."

Cirrostratus are sometimes referred to as "forerunners," since they sometimes warn us of impending precipitating, large-scale weather systems (see Chapter 5 for more on where forerunners come from). In addition, as we will discuss in a later chapter, deep convective clouds from storms produce anvils, which are composed of cirrostratus, and are also "forerunners" of sorts, but of a more-local phenomenon (a convective storm).

Smoke layers aloft sometimes look like cirrostratus (Fig. 2.40).

When cirrostratus clouds are relatively thin so that the sun is not completely blocked out, beautiful optical effects such as sun dogs (Fig. 2.41), halos (Fig. 2.42), and glories (Fig. 2.43) (Greenler 2020) may be seen.

Figure 2.19 *Drowning in a sea of fog as it creeps uphill. A sea of fog undercast, below, over Boulder, from the road up to Flagstaff Mountain, on Dec. 30, 2017; viewed to the south. The fog represents very cold, shallow, saturated arctic air behind a cold front. Time increases in each row to the right, and then down to the next row, and then to the right again. My separate time-lapse video of this event shows the cloud tops rising and falling like waves in the ocean, while the overall cloud top is slowly rising, like an incoming tide.*
Photos courtesy of author.

Cirrostratus have played a role in folk wisdom. There is an old saying: "red sky at night sailors delight; red sky in the morning sailors take warning." Similar words appear in the New Testament, in which Jesus says (in Matthew XVI: 2–3), "When in evening, ye say, it will be fair weather: For the sky is red. And in the morning, it will be foul weather today: for the sky is red and lowering." In midlatitudes, weather systems tend to propagate[4] from west to east. When a precipitating weather system is located to the *west*, then the cirrostratus emanating from it will appear in the western sky *in advance* of the precipitating weather system and it should be clear to the east. If this happens to be at

---

[4] Weather systems don't actually *move*; they "propagate." Air is continuously moving into and out from weather systems. The apparent motion of weather systems (for example, cyclones) is not the same as the motion of the air that makes up weather systems, which are curved paths around the cyclone. The center of cyclones appears to move continuously to the east (propagates). The winds at the center of the cyclone are actually calm with respect to the cyclone.

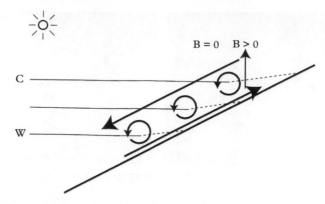

Figure 2.20 *Illustration of how air begins to spin about a horizontal axis along the slope of a mountain as the surface is heated by the sun's radiation. This vertical cross-section shows the slope of the ground by a thick, solid black line, which in this case slopes upward to the right; it is morning and the sun is rising on the left. Isotherms (lines of constant temperature[a]) are represented by the thin black lines away from the ground and by thin dashed lines near the ground where heat has been transferred from the ground to the air above. The temperature decreases with height (W—warm at low altitudes, C—cold at higher altitudes). Near the surface of the sloped ground, it is warmer (dashed-line area) than it is farther to the left (solid-line area). As a result, the air near the ground is warmer than its surroundings (to the left) at the same height, so it is buoyant (B > 0) and air is forced upward (upward arrow). (Buoyancy will be discussed in more detail in Chapter 4).[b] The air is therefore forced up near the ground, but not at all off to the left (where B = 0). If a paddlewheel were placed in the flow, it would rotate in the sense of the curved lines with the arrowheads. All the forced circulations add up to air flowing upslope near the ground and downslope above the ground (the rising and sinking branches from neighboring vertical circulations cancel each other out), as indicated by the long arrows. In this way, air is forced upslope at the surface.*

[a] In this figure, isotherms are shown, so that since the temperature decreases with height, the isotherms are raised near the mountain slope, indicating that is locally warmer than the environment. Meteorologists usually show potential temperature in a figure like this one, so that the lines of constant potential temperature lean downward, because potential temperature usually *increases* with height. Potential temperature is the temperature a parcel of dry air would have if brought down to 1000 hPa, near the surface, without any exchange of heat with its environment.

[b] Although the clouds discussed in this chapter are not buoyant, the airflow that generates some of them may be affected by buoyancy in unsaturated. In this case, buoyancy is important for considering how the airflow is affected by a mountain, but not for the dynamics of the cloud itself.

Figure courtesy of author.

sunrise, the clouds to the west will appear to have a reddish hue for a short time then: blue light is predominantly scattered away (filtered out) as the sunlight penetrates through a relatively thick slice of the atmosphere, leaving mainly the red-tinted light. Look out: There could be some adverse weather later. On the other hand, when a precipitating weather system is located to the *east*, then the back edge of the cirrostratus emanating from it will appear *after* the system has moved away. If this happens at sunset, it should be clear to the west, and the eastern sky will appear to have a reddish hue for the same reason the western sky appears to have a reddish hue at sunrise. This folk wisdom doesn't

Figure 2.21 *Fog near the summit of a mountain near Rotorua, New Zealand, on Feb. 13, 1984. The contrast between the clouds below illuminated by the sun and the rocky surface is almost blinding. The very ragged appearance of the cloud tops is indicative of vigorous vertical mixing.*
Photo courtesy of author.

work, however, if clouds completely block the sun or if the clouds aren't moving, just to note a few of a number of possible exceptions.

Cirrostratus, however, may originate very far away and precipitation may *not* be likely any time soon. I recall being amazed, when I was a beginning graduate student in Cambridge, Massachusetts, that after seeing cirrostratus overspread the New England sky, I looked at a weather satellite image (cloud images were just becoming available then and were quite a novelty in those early years) and found that the clouds seemed to have originated from a weather system thousands of miles away in the tropical Pacific. The ice crystals I surmised must have made it all the way, carried along by a jet stream, without producing any precipitation or disappearing by sublimating (turning back into water vapor directly, skipping the liquid phase entirely). It is more likely, however, that the ice crystals were reforming continuously at the upstream edge of the cloud in a region of rising motion and/or convective storms, moving along with the wind, and then sublimating at the downstream edge, in a region of sinking motion; the back and front edges of the cloud probably progressed farther and farther downstream with time. In recent years these long bands of clouds, since they often harbor copious amounts

Figure 2.22 *Sea, advection fog in a moist, marine layer of air, making landfall along the coast highway in California, south of Monterey, on Aug. 7, 1981. A view from near the top of the fog layer.* Photo courtesy of author.

of moisture at different altitudes (not just at high altitude), have been associated with "atmospheric rivers," which themselves tend to be located at middle and lower altitudes than the cirrostratus visible in satellite imagery.

Before we had access to the internet and cell phones, and hence real-time weather-radar observations and satellite cloud observations, we had to rely, while storm chasing, on the appearance of convective-storm anvils on the horizon as "forerunners" of convective storms. I (and many others) still use this low-tech technique to locate possible storms on the horizon before I check my cell phone for radar and satellite imagery. When driving, this technique is surely the safest one.

2.2   Orographic clouds: Clouds created by the unique airflow characteristics near mountains

Imagine that you could do an experiment in which you forced air to flow over a mountain range and view the sky's response from different vantage points, at different times of the day and year. You might expect there would be substantial variations in the shape and

Figure 2.23 *Marine fog banks offshore, along the California coast. (top) Offshore fog bank (compare to bottom panel) viewed head on at sea level, from Pacific Grove, California, on Oct. 28, 2015, and casting a shadowy gloom over the ocean. (bottom) Fog offshore south of the Monterey peninsula in California, on Oct. 23, 2006. Unlike the fog seen in the top panel, this fog is directly illuminated by the sun.*

Photos courtesy of author.

Figure 2.24 *Fog bank receding from San Francisco airport on July 12, 2019 as the ground is heated by the sun. This fog is totally cloaking a view of the hills to the east.*
Photo courtesy of author.

size of clouds that formed and in how the clouds looked to observers on the ground with different views of the clouds. You might also wonder what it might feel like to fly through the clouds from various angles and heights above the ground: Would the ride be smooth or bumpy? Would you feel the clouds' presence just before entering them, just afterward, or only while in them? What can one say about the weather from the appearance of the clouds?

Some of these thought experiments could be accomplished by traveling to a mountainous region, camping out, and monitoring the sky continuously during the daylight hours for many months. These experiments would also require simultaneously recording measurements of wind direction, wind speed, temperature, and humidity, all which are readily obtainable in the US from the National Weather Service, but only at certain widely spaced locations and only twice a day. For most of us, this kind of effort is not a practical exercise. On the other hand, if one lives in or frequently visits a mountainous area over for many years as I have, one might be able to piece together such an experiment. As one alternative to camping out for a few years, I will illustrate some of the variety of clouds one would see while on that hypothetical camping trip. My

Figure 2.25 *Fog above and below bridges near San Francisco. (top) Golden Gate bridge in low clouds and fog, San Francisco, on Aug. 12, 1981. This iconic scene is almost banal, having been seen and reproduced in magazines and travel brochures so often. Yet I still find it striking, owing to the majesty of the bridge, the mood induced by the fog, and the contrast between the crisply defined features of the bottom of the bridge and the fuzzy cloud at the top. (bottom) Ground fog flowing underneath the Oakland Bay Bridge on Dec. 29, 1998, as viewed from San Francisco.*
Photos courtesy of author.

Figure 2.26 *Airborne view of low clouds/advection fog over glaciers in southern Greenland, Sept. 7, 2019. It may be difficult to tell the difference between the clouds and fog: To locate the clouds, look for the diffuse boundary between the clouds and the exposed snow-covered peaks.*
Photo courtesy of author.

particular perspective is mainly from Boulder, Colorado, situated just east of the Continental Divide of the Rocky Mountains, which is oriented approximately in a north-south direction; the elevation increases to the west by about 7, 000 feet (~2, 100 m) in only about 20 miles (32 km). However, I will also show that similar clouds can also be seen in mountainous areas elsewhere and even on rarer occasions in non-mountainous areas.

The nature of clouds that form near a mountain is affected by the nature of the airflow near the mountain: In particular, we ask the following questions: Does air flow up and over the mountain or is it forced to divert around the mountain or does it flow partially up and around the sides of the mountain? In each case, how much is air lifted? When will it be lifted enough for a cloud to form? We therefore must consider what types of air motions are possible near a mountain.

Very smooth-looking clouds that appear to have been sanded down and buffed to perfection sometimes appear above and/or downstream (in the direction toward which the wind is blowing) from mountains. Clear (cloudless) air is brought to the cloud's location, where it suddenly turns into a visible cloud, and then exits the cloud into a

Figure 2.27 *Fog on top of snow and hail. (top) Advection ground fog created when moist air is cooled while flowing over colder, melting snow, Massachusetts Institute of Technology, Killian Court (formerly called the Great Court), Cambridge, Massachusetts, Jan. 27, 1976. (bottom) Hail fog in Boulder, Colorado, on July 30, 1997. The air near the ground is cooled to saturation by the recent hail fall as warmer, moist air flows over it.*

Photos courtesy of author.

Figure 2.28 *Advection fog at Chatham, Massachusetts, on Aug. 6, 2001, as relatively humid air from the south is cooled to saturation as it flows over the relatively cool ocean. In this case the ocean is cool, north of the Gulf Stream, for reasons other than upwelling, as off the Pacific coast. Note how soft the lighthouse and the dwelling and flag appear: The fog seems to have a softening effect, filtering out sharp outlines.*
Photo courtesy of author.

clear sky.[5] The cloud resembles a living organism, taking in air, using it to produce a beautiful sky sculpture, and then expelling the air after it has been used keep the cloud alive. But not all orographically produced clouds appear to be stationary; some actually do progress with the wind or rapidly change shape.

2.2.1   Airflow near mountains: Unblocked flow, blocked flow, and gravity-wave generation and propagation

Airflow near mountains can take a number of characteristics depending on the geometry of the mountains and the vertical profile of winds, temperature, and moisture in the vicinity of the mountains. The geometry of the mountains is described by the height of

---

[5] I and others have taken many time-lapse videos and movies exhibiting this phenomenon.

Figure 2.29 *A wraithlike web of scud clouds forming and moving upslope over Pawnee Peak (3, 945 m, 12, 943 ft MSL) in the Indian Peaks Wilderness, on July 26, 1994, after a thunderstorm. The turbulent mixing of two air masses is probably responsible for the highly intermittent appearance of the cloud fragments. The intermittent appearance was likely a result of sporadic bursts of air forced up the mountain.*
Photo courtesy of author.

the crests above the surrounding terrain, the slope of the mountain sides, and how neighboring peaks are aligned with respect to each other (e.g., if is there a one-dimensional, mountain range, an isolated peak, or an area of individual peaks, all closely connected by saddle points between neighboring peaks, etc.). Let's ignore moisture for the time being to keep the discussion as simple as possible (e.g., neglecting the effects of condensation, evaporation, and sublimation) and explore only the most fundamental processes that affect airflow.

Suppose there is a rapid decrease in temperature with height, air is driven by the horizontal wind up a mountain, and the air happens to cool at precisely the same rate at which the temperature decreases with height. Then the temperature of the air parcel is exactly the same as that of the air in the environment (away from the mountain), where there is no lifting going on: There is therefore no buoyancy (to be discussed in detail in Chapter 3) whatsoever. The atmosphere is "neutral," so that any air being lifted on

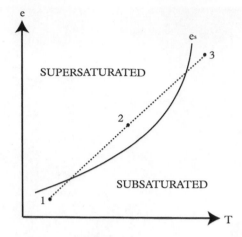

Figure 2.30 *Illustration of how air can become saturated by mixing two different* unsaturated *parcels of air together* without *any change in pressure. Variation of saturation vapor pressure (e$_s$) with temperature (T) (curved, solid line) and the profile of the temperature of a subsaturated air parcel at temperature T$_1$ and vapor pressure e$_1$ (point 1) that has been mixed with another subsaturated air parcel at temperature T$_3$ and vapor pressure e$_3$ (point 3), as a function of vapor pressure (e) (straight, linear dashed line), for various proportions of each air mass. Note the region where the vapor pressure is greater than the saturation vapor pressure, inside the concave curve (e.g., at point 2).*
Figure courtesy of author.

the windward side keeps ascending the mountain to its peak, after which it descends on the lee side. There is no force retarding or speeding up the air. As it descends, it warms at exactly the same rate at which the temperature in the environment increases in the downward direction. A little nudge will force air up and over the mountain and the air parcel always has the same temperature as that of its environment. (We will ignore for now the slowing down of the air as it rubs against the ground.) It turns out that for "dry" air, air with no moisture, this rate of decrease in temperature with height is about 10°C per kilometer,[6] the "dryadiabatic lapse rate." In this case the airflow is unblocked and the mountains do not have much of an effect (hence, a condition of "neutrality"): Air behaves almost as if the mountains were not there, except, of course, that air doesn't go into the ground and must follow the contours of the mountain.

On the other hand, suppose that the air temperature in the environment does not decrease so rapidly with height. Then air that is forced up a mountain still cools at the rate of 10°C per kilometer; but as the air is forced up the mountain, it finds itself being *cooler* than that of its environment (away from the mountain), so that it has negative buoyancy and there is therefore a resistance to it being forced upward further. It might not make it all the way up to the top if we were to stop forcing it up. The air then is

---

[6] This is a consequence of the thermodynamics of dry air. If the reader is technically inclined, please refer to any textbook on elementary atmospheric science. Otherwise, the reader is asked to just accept this assertion.

Figure 2.31 *Steam fog/arctic sea smoke over Lake Thunderbird, Norman, Oklahoma, on the bitter cold and windy morning of Dec. 22, 1989.*
Photo courtesy of author.

blocked and diverted around the side of the mountain. Said another way, if air is very heavy relative to the overlying air and given an impulsive push up a mountain, there may not be enough kinetic energy for the air parcel to make it all the way to the mountain-top, so it gives up and flows around the mountain. This situation sounds a bit like that of the mythological Greek, Sisyphus, who tries to roll a boulder up a mountain, but the boulder keeps rolling backward. How frustrating. However, even if the air is very stable, i.e., air that is lifted and becomes negatively buoyant, if we give the air enough of a push (if the component of the environmental wind in the direction of the uphill gradient in elevation is strong enough), then it might in fact make it all the way to the top. Kinetic energy of the airflow going up the mountain is converted into potential energy: The potential energy increases because the air gets higher and higher (potential energy is given by the mass of a parcel of air times the acceleration of gravity times the height of the center of gravity above some reference height, in this case the height of the surrounding terrain), but the kinetic energy (the mass of a parcel of air consisting of the mass of the same number of air molecules multiplied by ½ the square of the wind speed) decreases because the air motion (wind speed) decreases.

Now, suppose that we consider what happens if the atmosphere is stable, but the winds are strong enough to make it all the way up and over the mountain-top: The

Figure 2.32 *Low clouds/fog undercasts in mountainous areas. (top, left) Low clouds trapped below to the west, from the summit of Green Mountain, Boulder, Colorado, on Sept. 25, 2005. The Continental Divide is visible to the west, on the horizon to the right, with South and North Arapaho Peaks in the center, right. Cumulus clouds are visible to the left, above, and to the west of the fog layer, over higher terrain. (top, right) Sea-of-clouds undercast over Boulder, Colorado, viewed to the east, from the summit of Green Mountain, also on Sept. 25, 2005, but with a view higher above the stratus than in the top, left panel. I believe I first saw an undercast like this in a photograph of stratus seen from the Olympic Mountains in western Washington, and was immediately struck by the stark contrast between the dark vegetation above the clouds and the bright sea of clouds below. (bottom, left) Undercast of low clouds viewed from Niwot Ridge, in the Indian Peaks Wilderness of Colorado, on July 11, 2022. The view is to the southeast. Green Mountain (left) from which the photograph in the top panels was taken on another day, Green Mt., Bear Peak, and S. Boulder Peak (left to right), part of the Foothills to the Rocky Mountains, are visible poking up above the low clouds. (bottom, right) A sea of stratus being forced up a mountain in the Sangre de Cristo range, near Santa Fe, New Mexico, on March 19, 1999. Note the* fractus *fragment indicative of vertical mixing, as in Colorado in Figures 2.34, 2.35, and 2.36.* Photos courtesy of author.

kinetic energy of the air given by its initial push is sufficient to overcome its loss as it is converted into potential energy. Then, as the air goes over the top of the mountain, air is also displaced upward above its peak.[7] When this happens the air above the mountain-top cools and becomes cooler than its surroundings, where air is not being displaced

---

[7] In the neutral case, air above the mountaintop is also displaced, but it retains the same temperature as that of the environment.

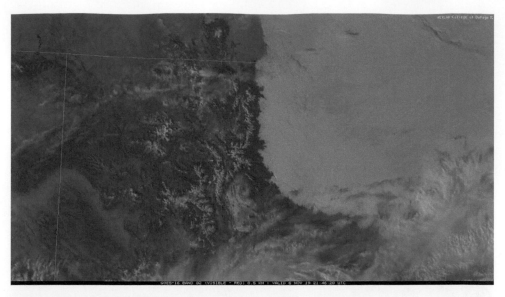

Figure 2.33 *Visible-channel National Oceanic and Atmospheric Administration (NOAA) satellite image of northern Colorado on Nov. 6, 2019, at 21:46 UTC. The snow-capped Continental Divide is seen extending in a north-south direction at center, top part of the image. The white clouds in northeastern Colorado are low clouds and fog, which intersect the higher terrain east of the Continental Divide. The air under the low clouds is relatively cold, in a shallow cold layer behind a cold front. Note all the fine-scale structure in the cloud edge as a result of variations in elevation.*
(Courtesy of College of DuPage Meteorology.)

vertically; in other words, the air above the top of the mountain becomes negatively buoyant. It therefore accelerates downward as it moves to the lee side. As it does so, it warms, overshooting its equilibrium level (where it has the same temperature as that of its environment, zero buoyancy and zero vertical acceleration) eventually becoming warmer than its environment, so it becomes positively buoyant and accelerates back upward, then downward, and so forth, producing a wavelike pattern in its motion (Fig. 2.44) (Federal Aviation Adminsitration 1997). It's the kinetic energy associated with the initial push that allows the air to keep overshooting its equilibrium level both when going upward and downward. Such a wave in the atmosphere is known as an atmospheric *gravity wave* because it is effectively a consequence of Earth's gravity [and not the same as "gravitation waves" highly sought after, and recently found for the first time, by physicists]. When air is lifted within the gravity wave, it might be cooled enough to achieve saturation, and thus produces a cloud; when it sinks, cloud droplets evaporate (or ice crystals sublimate) when it warms above the dew-point temperature. If you get a picture of air bobbing up and down like a cork pushed into water then you understand the basic idea of a lee (downstream from the mountain) mountain wave also known as an orographic gravity wave (which is just a gravity wave forced by flow over a mountain).

Figure 2.34 *Fog filtering in between higher terrain. (top) Airborne view of marine stratus near Mt. Rainier in western Washington, on 22 July 2013. (bottom) Stratus creeping upslope like a serpent (to the left, west), in local valleys, of a cold air mass behind a cold front, as viewed to the north, from Lost Gulch Overlook, west of Flagstaff Mountain in Boulder, Colorado, on Feb. 11, 2021.*

Photos courtesy of author.

Figure 2.35 *Mountains poking out above fog. (top) Valley fog in the Cascades of Washington State on Dec. 26, 1986. Note how stark the contrast is between the fog and snow-capped mountains, and the dark sky above, owing to a solid high overcast. The top of this cloud layer is much smoother than that seen in some of the other preceding figures. (bottom) Low clouds filtering around mountains in the Alps, viewed from near Garmisch-Partenkirchen, Germany, on Sept. 3, 2014. I can imagine hearing Rimsky-Korsakov's arrangement of Mussorgsky's* Night on Bald Mountain, *during which the mists are being summoned to swirl around the mountains during a witches' sabbath.*

Photos courtesy of author.

Figure 2.36 *Fog creeping upslope onto frozen Barker Reservoir, Nederland, Colorado, after a shallow cold front had passed by on the plains to the east of Boulder Canyon, on Feb. 22, 2019. The push of cold air below from the east associated with a cold front was strong enough to propel the fog layer all the way up to the reservoir, which is about 2, 000 feet (~600 m) above Boulder, Colorado. The wall of the fog bank looks grimly gray, a harbinger of the way the landscape locally will look very shortly. This photograph is truly "creepy."*
Photo courtesy of author.

2.2.2   Cap clouds, foehn clouds, and banner clouds

Suppose that air is forced upslope on the windward side of a mountain in an environment that is stable, and a cloud appears just upstream from the summit because there is sufficient moisture, while the air warms and the cloud droplets evaporate as the air sinks on the lee side, downstream from the summit. A smooth-looking cloud that envelopes the summit of a mountain that is isolated and relatively steep and appears not to move is called a *cap cloud* because it looks as if there is a nearly symmetrical cap on the mountain peak. Any hikers at the summit looking for a spectacular view may be disappointed if they are enshrouded in fog or underneath a low cloud base, keeping them shielded from the sun. The cap cloud often takes on a smooth lens shape, from which the moniker "lenticular" comes to describe it (Fig. 2.45). Later on, in Chapters 3 and 4, we will look at cap-like clouds that form over cumulus, cumulus congestus, and even some cumulonimbus clouds.

When strong winds impinge on a mountain range, a cloud may form not just at the summit, but also on the windward side (Figs. 2.46 and 2.47). As the air descends down the lee side, the cloud evaporates, leaving behind a *foehn* cloud, which can have a very sharp edge where the airflow begins downward and cloud droplets evaporate or snow crystals sublimate. The *foehn* is a dry, relatively warm wind that flows down the lee side

Figure 2.37 *Fog hitting mountain peaks. (top, left) Clouds/fog near the summit of Navajo Peak (13, 409 feet; 4, 087 m MSL), Indian Peaks Wilderness, viewed from Niwot Ridge, on July 16, 2013. (top, right) Fog alongside the upper portion of Niwot Ridge, in the Indian Peaks Wilderness, on Aug. 14, 2006. (bottom) Fog creeping up the side of Mt. Audubon (4, 032 m, 13, 229 ft MSL) and Paiute Pk. in the Indian Peaks Wilderness on July 10,[a] 2012, on an otherwise beautiful, sunny day. The leading edge of the fog is wedge-shaped in places, perhaps as a result of vertical shear at the edge of the advancing cloud mass (the wind has a component from right to left at the cloud-clear air interface, but from left to right inside the cloud).*

[a] This date is not entirely certain: The electronic files in the camera are dated July 10, while my diary shows July 9. I would defer to the documentation of the electronic files as the more accurate one.
Photos courtesy of author.

of mountains, particularly the Alps. We often see foehn clouds, which appear to be stationary, along the Colorado Front Range during the winter when relatively moist Pacific air is forced up on the windward side. The downslope winds on the lee side tend to be strong, gusty, warm, and dry. In the Rockies the foehn is locally known as a *chinook*. I have experienced many chinooks in Boulder, which to my consternation can cause deep snow cover to disappear almost overnight owing to the warm and dry wind, which induces melting and sublimation. The name "chinook" comes from the native American people known as the Chinooks, who described a warm, *moist*, southwesterly wind along the Columbia River in the Pacific Northwest. It is arguably a misnomer when applied to the Plains of the US because it originally referred to a warm, *moist* wind coming from

Figure 2.38 *Different patterns and textures in cirrostratus at sunrise or sunset. (top, left) Sun setting over the Adriatic Sea in broken cirrostratus clouds over Pula, Croatia, on Sept. 22, 2017. (top, right) Cirrostratus clouds brilliantly illuminated at sunrise over Boulder, Colorado, on July 17, 2020. (bottom, left) Bands in cirrostratus at sunrise, over Boulder, Colorado, on Jan. 11, 2022. (bottom, right) Sun brilliantly illuminating the underside of cirrostratus on Jan. 27, 2010, over Norman, Oklahoma, at sunset. The cirrostratus in these images were true "forerunners," as there were freezing rain and ice pellets the next day.*
Photos courtesy of author.

the *ocean*, not a *dry* wind involving flow over mountains. It has been reported that the Blackfoot Native Americans used the term "snow eater" to describe the downslope, dry, winds in the lee of mountains in the interior, especially over the Plains of the US.[8] When a broad north-south oriented channel or jet of high winds impact on the Rockies the long wall of cloud[9] produced is called a *foehn wall* (Fig. 2.47, top panel). I have many times penetrated the foehn wall on cross-country skis. Doing so is like entering a portal to another dimension: You are skiing along under sunny skies, with a strong headwind, but with relatively warm temperatures . . . a lovely day. Suddenly it begins to snow, while the sun still shines through the snow. You then quickly enter the cloud, the sun disappears, the sky darkens, and it snows heavily, with the snow driven horizontally in your face by the wind. There is thus a sudden reversal in the weather. When you turn around

---

[8] I have for years mistakenly believed that the word "chinook" means "snow eater" in a Native American language. I thank Dale Durran for pointing out this error to me.
[9] A "wall of cloud," which is different from a "wall cloud," to be discussed in a later chapter on supercells.

Figure 2.39 *Cirrus cloud bands with some upward growth. (top) Cirrus and cirrostratus clouds exhibiting some indication of buoyancy in a few bands, colorfully illuminated at sunrise over Boulder, Colorado, on Dec. 20, 2021. (bottom) Cirrostratus cloud bands showing some evidence of buoyancy in bands, over Norman, Oklahoma, on Nov. 27, 2021.*
Photos courtesy of author.

Figure 2.40 *Stratiform base of "cloud" of smoke from a forest fire. (top) A stratiform high or midlevel-level cloud of smoke from a forest fire northwest of Boulder, Colorado, on Sept. 27, 2020. The view is to the west. The cloud looks like a cumulonimbus anvil or cirrostratus. (bottom) Vivid colors and some small mamma from the anvil of smoke emanating from a forest fire, as seen in the above panel, but shortly later.*

Photos courtesy of author.

Figure 2.41 *Optical phenomena in thin, cirrus clouds. (top) Sun with a sun dog (or "mock" sun; right) over Norman, Oklahoma, on Sept. 4, 2015, without a halo, owing to the lack of cloud surrounding the sun. (bottom) A "crystal"-clear airborne view of the sun's reflection and 22° rainbow arc/sundog (right) inside cirrus clouds on Feb. 25, 2004, at an unknown location over the western US, between Denver and Seattle.*

Photos courtesy of author.

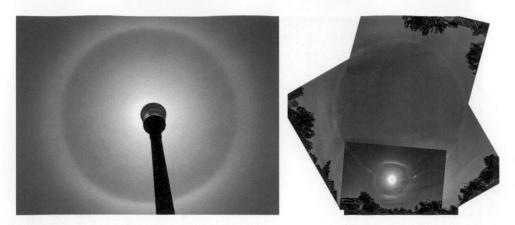

Figure 2.42 *Solar halos in cirrus. (left panel) Full halo in a thin, cirrostratus overcast, in Norman, Oklahoma, on April 9, 2020, with the sun strategically eclipsed by a street lamp. (right panel) From Norma to Norman: Composite photograph of a 360 deg "parhelic circle" over Norman, Oklahoma, on 21 Oct. 2023. The cirrus cloud material originated from Hurricane Norma, which was over Mexico.* Photos courtesy of author.

and return, you exit the foehn wall and it is once again a bright, sunny day, all in a matter of a few minutes: two worlds, side by side, remarkably accessible on foot.

Banner clouds, which appear streaming from the summit to the lee side of a mountain (like a flag from the top of a pole), are more difficult to explain (Voigt and Wirth 2013). They tend to occur in the lee of very steep, isolated mountains with sharp edges, when the moisture conditions are just right: If there is too much moisture, then the entire area around the mountains is cloudy; if there is not enough moisture, then obviously no such clouds are visible. Banner clouds have been frequently photographed over the Matterhorn in the Alps and also around Mt. Everest. I am always on the lookout for a banner cloud in the Colorado Rockies where I spend much more time, but most of the peaks may not be steep enough or have sides sharp enough. It may seem odd at first thought to find a cloud on the lee side of a mountain, where we expect air to be flowing downward, warming and evaporating. Upon closer analysis, though, it turns out that the air, counterintuitively, is in fact actually ascending there, consistent with the cloud we see. The subsequent brief discussion is intended to understand why upward motion and accompanying banner-cloud formation occur where they do and may be skipped over by the more casual reader.

To address the problem of how there can be ascending motion in the lee of a mountain (and of other problems in subsequent chapters), we must understand how the pressure distribution (the way it varies in space) affects the movement of air.

Let's now go back to the issue of what air does when it encounters a mountain. What happens is this: Air does not have enough kinetic energy (the winds are too weak and the environmental air is too stable) to make it up on the windward side all the way to the top

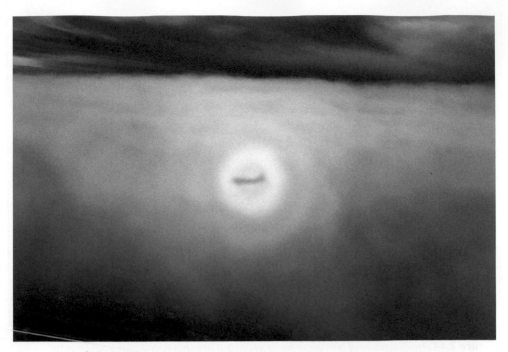

Figure 2.43 *An airborne view of a "glory" in a deck of altostratus probably composed of water droplets, Sept. 23, 1984, off the west coast of Mexico. The shadow of the airplane is seen surrounded by a ring of rainbow-like colors. Glories are seen with the sun located directly behind the observer. I was aboard a NOAA P-3 aircraft on the way back from a hurricane mission in the eastern Pacific when I took this picture.*
Photo courtesy of author.

of the mountain. The air is therefore deflected around the mountain, where it becomes cooler than that of the environment owing to the upward motion on the windward side (Fig. 2.48). The slowdown and deflection are related to a horizontal pressure-gradient force that acts in the direction opposite to that of the wind, with a region of relatively high pressure on the windward slope.

A consequence of the horizontal temperature gradient mentioned just a few paragraphs ago is that there is also a horizontal gradient in buoyancy. The ascending, shunted-aside air, since it is locally negatively buoyant (it is cooler than its environment), experiences a downward-acting force. Away from the mountain, however, there is no buoyancy in the environment. Thus, if you imagine that if a paddle wheel were placed in the vicinity of the mountain, it would begin to spin about a horizontal axis, downward near the surface of the mountain and upward farther from the surface (Fig. 2.48), owing to the horizontal gradient in buoyancy. If we were to follow this parcel of air around to the right of the mountain, it would keep spinning, but as gets around to the other side of the mountain, it starts descending. As it descends, its axis

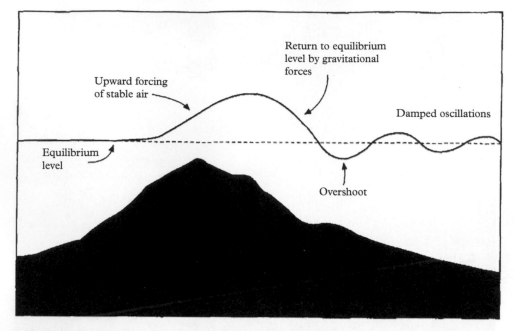

Figure 2.44 *An illustration of how air is forced up, over a mountain, and then overshoots its equilibrium and oscillates downstream (bobs up and down). Solid line is a streamline, from west to east (left to right).*

(From *Hazardous Mountain Winds and Their Visual Indicators*, 1997, Federal Aviation Administration, Fig. 4–3; Fig. 4 in my older, undated, hardcopy version.)

of rotation shifts from horizontal to vertical, so that a counterclockwise-rotating vortex forms on the lee side. Similarly, air that is diverted around to the left of the mountain eventually becomes part of a clockwise-rotating vortex. Thus, there is a pair of counter-rotating vortices on the lee side of the mountain. Counter-rotating *wake* vortices such as these are often found in the lee of volcanoes in Hawaii (Smith and Grubišić 1993), where the trade winds rather consistently impinge up the volcanoes and the low-level air is very stable.

For a banner cloud to form, the mountain slope must be very steep and the edges sharp. In this case, the flow approaching the mountain splits immediately around the mountain and counter-rotating wake vortices appear on the lee side. The dynamics of these wake vortices appear to be different from that associated with counter-rotating vortices in a stable atmosphere. The vortices become more horizontally oriented at higher altitude and connect via a horizontal vortex on the lee side, so that what looks like the shape of a Slinky appears, which is also known as a "bow vortex."[10] The net result of all this vortex formation is that there is upward flow on the leeside, near the top of the

---

[10] Not to be confused with a "bow echo," to be discussed in Chapter 4.

Figure 2.45 *Cap clouds. (left) Two-tiered cap cloud enshrouding the top of Mt. Rainier, as seen from the Space Needle in Seattle, Washington, on Jan. 23, 2011. Some lenticular clouds associated with vertically trapped gravity waves in the lee of the mountain (see Section 2.2.3) are seen to the left, east of Mt. Rainier. (top, right) Lens-shaped cap cloud above Mt. St. Helens, Washington, viewed to the north from Portland, Oregon, on Nov. 23, 2016. (bottom, right) Irregular cap cloud above mountains northwest of Nederland, Colorado, on Jan. 12, 2016, with a wavy wisp on the downstream end as waves are generated downstream as air parcels overshoot and undershoot their equilibrium levels.* Photos courtesy of author.

mountain, which is counterintuitive, since flow on the lee side of mountains is usually downslope. The nature of the airflow responsible for producing a banner cloud is one that is not immediately obvious and is an example of how fascinatingly complicated the atmosphere can be. In this case, the air flowing up and over the mountain "separates" from the prevailing wind and flows backward to where it had originated.

2.2.3 Vertically propagating and vertically trapped gravity waves in the lee of mountains

Let us return to the formation of gravity waves as air flows up and over a mountain range to its lee side and how they can produce absolutely gorgeous, smoothly sculpted clouds that appear nearly motionless in the sky. In a sense, a mountain reminds me of the 1963 Beach Boys' song "Catch a Wave," in which a surfer who catches a wave is sitting on top of the world. How different types of gravity waves are produced is rather complicated and involves some mathematics that is probably out of the reach of many readers, so I will try to simplify what happens as much as possible, without tossing out the most significant features (without "throwing the baby out with the bath water") and

Figure 2.46 *Foehn clouds. (top) Foehn cloud over the Continental Divide onto the Indian Peaks of northeastern Colorado, on Feb. 9, 2021. On this day the tops of the clouds look smooth, suggestive of laminar, stable ascent with snow leaking downstream over the eastern side of the Continental Divide. (bottom) Foehn cloud with snow showers leaking over the Continental Divide over the Indian Peaks, Colorado, on Feb. 11, 2021. Both photographs were taken from the Lost Gulch Overlook, west of Flagstaff Mt. in Boulder. The tops of the clouds are bubbly looking, suggestive of a convective environment.*

Photos courtesy of author.

Figure 2.47 *Foehn and upslope clouds. (top) Foehn wall stretching from the southwest to the northwest along the Continental Divide, March 14, 2016, viewed from Boulder, Colorado. A prairie dog village is seen in the foreground. (bottom) Clouds forming due to moist upslope winds west of the Continental Divide during the summer, on Aug. 1, 2011. View to the west over the Indian Peaks, Colorado.*
Photos courtesy of author.

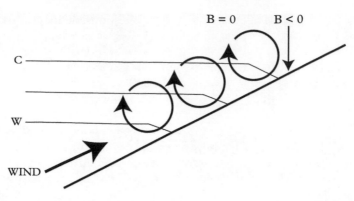

Figure 2.48 *Illustration of how air begins to spin about a horizontal axis along the slope of a mountain as air is forced up the slope and cools right near the ground. This vertical cross-section shows the slope of the ground by a thick, solid black line, which in this case slopes upward to the right. Isotherms (lines of constant temperature) are represented by the thin, gray lines. The temperature decreases with height (W—warm at low altitudes, C—cold at higher altitudes). Near the surface of the sloped ground, it is cooler than it is farther to the left. As a result, the air near the ground is cooler than its surroundings (to the left) at the same height, so it is negatively buoyant (B) and air is forced downward. (Buoyancy will be discussed more in Chapter 4) The air is therefore forced downward near the ground, but not at all off to the left. If a paddlewheel were put in the flow, it would rotate in the sense of the curved line with the arrowhead. This situation is the opposite of that depicted in Figure 2.20, in which heating at the ground acts to produce a buoyancy gradient and spin in the opposite direction. If this figure were made three-dimensional, then, if were an isolated mountain peak and if the kinetic energy of the air was not sufficient to reach the summit, the spinning air would be diverted around both to the left and right of the peak and then downward to produce counter-rotating vortices in the lee of the mountain.*
Figure courtesy of author.

without being misleading . . . and, perhaps most pleasing to the reader, without showing any mathematical equations.

Most analyses of gravity waves are done for waves whose amplitudes (the difference between the peaks and the valleys of vertical motion—the difference between the maximum upward motion and the maximum downward motion) are not "too large." Small-amplitude waves are "linear" because the perturbations in the wind and temperature fields, that is, the deviations from the environmental conditions, are so small that we can neglect any mathematical terms that represent the *products* of two or more quantities in the equations that describe the wave motion. If a perturbation is represented in a term in which there is only one perturbation as a factor, then the larger the quantity multiplied by the perturbation, the larger the term is: A perturbation that is twice as large as another one, for example, produces a term that is twice as large, not four times or ten times as large. If we were to graph the variation of the term as a function of the perturbation, it would follow a tilted, straight line. When the amplitude is large enough that nonlinear effects are important, the term would increase more rapidly than it would if it were linear, explanations are more complicated mathematically, and a computer simulation may be needed. It also turns out that nonlinearity is responsible for our inability to make accurate weather forecasts more than a certain number of days for midlatitude cyclones, anticyclones, troughs, and ridges, features which are responsible for everyday weather; nonlinearity is also responsible for our inability to make accurate weather forecasts more than a certain number of hours for smaller-scale, more rapidly varying features such as thunderstorms, owing to greater sensitivity to the nonlinear terms. Atmospheric scientists try to understand the simple, linear effects first before tackling the more complicated nonlinear effects. Often, many phenomena can be understood with only linear analyses that most of us modestly mathematically inclined mortals can understand.

The linear theory of mountain gravity waves was first advanced significantly by the late Richard Scorer, a British meteorologist in the late 1940s (one whose book on clouds in part inspired me to write this book), and later further advanced by scientists such as Dale Durran (e.g., Durran 1990; Durran and Klemp 1982) at the University of Washington, among others. It turns out that it is easier mathematically to consider first an infinite, periodic set of parallel mountain chains, which resemble the ruffles in corrugated cardboard, because their distribution of heights above some reference level, such as sea level, can be described by simple sinusoidally varying functions (sines and cosines) (Fig. 2.49). These functions have the nice property that their derivatives are also sinusoidal. Simply put, linear responses tend to be roughly sinusoidal. Isolated ranges and isolated peaks are a bit more difficult mathematically to explore because either the sum of many different sines and cosines having different wavelengths and amplitudes are required (Fourier series[11]) or non-sinusoidal functions must be used and these functions are much more difficult to manipulate. A lot can be understood by considering only the simplest models of the atmosphere.

Suppose the width of a mountain range is not so wide that the air is disturbed mainly in the vertical direction only; if a mountain is too wide, then the airflow is disturbed mainly in the horizontal direction through action of the Earth's rotation about its axis (the Coriolis force) rather than in the vertical direction. If the mountain range is relatively narrow, then it takes air a relatively short amount of time to pass across the mountain range, so that the effects of Earth's rotation, which act to deflect air horizontally, are minimal. "Not too wide" means roughly less than several hundred kilometers.

Figure 2.49 *A periodic, sinusoidally varying mountain range; the wavy line indicates the changes in elevation of the surface.*
Figure courtesy of author.

---

[11] Fourier series is a higher-level mathematical topic, so don't worry if you don't have a mathematical background and don't fully understand what it means.

It is found from a mathematical analysis of Newton's second law of motion applied to the atmosphere and a statement of the law of thermodynamics, which specifies what happens in the interchange between heat energy, temperature, and kinetic energy, what conditions permit gravity waves. Stationary[12] gravity waves (like those in Fig. 2.44) appear on the *mesoscale* (having horizontal wavelengths of tens of kilometers to several hundred kilometers), and make their effects on the vertical currents of air felt far above the mountain-top level: When the atmosphere is very stable and there is no vertical wind shear (the winds don't vary with height), then in the reference frame of the airflow (imagine yourself moving along with the wind), waves having relatively long (with respect to a measure of the stability) horizontal wavelength propagate vertically, while waves having relatively short horizontal wavelength decay in amplitude with height above the mountains; or, to paraphrase, "changes in altitudes, changes in amplitudes," with apologies to the singer-song writer Jimmy Buffett who wrote and sang a song in the mid-1970s with a similar name. An atmospheric inversion (a layer in which the temperature increases with height) is characterized by high stability, so only waves with relatively long wavelengths, on the "mesoscale," can propagate upward, making their effect on vertical motion felt. For all practical purposes, "mesoscale" waves (those tens to hundreds of kilometers in width and lasting for less than a day, the width being comparable to the width of the mountains) triggered by flow over the mountain range don't make themselves known much higher up because their wavelengths are just too short and are effectively prevented from affecting the airflow well above the mountains.

2.2.3.1 *Stationary, vertically propagating gravity waves*

Let's now include the effects of a vertical variation of the wind with height, the *wind shear*. Stationary gravity waves can exist and propagate vertically[13] when a quantity named after the pioneer Richard Scorer bears a specific quantitative relationship with the spacing between neighboring mountain ranges. Vertically propagating, low-amplitude waves are most likely when the distance between neighboring mountain ranges (the wavelength of the sinusoidally varying orography) is greater than the inverse of (one divided by) the "Scorer parameter." This is probably confusing at first glance. More simply stated, the most likely circumstances under which we can expect vertically propagating gravity waves to be produced are when mountain ranges are far apart and the Scorer parameter is large.

The *Scorer parameter* increases with the amount of stability in the environment, that is, the rate at which temperature changes with height. The more slowly temperature decreases with height or the more rapidly it increases with height, the greater the stability. The Scorer parameter also decreases with the component of the windspeed normal

---

[12] Stationary in the sense that the waves do not appear to be moving in the cross-mountain direction; i.e., the crests and troughs all remain in the same place.

[13] We consider only stationary waves, not horizontally propagating waves, because they are simpler to describe and also fit what we often observe: clouds often appear not to be moving along with the wind and remain somewhat fixed in space with respect to the mountains, but they do sometimes move slowly away from the mountains.

to the mountain range (it is inversely proportional to the windspeed: so, when considering how the Scorer parameter varies with height, vertical shear must also be taken into account). It follows that stationary, small-amplitude, vertically propagating gravity waves are most likely when the air is relatively stable and the winds relatively weak (the latter condition seems counterintuitive because wind is needed to trigger waves). If the air is just weakly stable and the winds strong, only extremely long-wavelength wave clouds are possible; for all practical purposes, under these conditions air flow over mountains does not produce wavelike patterns of rising and sinking motion, and wave clouds. [The Scorer parameter also depends weakly on the shape of the vertical profile of the windspeeds normal the mountain range (e.g., on whether or not there is a local maximum or minimum in windspeed), much less so than on the stability and windspeeds themselves.] When air flows over an *isolated* mountain peak (as opposed to a periodic set of mountain ranges), an upstream environment at and above mountain-top level of *constant Scorer parameter or one that decreases only very slowly with height* supports vertically propagating waves.

Figure 2.50 (left panel) shows schematically what the airflow over a single mountain in a vertically propagating gravity wave looks like. First, the waves tend to lean with height in the direction opposite to that of the airflow normal to the mountains. So, the air hitting the north-south oriented Rocky Mountains in Colorado from the west may produce stationary waves that lean to the west with height. The phase of the wave at high altitude may be as much as 180° less than that of the wave near the surface. For example, in Figure 2.50, in the left panel, the airflow is downward just east of the mountain peak at low altitudes, but is upward just east of the mountain peak at high altitudes. If there is enough moisture flowing in from the Pacific and if the amplitude of the waves is high enough, then clouds or layers of clouds may form that lean upwards to the west. Moisture in the atmosphere tends to be non-uniformly distributed. Low-level moisture is more likely to be limited near the Rocky Mountains because they are far from oceans, the sources of low-level moisture. High-level moisture tends to originate from air that has been lifted upstream by large-scale weather systems.

2.2.3.2 *Vertically trapped gravity waves*

When the Scorer parameter decreases *rapidly* with height in the environment upstream from a mountain range and at and above mountain-top level, *vertically trapped gravity waves* are possible in the lee of mountains. This sets them apart from vertically propagating gravity waves in the lee of mountains, which are favored by a nearly constant or only slowly decreasing Scorer parameter with height. Vertically trapped gravity waves are found only on the lee side of a mountain range and mainly at lower altitudes, and not on both the windward and lee sides and at all altitudes, like vertically propagating waves. Unlike vertically propagating gravity waves, though, trapped waves do not tilt with height, but rather are vertically stacked; their phases are independent of height (i.e., crests are aligned vertically with crests and valleys are aligned vertically with valleys) (Fig. 2.50, panel at the right). If the air is more stable at low altitudes than at high altitudes and/or the wind speeds normal to the mountain increase in speed with height, then the Scorer parameter decreases with height. The latter tends to be the case in the winter along the Rocky Mountains. When there is a rapid increase in windspeed with height with winds from the same direction (typically from the west), there tends to be a strong horizontal temperature gradient (typically, colder to the north, warmer to the south) vertically averaged through the troposphere. Often the stability of the air does not vary much with height, but simply having the wind from the same direction and increasing in speed markedly with height is sufficient to have the Scorer parameter decrease with height and permit vertically trapped gravity waves.

What prevents trapped gravity waves from propagating vertically? Vertically propagating waves are possible when the Scorer parameter is relatively large. Suppose that stability is high and the windspeed low at low altitudes: then the Scorer parameter is large. If the Scorer parameter is too small, then vertically propagating waves are not possible. Suppose that at higher altitudes the stability is low and the windspeed high. In this case, the Scorer parameter decreases with height. For this scenario, gravity waves triggered at low altitude by mountains propagate upward, but reach a level at which they no longer can no longer propagate upward. Because energy must be conserved, the waves are reflected back downward and interact with upward-traveling waves. This can

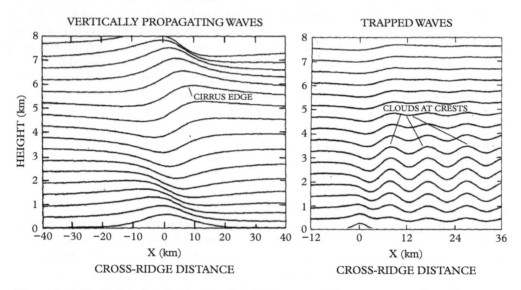

Figure 2.50 *Streamlines (based on a numerical simulation) in west-to-east airflow (from left to right) which is in steady state (doesn't change with time at all locally, at a fixed point on the Earth) over an isolated mountain shaped like a bell, when the Scorer parameter does not vary with height, for (left) vertically propagating waves and (right) trapped lee waves, when the Scorer parameter decreases abruptly with height. Cirrus clouds are more likely to form where air is rising, near the crests.*

(Adapted from Durran and Klemp 1982, their Figs. 1b and 4a; © American Meteorological Society. Used with permission.)

result in an interference pattern of waves that leads to a vertically stacked (no tilt with height) pattern of wind.

2.2.4 Examples of orographic wave clouds

Why do we care whether or not gravity waves propagate vertically or are trapped? We care because the lifting of air at high altitudes, far above the mountains, is required for cloud formation there, where we often see beautiful orographic wave clouds. It's as if the mountains are magically responsible for upward air motion and clouds far removed, above, though not necessarily directly above them, which is "action" at a far "distance."

When one looks at an orographic gravity wave that produces clouds, some appear to have many layers Figure 1.1 (bottom panels and right panel, third row) and elsewhere, and if the layers are closely spaced in the vertical, are seen as striations. Richard Scorer postulated that the layers are due to the fine-scale structure of the vertical distribution of moisture, which is thought to be in part a result of vertical wind shear, which can stretch out moisture into thin, concentrated filaments (Fig. 2.51). Directional shear, not illustrated in Fig. 2.51, might be even more important than the speed shear illustrated in Figure 2.51 because with directional shear, the filaments of enhanced moisture become even more removed from the sources of the moisture, humid plumes. It has

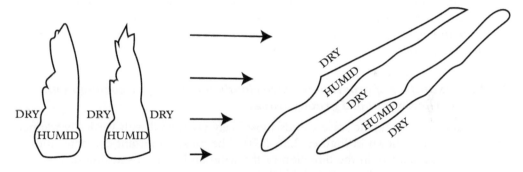

TILTING OF VERTICAL PLUMES OF HIGH HUMIDITY AND
STRETCHING INTO THIN FILAMENTS
BY VERTICAL SHEAR

Figure 2.51 *Illustration of how a series of alternating dry and humid air columns (possibly associated with a series of cumulus clouds that have dissipated) can be tilted and stretched out into elongated, narrow layers. Eventually (not shown) the layers may become nearly horizontal, resulting in very strong vertical gradients in moisture. (a modified version of Fig. 5.1.iii in Scorer's 1972 book).*
Figure courtesy of author.

been demonstrated, using a computer—model simulation that relative humidity differences as small as ±0.25% can replicate the striated, layered structures we see in many orographic wave clouds.[14] This was surprising to me and to many of my colleagues, who first thought that the vertical structure of orographic wave clouds must be a result of vertical variations in vertical air velocity, owing to some type of gravity wave in the atmosphere. Current instruments are not sensitive enough or struggle to detect such small variations in relative humidity.

The highest orographic wave clouds can be the most colorful, since at sunset they can exhibit bright red or orange coloring. Orographic wave clouds tend to be composed of supercooled water droplets. When they are very high up, they may contain ice crystals. Orographic wave clouds may be classified as altostratus or cirrostratus and sometimes as altocumulus.

Although we are guided by Figure 2.50 in distinguishing orographic wave clouds that are forced by vertically propagating waves from those forced by vertically trapped waves, the real world is more complicated. Mountains are not necessarily shaped symmetrically, so that the slope on the windward side is not necessarily the same as the slope on the lee side; this asymmetry has an effect. When there is more than one mountain range separated along the direction of the ambient wind flow, then waves generated by the first mountain range to encounter the flow of air combine with the waves generated by the second mountain range, to generate more complex patterns. And most noticeably, the elevations of individual mountain peaks *along* the ranges are not the same.

2.2.4.1 *Examples of orographic wave clouds most likely associated with vertically propagating gravity waves*

Photographs of orographic wave clouds most likely associated with vertically propagating gravity waves are shown in Figures 2.52–2.55. The most important thing to look for is a tilt with height upstream, i.e., from which the direction the wind is coming. Also, they tend to produce clouds at high altitudes.

2.2.4.2 *Examples of orographic wave clouds most likely associated with vertically trapped gravity waves*

Photographs of orographic wave clouds most likely associated with vertically trapped gravity waves are shown in Figures 2.56–2.70. The most important characteristics to look for are periodicity in the direction of the wind and vertical stacking of plates or striated cloud edges, with no tilt with height.

From the preceding orographic wave-cloud photographs, we find that it is often difficult to distinguish between orographic wave clouds that are formed by areas of rising motion in vertically propagating lee waves from those formed by areas of rising motion in vertically trapped lee waves. One reason for this predicament is that there may not be sufficient moisture available to produce clouds everywhere there is rising motion. There

---

[14] Hills and Durran, 2014.

Figure 2.52 *Orographic clouds probably forced by vertically propagating gravity waves. (top, left) Piles-of-plates and everything else: Complex back side to an orographic wave cloud at sunset over Boulder, Colorado, on Jan. 3, 2021. The view is to the north-northwest. Since the cloud at the top seems to be leaning to the west (left) with height, it may be caused by vertically propagating gravity waves. (top, right) An orographic wave cloudx viewed to the northwest from Boulder, Colorado, on Feb. 17, 2023. (middle. left) A chaotic-looking sky at sunset on Dec. 24, 2007, looking north from Boulder, Colorado. There are several striations on the nearest cloud. Since most appear to lean to the west with height, they are probably due to vertically propagating waves, but they also exhibit some periodicity normal to the wind and vertical stacking, which would argue for vertically trapped waves. The mid-section of this cloud looks like an apartment complex. (middle, right) A swarm of orographic wave clouds over Boulder, Colorado, on Dec. 12, 2021, probably associated with vertically propagating waves because some appear to lean upstream (to the west) with height. (bottom, left) Wave clouds on Jan. 5, 2023 over Boulder, Colorado, with a view to the north. These clouds appear to tilt with height to the left, or northwest. (bottom right) Wave cloud on Dec. 19, 2022 at sunset and viewed to the south-southwest from Boulder, Colorado. The cloud appears to tilt to the right, or west or northwest, with height.*

Photos courtesy of author.

Figure 2.53 *Orographic wave clouds possibly associated with vertically propagating waves. (top) after sunset on Jan. 5, 2012, over the Foothills of Boulder, Colorado. Note the elliptical pattern associated with inhomogeneities in the north-south direction. There is a sharp western edge, but there is some along-the-mountain variability and some leaning to the west with the height of the cloud bottom. (bottom) After sunset on March 31, 2003, over the Foothills of Boulder, Colorado. Note the downward protuberances, which look like mamma under anvils of cumulonimbus clouds (see Chapter 4). These clouds are reminiscent of the clouds during the parting-of-the-Red-Sea scene in the epic and iconic 1956 movie* The Ten Commandments.

Photos courtesy of author.

Figure 2.54 *Circular/elliptical orographic wave cloud over the Foothills in Boulder, Colorado, on Jan. 10, 2010, after sunset. Since it might be leaning with height to the west, it may be associated with vertically propagating gravity waves.*
Photo courtesy of author.

could be periodicities that are not evident and stacking, which is not evident. In addition, it is sometimes difficult to determine if an orographic wave cloud that is parallel, and near to another one, lies downstream or above or below its neighbor. So, in my photographs I tried first to identify those cases in which there is the least ambiguity. For many cases, however, there is sufficient ambiguity that I refer to the clouds simply as "orographic wave clouds," with no mention of the types of invisible wave motion going on, which we cannot verify anyway without any actual wind measurements. Furthermore, there are instances when there might be clouds formed by both types of waves (e.g., in Fig. 2.59, lower left), but especially in different sections of the sky or at different altitudes. To many readers the distinction between propagating and vertically trapped waves should not affect their opinion of the *beauty* of the cloud. However, the point I want to make is that there are clouds that have similar appearances, even if the mechanism for their formation is not certain and the mechanisms may even be different for each one.

Figure 2.55 *Edge of an orographic wave cloud on Oct. 8, 2002, near Ward, Colorado, viewed on end, to the south. There is some purple and green iridescence along the cloud edge, which is. indicative of uniform-size, small water droplets. Note the silhouette of the "flag tree" or "banner tree" at the lower left, in which only vegetation on the sheltered lee side (left) survives under the prevailing, strong westerly flow (from right to left) at this high elevation. Since there is some leaning of the cloud in the upstream direction (to the right, west) with height, the cloud may be caused by a vertically propagating gravity wave.*

Photo courtesy of author.

Figure 2.56 *Orographic wave clouds probably associated with vertically trapped gravity waves. (top left) on Jan. 18, 2016, over Boulder, Colorado, owing to the periodicity in the downstream direction (view is to the east) and vertical stacking. (top right) over Boulder, Colorado, Dec. 8, 2017. Note all the*

2.2.5 Hydraulic jumps

There is another set of circumstances under which clouds may appear at high altitudes near mountain ranges; these foster what appears to be the atmospheric analog to what is known in water as a *hydraulic jump* (Fig. 2.71) in models of fluid flow characterized by two layers, consisting of a dense, lower layer and a less-dense, upper layer (like water below and air above). Rich Rotunno, an atmospheric scientist at the National Center of Atmospheric Research (NCAR) in Boulder, Colorado, has pointed out hydraulic jumps to me in water streams in wilderness areas of Colorado. Water flows downstream and up and over obstacles such as rocks, followed by a descent into a region of turbulence (choppy, chaotically flowing water), followed by an increase in the height of the water level just downstream from the turbulent region that is not turbulent (Fig. 2.71). Hydraulic jumps in the atmosphere were first studied using bodies of water as an analog: When water flows up a mountain range and flows more slowly than the speed of "shallow-water" gravity waves, it speeds up as it reaches the summit and continues to increase in speed as it descends on the lee side where it flows more quickly than the speed of gravity waves, resulting in very strong downslope flow and turbulence. In the atmosphere, a pressure-gradient force acting in the same direction as the wind on *both* sides of the mountain is what keeps the air accelerating. The height of the interface between the denser fluid below and the less-dense fluid above decreases as the wind speed keeps increasing, in order to conserve energy (potential energy is being converted into kinetic energy). However, at the base of the mountain the wind speed decreases substantially owing to the vertical mixing and end of the mountain slope, and the height of the interface between the denser fluid below and the less-dense fluid above jumps up in height abruptly. This jump is necessary to conserve energy and to produce a pressure-gradient force opposite to the wind direction so that the speed of the airflow slows down.

Figure 2.56 (Continued) *layers and striations separating the layers. These clouds are likely caused by vertically trapped gravity waves owing to the periodicities and vertical stacking. View is to the east. (second row, left) Stacks of orographic wave clouds over Boulder, Colorado on Jan. 21, 2021. Otherwise as in the previous panels (second row, right) near Nederland, Colorado on Jan. 26, 2016. There is vertical stacking, but less evidence of periodicities. View is approximately to the south or southwest. (third row, left) with many plates, over the Foothills of Boulder, Colorado, on March 17, 2017. Note NCAR faintly glowing at sunrise below the Flatirons on the bottom right of the photograph. This cloud was probably associated with vertically trapped gravity waves, but from this view there is no sign of periodicity in the east-west direction. (third row, right) Series of orographic wave clouds over Boulder, Colorado, on Dec. 17, 2020, probably a result of vertically trapped gravity waves, owing to the periodicities and vertical stacking. View is to the northwest. (fourth row) A number of vertically stacked orographic wave clouds over Boulder, Colorado, on Jan. 21, 2021, viewed to the west through northwest. (bottom, left panel) Vertically stacked plates of clouds over the Foothills of Boulder, Colorado, on Dec. 23, 2022. (bottom, right panel) Vertically stacked plates of clouds over Boulder, Colorado, on Jan. 8, 2023.*
Photos courtesy of author.

Figure 2.57 *Airborne view of a series of vertically trapped gravity wave clouds over the intermountain western US on Jan. 10, 2004. These clouds are periodic and vertically stacked. View is approximately to the south.*
Photo courtesy of author.

Results from a recent study based on computer simulations in air whose density decreases *continuously* (not abruptly as in a two-layer model), which is a more realistic scenario, have indicated that when the mountain–normal flow is strong enough to make it up and over a mountain range, the air becomes warmer to the east of the mountains as it descends on the lee side. As mentioned earlier, since there is a horizontal gradient in buoyancy, an imaginary paddle wheel would rotate such that the axis of rotation lies in between the ground and upward buoyancy forces to the right in the warmer region. Horizontally oriented rotating columns of air descend and some similar rotating air columns downstream cause a reversal in wind direction aloft, with air moving backward, *toward* the mountain, rather than away from it. This air is relatively cool compared to the warmer air below, leading to a steep, nearly vertical wall of air characterized by a super-dryadiabatic[15] lapse rate and turbulence. Hydraulic jumps may be thought of as vertically propagating waves that form above and downstream from a mountain and become unstable with respect to vertical displacements in the midtroposphere; there is resulting turbulence and vertical mixing, and a steepening of the slope of air currents.

---

[15] To be discussed in Chapter 3. The importance of a super-dryadiabatic layer is that there can be instability with respect to vertical displacements of air, which promotes turbulence and vertical mixing.

Figure 2.58 *Orographic wave clouds probably associated with trapped waves. (top, left panel) A series of vertically trapped lee waves are responsible for this series of orographic, stacked, wave clouds, viewed to the south from Boulder, Colorado, on Feb. 23, 2021. (bottom, left panel) Widely separated pile-of-plates-type orographic wave cloud over Boulder, Colorado, on Feb. 23, 2021. Some cumulus* fractus *rotors (discussed later) are also seen on the left (east). This cloud is probably evidence of a vertically trapped gravity wave, owing to the vertical stacking, and periodicities seen in the top, left panel, at approximately the same time, but a zoomed-in view. (right panel) Orographic wave clouds in bands streaming downstream and an isolated, vertically stacked orographic wave cloud, on Jan. 23, 2012, in Boulder, Colorado. The latter is evidence of vertically trapped gravity waves. Any leaning with height upstream (west, left) is evidence of vertically propagating waves.*
Photos courtesy of author.

The vertically propagating waves are said to be "breaking," in analogy to what happens when an ocean wave breaks as water curls around air, and thus falls over, in effect breaking the wave and resulting in turbulent flow and whitewater conditions. In Figure 2.72 we see what the clouds can look like in a hydraulic jump as depicted by Figure 2.71, top. The resulting configuration of wind and temperature results in a strong downslope flow at the surface on the lee side, and possibly a column of rising motion aloft associated with a rotor (Fig. 2.71, top panel).

There are often orographic wave clouds that have a relatively sharp western (upstream) edge and are solid in the north–south (along-the-mountain) direction (Figs. 2.73 and 2.74), and stream far off downstream. Sometimes there are variations in the

Figure 2.59 *Orographic wave clouds that exhibit some periodicity. (top, left) Orographic wave clouds near Nederland, Colorado, on Jan. 11, 2014. The waves might be vertically trapped gravity waves since they exhibit periodicity and vertical stacking. There is what appears to be a ragged-looking rotor cloud in the lower right of the photograph. View is to the south or southwest. (top, right) Waves in cirrus clouds over Boulder, Colorado, on Jan. 1, 2017. The waves might be caused by vertically trapped lee waves, but they appear to be at rather high altitude, where vertically trapped waves don't usually occur (see Fig. 2.67). (bottom, left) Orographic wave clouds over Boulder, Colorado on Dec. 30, 2016, possibly associated with vertically propagating gravity waves since some of the clouds lean with height upstream (to the left), but they also exhibit some characteristics of vertically trapped gravity waves, owing to their periodicity. View is to the north. (bottom, right) Vertically trapped gravity wave clouds with iridescence, and the sun, over Boulder, Colorado, Feb. 18, 2018. The iridescence is indicative of uniform-sized small water droplets. These clouds exhibit some periodicity and also vertical stacking.* Photos courtesy of author.

cloud structure in the along-the-mountain-range direction (Fig. 2.75). A column of rising motion may produce cloud material which flows/spreads downstream aloft, like a thunderstorm anvil. In fact, sometimes the cloud material and water vapor produced by evaporating or sublimating cloud particles can be found far downstream: In Figure 2.76 (top and middle panels) there is a nearly continuous plume of cloud material or water

Figure 2.60 *Isolated orographic wave cloud, south of Boulder, Colorado, on Feb. 12, 2016. This cloud appears to be tilted downward in the downstream (lee, east) direction, but this tilt may be an illusion as a result of perspective, as the cloud may be oriented along the mountains. It appears to lean with height in the upstream direction (right, west), but not with respect to the orientation of the mountain range, which is nearly perpendicular to the axis of the cloud. Since, however, it is at a relatively low altitude with respect to the mountain crests to the west, it is more likely due to a vertically trapped wave.*
Photo courtesy of author.

vapor all the way from the lee of the Front Range in Colorado, southeastward into Oklahoma. Just as I was amazed at seeing evidence of cloud material or water vapor from the Pacific making it all the way up into New England, I have been amazed to see a mesoscale event in Colorado make itself known all the way to the Southern Plains. One can look up in the sky in Norman and exclaim "Wow! That cloud material was forced by upward motion associated with an orographic gravity wave west and northwest of Denver." Photographs of the back edge of the clouds seen in the top two panels, but taken from the ground, are seen in the bottom panel.

The origin of these types of orographic wave clouds is not entirely obvious. They are probably too high up to be caused by vertically trapped gravity waves, but the air may be so humid that all the cloud material does not disappear in subsiding branches of the downstream waves if there were trapped waves present. They may be caused by vertically propagating waves, but a tilt with height upstream of the clouds is not visible because very humid air is confined to only a thin layer. They are forced by the upward motion along a hydraulic jump, but they don't look at all like the photograph of the hydraulic

Figure 2.61 *Colorful orographic wave clouds at sunset. (top) over the Foothills, Boulder, Colorado, on Jan. 4, 2021. The thin cloud band partially hidden by the mountains may be a rotor or another in a series of vertically trapped gravity waves. (top) over the Foothills of Boulder, Colorado, at sunset on Jan. 4, 2019. Note the transverse (normal to the mountains) cloud bands at the upper left. There is some vertical stacking, which is characteristic of vertically trapped gravity wave clouds, but little evidence of periodicity in the direction normal to the mountains.*
Photos courtesy of author.

Figure 2.62 *Stacked-plate-looking stationary orographic wave clouds over the Foothills of Boulder, Colorado. (top panel) Clouds with a pinkish glow at sunrise, on Feb. 6, 2016, possibly associated with vertically trapped gravity waves, owing to the vertical stacking. (bottom panel) As in the top panel, but during the late morning on Jan. 8, 2023.*

Photos courtesy of author.

Figure 2.63 *Series of orographic wave clouds probably associated with vertically trapped gravity waves during the daytime. (top, left)) over Boulder, Colorado, on June 21, 2020. View is to the northwest. (top, right) As in the top, left panel, but viewed to the northwest, on Jan. 3, 2018. (middle, left) Striated flying saucers spaced apart, in the form of altocumulus lenticularis, invade Boulder, Colorado, on April 24, 2021. View is to the northwest. (middle, right) A series of striated lenticular clouds, viewed to the east, down Boulder Canyon, from Nederland, Colorado, on March 9, 2007; like the top two panels, but viewed from the opposite direction, looking downstream, not upstream. (bottom) A series of orographic wave clouds widely spaced apart, over Boulder, Colorado on Dec. 11, 2021. View is to the west.*
Photos courtesy of author.

Figure 2.64 *Stationary parallel cloud bands. (top) Parallel bands of relatively low clouds possibly associated with stationary, vertically trapped gravity waves over and west of Boulder, Colorado, on Feb. 20, 2003. Since there is no evidence that there was vertical stacking of the clouds, however, the classification of the clouds is uncertain. (middle) Parallel bands of relatively low clouds, similar to those seen in the top panel, possibly associated with stationary, vertically trapped gravity waves over and west of Boulder, Colorado, on Jan. 15, 2016. If there was evidence that there was vertical stacking of the clouds, then the cloud classification would be more certain. (bottom) Parallel bands of relatively low clouds, like those seen in the first two panels, possibly associated with stationary, vertically trapped gravity waves over Boulder, Colorado, on March 10, 2003. South and North Arapaho Peaks are seen on the horizon; the street is aptly named Arapahoe Ave., but the spellings are different. The top of the main cloud is smooth and laminar, probably because the air was very stable. The second in a series of cloud lines is barely seen at the top of the photograph. If there were less—ambiguous evidence of vertical stacking of the clouds, then their classification would be more certain. It is also possible that these clouds could be associated with boundary-layer rolls, as described in Section 2.3 and not related to mountains at all.*

Photos courtesy of author.

Figure 2.65 *Parallel cloud bands with the sun or the moon. (top) The moon hiding behind parallel bands of higher orographic wave clouds over the Foothills, Boulder, Colorado, on March 28, 2021, well before sunrise. This is a scene that the late nineteenth- and early twentieth-century American painter Albert Pinkham Ryder, who produced some incredibly moody moonlight scenes, would have loved. Since there are a number of parallel bands, this might be caused by a stationary, vertically trapped*

jump in Figure 2.72, and there are not ragged-looking rotor clouds created as a result of turbulent mixing. Orographic wave clouds, whose origins are not obvious, are shown in Figures 2.77–2.81; some were taken either at sunrise or sunset, when brilliant colors could be seen. Similar-looking clouds are grouped in panels in some of the figures.

Orographic wave clouds may be significant not only for esthetic reasons, but also because they may block out views of the sky locally, possibly blocking significant astronomical phenomena.[16] Forecasting whether or not a local, orographic wave cloud in an otherwise quiescent atmosphere may hinder viewing of the sky is very difficult. In addition, a local orographic wave cloud may reduce the amount of solar radiation reaching the surface during the day and reduce the amount of infrared radiation passing out into space at night, thus making it cooler during the day and warmer at night than that which otherwise be anticipated.

The hydraulic-jump-like condition (seen in Fig. 2.71) is thought to be responsible for some downslope windstorms, which sometimes have winds as strong as hurricanes. There are several other theories for why downslope windstorms form, which involve the interactions between upward-propagating waves and downward-propagating waves that are partially reflected, but they are mathematically complicated. Unlike vertically propagating, stationary gravity waves, the hydraulic jump produces a cloud that does not necessarily lean upstream with height. It is often challenging, however, to detect any tilt with height owing to one's visual perspective or the particular vertical distribution of moisture. Another difference between purely vertically propagating, stationary gravity waves and the special-case, hydraulic jump is that the former can be described by linear theory while the latter is distinctly nonlinear.

Orographic wave clouds, however, can be especially beautiful, especially at sunrise or sunset, when they are found at a relatively high altitude and have sharp cloud edges. Just as it is sometimes difficult to distinguish clouds generated by vertically trapped gravity waves from clouds generated by vertically propagating gravity waves, it is also frequently difficult to tell how a cloud produced by vertically propagating waves differs in appearance from a cloud produced by a hydraulic jump, especially since the latter is a special

---

Figure 2.65 (Continued) *gravity wave, but there is not much evidence of vertical stacking. (middle) Parallel bands of higher orographic wave clouds over Boulder, Colorado, on Jan. 17, 2021, near sunset. Note the similarity of this photograph to that in the upper panel which shows wave clouds before sunrise, but illuminated by the moon. Since there are parallel bands, this cloud might be caused by stationary, trapped gravity waves, but there is not much evidence of vertical stacking as suggested in Figure 2.85 and elsewhere. (bottom) Orographic wave clouds over Boulder, Colorado, on Dec. 11, 2021.*
Photos courtesy of author.

[16] I photographed the conjunction of Jupiter (left) and Saturn (right) during the evening of Dec. 21, 2020 just underneath the back edge of an orographic wave cloud, viewed to the southwest, from Boulder, Colorado. Some in other locations were probably not as fortunate, in that what should have been a clear evening was ruined by an orographic wave cloud.

Figure 2.66 *Parallel bands of orographic wave clouds over the Foothills of Boulder, Colorado, on June 20, 2020. These two cloud bands were possibly forced by vertically trapped gravity waves, but there is just some evidence of vertical stacking (the more distant cloud at the bottom has some vertical stacking and striations). The closer band is fuzzier, exhibits no evidence of stacking, and exhibits variations in the north-south direction, parallel to the Continental Divide.*
Photo courtesy of author.

case of the former. When no tilt with height backward against the airflow is readily apparent and when there is one sharp back edge to the cloud, then I would guess that the cloud is most likely caused by a vertically propagating gravity wave without a hydraulic jump in an environment in which there is only a thin layer of high humidity; thus the air might be too dry to produce a rear edge to the cloud at more than one level, so that no tilt with height can be discerned. The sharp back edge, however, could also be created by airflow in a rotor in a hydraulic jump. In many instances of high-altitude cirrus clouds exhibiting a sharp back edge, however, strong surface winds and turbulence may not be present near the ground; these features are hallmarks of hydraulic jumps (as in Fig. 2.71; note also that there are no cirrus clouds at all in this unambiguous example of a hydraulic jump). It is likely, though, that broad areas of cirrus clouds having a sharp western edge are not likely due to vertically trapped gravity waves, since they appear at high altitude and do not exhibit evidence of periodic, wavelike structures. As noted earlier, some people side-step the problem of associating a given cloud with a given type of gravity wave by simply referring to the cloud ambiguously as an "orographic" or "mountain" wave

Figure 2.67 *Parallel cloud bands. (top) Parallel bands of higher clouds possibly associated with stationary, vertically trapped waves, over Boulder, Colorado, on Feb. 24, 2016. If there were evidence that there was vertical stacking of the clouds, then their classification would be more certain. (bottom) Orographic wave clouds over the Foothills of Boulder, Colorado, on Sept. 11, 2020, just several days after a rare, early-Sept. snowstorm. Since they appear as parallel bands, they are possibly associated with vertically trapped gravity waves, but there is little evidence of vertical stacking. They look somewhat like the clouds seen in the top panel, but at sunset, and a bit less laminar.*
Photos courtesy of author.

Figure 2.68 *Wave-like clouds possibly associated with vertically trapped gravity waves, over Norman, Oklahoma, on May 4, 2013. One cannot use the adjective "lee" to describe the waves, however, since there are no mountains from which the clouds are in the lee. It is also possible that these clouds could just be the tops of buoyant, boundary-layer plumes (see Chapter 3) pushing upward against a stable layer/inversion.*
Photo courtesy of author.

cloud, without specifying what type of wave created the cloud. I have often submitted to this cowardly way out from time to time (e.g., Fig. 2.81).

2.2.6 Bar clouds

In the lee of the Front Range of the Colorado Rockies, there is at least visually a unique type of orographic wave cloud that is colloquially and locally referred to as a "bar" cloud, the name first used by Dale Durran, John M. Brown, and colleagues at the NCAR and the National Oceanic and Atmospheric Administration (NOAA) in Boulder, Colorado, in the late 1970s and early 1980s. Bar clouds are relatively narrow and have definite northern and southern edges (Figs. 2.82–2.88). The first time I saw such a cloud, I gazed in wonderment, only to be told by my friend and colleague John Brown that, "Oh, we see these here all the time!" In the following figures, we see bar clouds under various types of lighting, from sunrise, sunset, and during the middle of the day under a deep blue sky, as well as from different vantage points. Perhaps the bar cloud under

Figure 2.69 *Possible vertically trapped gravity wave clouds over Norman, Oklahoma on Nov. 12, 2011, owing to their periodicity, but there is no evidence of vertical stacking. These parallel cloud bands are* not *contrails. They are probably too high to be considered bores (see Section 2.3) and appear to be sui generis.*
Photo courtesy of author.

this condition should be called the "sky bar," in eponymous recognition of the candy bar formerly made by the Necco candy company, popular during the mid-1950s in the US.

Bar clouds tend to occur by themselves. They are often later accompanied by or replaced by vertically propagating wave clouds or hydraulic-jump clouds. In time-lapse movies and in videos we can see cloud material forming on the northern edge and dissipating on the southern edge of the bar cloud. In a climatological study done by two undergraduate students under my supervision, and that of my friend and colleague John Brown, we found evidence that bar clouds are associated with substantial along-the-mountain-range flow, from north to south, consistent with the time-lapse studies. We also found that they are seen typically when higher values of moisture associated with

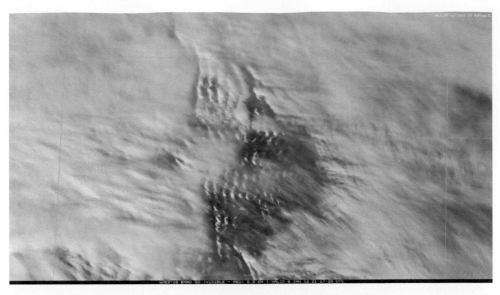

Figure 2.70 *NOAA satellite visible-channel images of highly probable vertically trapped gravity waves over/in the lee of the Rocky Mountains in north-central Colorado on Jan. 6, 2018. Note the north-south oriented, slightly wavy, parallel, narrow cloud bands. When viewed as an animation, the bands remain stationary, while the other, higher clouds stream on by to the east. Since the traveling high clouds do not show any banding, while the stationary, lower clouds do, it is highly likely that the lower level cloud bands represent vertically trapped gravity waves.*
(Courtesy of College of DuPage Meteorology.)

rising motion in an upstream trough[17] in the wind field at middle and upper levels in the troposphere first makes it around a downstream ridge[18] in the wind field and then encounters mountains, that is when the airflow is from the northwest, i.e., as just noted, with a large component of the wind *along* the mountain range.

We don't know whether these clouds are associated with vertically propagating waves or trapped waves or hydraulic jumps. They are likely influenced by the three-dimensional aspects of the airflow because there is a component of wind both perpendicular to (from the west) and along (from the north) the mountains. They could be result of waves generated upstream from a higher mountain, but also affected by waves generated by westerly flow along the Continental Divide where the terrain is locally higher. Whatever

---

[17] Troughs are regions of relatively low pressure and counterclockwise curvature in the winds, in the Northern Hemisphere, and are accompanied by unsettled weather (rising motion, clouds, and precipitation) downstream from the trough axes.

[18] Ridges are regions of relatively high pressure and clockwise curvature in the winds, in the Northern Hemisphere, and are accompanied by fair weather (sinking motion, clear skies, and dry conditions).

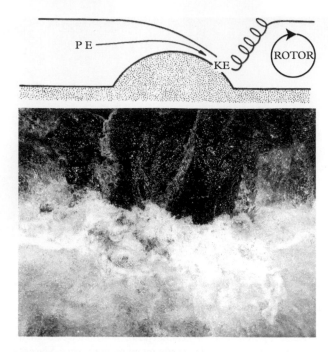

Figure 2.71 *An idealized and a real hydraulic jump. (top) Side view of the airflow across a mountain accompanied by an idealized hydraulic jump. The curved line represents the boundary between dense air near the ground and less dense air above. The air accelerates* both *as it ascends the mountain* and *as it descends the mountain, resulting in a continuous conversion of potential energy (PE) to kinetic energy (KE) regardless of whether the air is ascending or descending (which may seem counterintuitive), until the airflow becomes turbulent (indicated by the coils) and after vigorous vertical mixing through which the strong downslope winds disappear, the height of the boundary jumps up to what it had been upstream from the mountain. (from Durran 1990, their Fig. 4.5c; © American Meteorological Society. Used with permission.) (bottom) Photograph of a hydraulic jump in a stream in the Indian Peaks Wilderness of Colorado, below Mitchell Lake, June 13, 2022. The water flows smoothly over a rock (at the top) and downward, becoming turbulent with white water below. A cloud may form near the top of the jump in height of the hydraulic-jump boundary as a result of vertical mixing or along the ascending branch of a rotor (to be discussed later) which may form downstream and underneath the jump.*
Photo courtesy of author.

the reason for their existence, they certainly make for some very colorful sunrises and sunsets along the Front Range in Colorado.

Other miscellaneous orographic wave clouds are seen in Figures 2.87–2.89.

2.2.7   Rotors

Horizontal vortices called *rotors* are sometimes found near the ground, underneath vertically propagating gravity waves, vertically trapped waves, and hydraulic jumps. They

Figure 2.72 *Line of cumulus* fractus *(ragged-looking, small cumulus clouds) extending from north (left) to south (right) on Dec. 30, 2021, viewed to the east from Boulder, Colorado. The clouds are most likely associated with the upward motion in a hydraulic jump. At the time this photograph was taken, the downslope winds were blowing around 70 mph, with gusts as high as 80–100 mph, from the southwest (right to left and into the page). The smoke plume was the beginning of the "Marshall" fire, which tragically destroyed over 1, 000 homes and other structures. The ragged-looking cumulus clouds were* not *forced by buoyancy created by the fire as in pyrocumulus clouds (see Chapter 4), but rather by the upward branch of the hydraulic jump circulation via vertical mixing or a downstream rotor (Fig. 2.71, top panel). A classic example of a rotor cloud in the lee of the Sierra Nevada above the Owens Valley in 1950 may be found in Fig. 2 in Doyle and Durran (2002); the similarity between this photograph (not shown) and mine is striking.*
Photo courtesy of author.

can be seen as bands of ragged-looking, cumulus *fractus* clouds (Figs. 2.90–2.92). They are seen in time-lapse movies and videos as nearly stationary segments of clouds that roll around a horizontal axis. Rotors can present a substantial aviation hazard since aircraft may experience sudden, significant changes in headwind or tailwind, and therefore potentially dangerous, rapid changes in lift, which could cause an aircraft engine to stall. If this were to happen, because the aircraft is so close to the ground in the rotor, there would be little time for the pilot to recover from the stall. Sometimes rotors are made visible as long bands of isolated ragged-looking clouds, so it is not always obvious whether the rotor is from vertically trapped or vertically propagating waves, or from a hydraulic jump. In other instances, rotors appear underneath isolated, small orographic

Figure 2.73 *Large-scale orographic wave clouds with a sharp western edge and streaming to the southeast, are seen by a NOAA satellite, over northeastern Colorado on April 14, 2019, at 1606 UTC. The clouds may be associated with vertically propagating waves (we can't see underneath the back edge of the clouds to determine whether or not it tilts with height in the direction from which the wind is blowing) or a hydraulic jump. The snow-covered Continental Divide is clearly seen extending from north to south, in the center portion of the frame of the image. The sharp western edge remained nearly stationary, while, like deep convective storms (Chapter 4), anvil-like cirrostratus and altostratus flow downstream (to the southeast, in this case) from a source; in the case of convective storms, the source is the updraft portion of the storm, while in the case of wave clouds, the source is a narrow band of rising motion associated with an orographic gravity wave. See Figure 2.74 for examples of what the back edge of this type of cloud looks like from the ground.*
(Courtesy of College of DuPage Meteorology.)

wave clouds (Fig. 2.92); isolated lenticular clouds are the focus of the next section, Section 2.2.8. From computer-simulation experiments, we know that the drag of the wind against the ground plays an important role in producing some rotors, especially those caused by trapped lee waves. Rotors adjacent to hydraulic jumps may be created on the lee side of mountains, as in Figure 2.71, top panel.

2.2.8   Flying saucers: Altocumulus lenticularis

Isolated wave clouds sometimes look like flying saucers (Figs. 2.93–2.106); during the late 1940s and the early 1950s, when there was a flying-saucer craze in the US, many wave clouds may have been reported as unidentified flying objects (UFOs).[19] According

---

[19] Reed, R. J., 1958, "Flying saucers over Mount Rainier." *Weatherwise*, **11**, 43–45, 65–66.

Figure 2.74 *Clouds with a sharp western edge. (top, left) Western, sharp edge of a large-scale orographic wave cloud possibly associated with a hydraulic jump, on Dec. 20, 2016, over Boulder, Colorado, owing to the sharp, stationary, western, upstream edge and the relatively long downstream length. View is to the west-northwest. (top, right) Western, sharp edge of a large-scale orographic wave cloud probably associated with either a vertically propagating wave or a hydraulic jump, at sunrise, on March 16, 2016, over the Foothills of Boulder, Colorado. View is to the west. (bottom, left) Back edge of an orographic wave cloud over the Foothills, Boulder, Colorado, viewed to the west, on Jan. 21, 2021. There is also some evidence that the cloud leaned with height upstream (left, south) and may therefore be due to vertically propagating gravity waves. (bottom, right) Back edge of an orographic wave cloud in the shadows, over Boulder, Colorado, on June 21, 2020, with the Foothills to the west brightly illuminated, just after sunrise. It appears as if the mountains are about to be eaten up as the shadowy jaws overtake them.*

Photos courtesy of author.

to Dick Reed, a prominent, synoptic meteorologist from the University of Washington in Seattle, in quoting a 1953 book on flying saucers by DH Menzel,[20] the flying-saucer "craze" began on June 24, 1947, when Kenneth Arnold was flying in a small plane near Mt. Rainer and spotted UFOs close to the mountain. Reed surmised that what was spotted were likely lenticular clouds.

[20] Menzel. D. H., 1953, *Flying Saucers*, Cambridge, Harvard Univ. Press, 319 pp.

Figure 2.75 *Wave clouds with protuberances or inhomogeneities on the lee side of mountains. (top, left) Orographic wave clouds on Feb. 21, 2005, over Boulder, Colorado, as viewed to the south. Note the inhomogeneities in the north-south direction, possibly reflecting inhomogeneities in the mountain tops in the north-south direction. (top, right) Orographic wave clouds viewed to the south from Boulder, Colorado, on Dec. 12, 2021. (bottom, left) A mixture of orographic wave clouds possibly due to vertically propagating waves (cloud edge on right, leans to the west with height) and also vertically trapped gravity waves (there is some periodicity in the east-west direction and vertical stacking of the lower clouds), in Boulder, Colorado, viewed to the south, on May 9, 2006. Each individual scalloped cloud element might represent the effects of individual mountain peaks. (bottom, right) Orographic wave clouds, viewed to the south, from Boulder, Colorado, on Dec. 22, 2021. There are individual lobes, with some iridescence.*
Photos courtesy of author.

Sometimes the sky appears to be invaded by many flying saucers (Fig. 2.99). When this happens, it appears as if the atmosphere is hosting a large dinner party since there are plates all over the place. These wave clouds are technically known as altocumulus *lenticularis* because they are shaped like lenses (lenticularis refers to a lens shape or that of a lentil bean). The abbreviation for them used by cloud observers is ACSL (AltoCumulus Standing Lenticularis). These clouds, also called lenticular clouds, are particularly mysterious and beautiful, owing to their appearance, particularly when isolated, as sculptures in the sky apart from anything else. Why are they where they are and not anywhere else? They are particularly prone to exhibiting striations, probably due to tiny vertical variations in moisture, as has been noted earlier in other orographic wave clouds. Sometimes isolated lenticular clouds appear layered like a stack or pile of plates (e.g., Figs. 2.96, 2.98, and 2.100). When looking at the striations or even watching time-lapse videos/movies of them, they sometimes appear to be rotating. This effect is

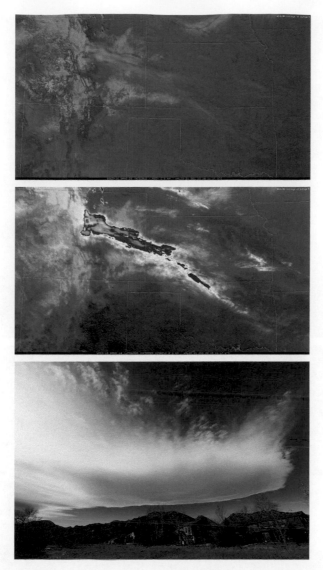

Figure 2.76 *Satellite and ground-based views of an orographic wave cloud. (top) A GOES-16-East visible-channel view of cirrus from an orographic mountain wave over the Colorado Front range stretching all the way down into Oklahoma, Dec. 21, 2020. (courtesy of College of DuPage Meteorology) (middle) As in the top panel, but in the longwave-infrared channel, which shows the stream of clouds more clearly. The coldest, highest top (orange) is seen right in the lee of the Rockies, while the warmer, lower tops (purple) are seen in a narrow band streaming into north-central and eastern Oklahoma on a northwesterly wind current (supported by an analysis of the wind field, but not shown here). The cloud material and water vapor are probably gently subsiding as they are carried southeastward by the wind. If it had been the late spring or summer, I would have first suspected that this cloud feature was an anvil from a convective storm. (bottom) Orographic wave cloud over the Foothills of Boulder, Colorado, on Dec. 21, 2020. A satellite view of this cloud is seen in the top and middle panel. Boulder is around 40° N, which is marked by the border between Kansas and Nebraska in the satellite image. View is to the west.*

Photo courtesy of author.

Figure 2.77 *Orographic wave clouds over the Foothills of Boulder, Colorado. (top panel) Wave cloud on June 10, 2020. View is to the west. (bottom panel) Wave cloud after sunset on Dec. 17, 2022. View is to the west.*

Photos courtesy of author.

Figure 2.78 *Orographic wave clouds streaming toward the viewer with pronounced inhomogeneities in the north-south direction, along the mountains. Viewed to the west from Boulder, Colorado. (top panel) On April 11, 2021. (bottom panel) On Jan. 5, 2023, at sunrise.*
Photos courtesy of author.

Figure 2.79 *Wave clouds at sunrise, that are partially brilliantly illuminated. (top) The underside of an orographic wave cloud, with a mostly well-defined back edge, viewed to the west over the Foothills of Boulder, Colorado, during sunrise on Jan. 3, 2012. It was possibly forced by a hydraulic jump. The vertical striations at the cloud's edge and the curtain-like effect make it look like an aurora. (middle)) Sunrise over the Flatirons of Boulder, Colorado, viewed to the west, with the underside of an orographic wave cloud brilliantly illuminated at sunrise, on Oct. 24, 2020, revealing complex, wavy structures. The lighting in this cloud is similar to that in the top panel, but the back edge is much more ragged-looking and the texture of the underside more wavelike. (bottom) Orographic wave cloud over Boulder, Colorado, on Dec. 19, 2021, at sunrise.*
Photos courtesy of author.

Figure 2.80 *Wave clouds at sunset with striations on their bottom side. (top, left) The colorful underside of an orographic wave cloud, exhibiting complex structure, near sunset, over the Foothills of Boulder, Colorado, on Jan. 7, 1996. This cloud might be triggered by a hydraulic jump. There is no evidence of the cloud base leaning upstream with height, so it is not likely associated with vertically propagating gravity waves. (top, right) Colorful orographic wave clouds at sunset over the Foothills of Boulder, Colorado, on Nov. 27, 1984. The cloud may be a bar cloud, but the northern and southern extents of the cloud are not visible in the frame. There is also some higher cloud above, which if it lies to the west of the lower clouds, is evidence of vertically propagating gravity waves. (bottom, left) The rear edge underneath an orographic wave cloud at sunset over the Foothills at Boulder, Colorado, on Jan. 30, 2021. Note the wavy striations and highly textured underside of the cloud. Clouds such as these are most colorful at sunset when there are no intervening clouds to the west, and about 10–20 minutes after sunset, depending on the height of the cloud base and the sun angle (related to the time of year). At sunrise, the colors are also most vivid about 10–20 minutes prior to sunrise, and when there are no intervening clouds to the east. (bottom, right) Underside of an orographic wave cloud over Boulder, Colorado, on Jan. 2, 2022, viewed to the west.*

Photos courtesy of author.

Figure 2.81 *Series of photographs (left to right, then down a row) of an orographic wave cloud over Boulder, Colorado, on Jan. 2, 2022, taken every few minutes beginning before sunset and continuing on until after sunset. These photographs are reminiscent of Claude Monet's drawings of Rouen cathedral under different lighting conditions. The colors here progress from a dominant blue-white-gray pre-sunset to yellow-dark blue to red-orange during sunset, to gray-white/light blue after sunset.* Photos courtesy of author.

Figure 2.82 *Bar clouds. (top, left) Bar cloud over the Foothills of Boulder, Colorado, on Jan. 16, 1989, with its reflection in a pond. It exhibits some stacking like a vertically trapped gravity wave cloud. (top, right) Bar clouds over the Foothills of Boulder, Colorado on Feb. 6, 2016. It exhibits some stacking like a vertically trapped gravity wave cloud. (middle) Bar clouds on April 2, 2009 (left, sunrise) and (right, later in the morning) over the Foothills of Boulder, Colorado. It exhibits some stacking like a vertically trapped gravity wave cloud. (bottom) Bar cloud, Boulder, Colorado, on Feb. 24, 2016. It exhibits some stacking like a vertically trapped gravity wave cloud.*

Photos courtesy of author.

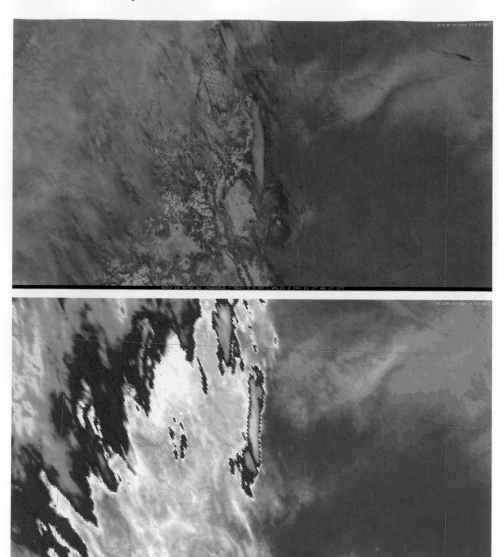

Figure 2.83 *Satellite view of a bar cloud. (top) A bar cloud (center) just east of the Continental Divide (the snow-capped Divide is seen as branches of white just to the west of the bar cloud) in northeastern Colorado as seen in the visible range by the NOAA GOES-East satellite on Jan. 3, 2021. (bottom) As in the top panel, but an infrared image, in which the yellow and orange shading indicates colder temperatures and higher altitude.*

(Courtesy of College of DuPage Meteorology.)

Figure 2.84 *Panoramic views of bar clouds and other wave clouds. (top) Bar cloud on Feb. 11, 1985 over the Foothills of Boulder, Colorado. There are also some higher-level clouds probably associated with vertically propagating or vertically trapped waves. However, the upper-level cloud bands do not exhibit any stacking, necessary for trapped waves nor any leaning with height upstream, necessary for vertically propagating waves, and so its classification is ambiguous. We would not see banding, however,*

likely an illusion, as cloud particles form on one side, move along, and dissipate at the other edge. I think that seeing curved lines fools one into believing that air motion follows the lines and that the lines curl back behind the cloud, which is not necessarily true (Figs. 2.98, 2.103, and 2.104). Later we will see that this phenomenon is also observed in some supercells around their updraft bases. In some instances, the wave clouds are asymmetric, having a wavy tail, like a curved panhandle, on the downwind side only (Fig. 2.97). They may appear to be tilted (Fig. 2.101), perhaps owing to the perspective of the viewer. In many instances only two plates can be seen (Fig. 2.102).

Are lenticular clouds initiated by vertically propagating waves or by vertically trapped waves? It is difficult to tell based only on their appearance, because when they appear totally isolated, it seems as if they cannot be due to vertically trapped waves because multiple clouds periodically distributed are not seen. It might simply be that there is not enough moisture to form more clouds elsewhere. When they occur along with or along bands of cloud, it is likely they are associated with trapped waves. If they appear just downstream from mountains then they could be due to vertically propagating waves, but there is not enough moisture available aloft for other layers of cloud to form. Another possibility is that when the mountains are particularly steep, that *nonlinear* gravity waves form and that waves of many different wavelengths are triggered, which when all combined are manifest as a single bump upward. When isolated lenticular clouds are seen in the lee of isolated mountains, such as the volcanic peaks in the northwestern part of the US, it is most likely that these clouds are triggered by vertically trapped gravity waves in the lee of the isolated mountains (Fig. 2.94). At other altitudes, standing lenticular clouds are called stratocumulus standing lenticular (SCSL) (low altitudes) and cirrocumulus standing lenticularis (CCSL) (high altitudes).

In some wave clouds, the spacing between plates may nearly vanish, so that the clouds appear as nearly vertically oriented cylinders (Fig. 2.103). In this case moisture variations in the vertical may be very small.

Figure 2.84 (Continued) *at high levels in vertically trapped waves. (Composite panoramic photograph) (second row) View to the west from Boulder on Dec. 23, 2022. (third row) Panoramic view, from south-southeast to north-northwest of an orographic wave cloud over the Foothills of Boulder, Colorado, on Jan. 3, 2021. This is the stationary, orographic wave cloud seen by satellite in Figure 2.83. It appears to be a bar cloud that is slightly longer than the typical bar cloud. The cloud appears artificially curved, owing to the camera's internal software for stitching together of frames. (bottom row) Panoramic view of a bar cloud and a cloud above, possibly associated with vertically propagating gravity waves (if the cloud leans to the west with height) or vertically trapped gravity waves (if the upper-level bands parallel to the mountains are periodic in the east-west direction), on Dec. 23, 2013, in Boulder, Colorado. The bar cloud exhibits some vertical stacking. Like the previous panel there is some artificial curvature.*

Photos courtesy of author.

Figure 2.85 *Bar clouds as viewed both on edge and head on. (top, left) Probably a bar cloud south of frozen Barker Reservoir, Nederland, Colorado, on Dec. 30, 2013. The view is to the south, nearly along the axis of the cloud. The cloud looks compressed in the north-south direction owing to the viewing angle: I was almost underneath the cloud, looking down the length of it. (top, right) Probably a bar cloud over Nederland, Colorado, on Dec. 20, 2018, looking somewhat like a comet streaking across the sky. It is similar to the cloud seen in the top, left panel, with the same angle of view. There are, however, also isolated lenticular (to be discussed shortly) clouds visible near the horizon. (bottom) Bar cloud seen in the top, left panel, but about 40 minutes later, near the end of the day, over the Foothills of Boulder, Colorado, on Dec. 30, 2013. In this view we are looking perpendicular to the axis of the cloud, thus giving credence to my interpretation of the cloud shown in the top, left panel as being a bar cloud.* Photos courtesy of author.

The edges of some wave clouds are very fuzzy looking (Figs. 2.105 and 2.106), possibly because the diameter of the cloud droplets varies widely, as noted earlier. In this case, the smaller droplets evaporate before the larger droplets do, thereby making the boundary between clear air and cloud appear less distinct.

Figure 2.86 *More bar clouds on edge. (top, left) Probably a bar cloud viewed on end, on Feb. 25, 2016, from near Ward, Colorado. The view is unusual, in that most bar clouds in Boulder are viewed perpendicular to the major axis of the cloud, not along the major axis. The cloud looks deceptively as if it is building upward, which it is not. (top, right) Orographic cloud viewed on end, on Nov. 2, 2001 near Ward, Colorado, viewed on end. It is possibly a bar cloud, but the northern end of the cloud is not visible in this frame. This cloud also looks deceptively as if it's building upward, but like the one seen in the top, left panel, it is not. (bottom, left) Orographic wave cloud on Dec. 27, 1988, near Ward, Colorado, as it stretches off to the south. It might be a bar cloud but the southern end is not visible. Wisps of cloud are seen mainly on the windward side (right) and to a much lesser extent, on the lee side (left). It is not known why there is this asymmetry. (bottom, right) Isolated wave cloud over Boulder, Colorado, on Jan. 4, 2021. This cloud is probably aligned at a large angle from the Continental Divide and is therefore not a bar cloud, though it looks like the same type of cloud seen in the other panels.* Photos courtesy of author.

Figure 2.87 *Long orographic wave cloud on Oct. 31, 2015 near Ward, Colorado, viewed looking almost down the length of the cloud as it stretches to the south. It might be a bar cloud, but the northern and southern ends are not visible in this frame.*
Photo courtesy of author.

2.2.9 Kelvin—Helmholtz billows and whale clouds

When altocumulus clouds are present at the interface between relatively dense air below and less-dense air above, and when the wind speed changes rapidly with height in the cloud layer, wavelike motions that are induced may wrap up into curls or cats'-eye-like features (Figs. 2.107, 2.108, and 2.112). These curls are called Kelvin–Helmholtz (K-H) waves or billows (Browning 1971) or rolls, in honor of the two scientists who first explained them. They occur not only in clouds, but also in clear air. They have also been seen in clouds on Jupiter and Saturn, and, in a singular sense, in woodblock prints of a breaking wave by the Japanese artist Hokusai.[21] Evidence of K-H waves was first detected by radar many decades ago. Figure 2.113 shows a radar depiction of K-H waves in a winter storm in Oklahoma (see also Houser and Bluestein 2011). While the most impressive-looking K-H waves I have seen are along portions of larger-scale wave

[21] The prints are actually of an isolated, breaking ocean wave, not specifically of a train of waves.

Figure 2.88 *A possible, very narrow bar cloud forming over the Foothills of Boulder, Colorado, on Feb. 1, 2009, at sunset. There were northern and southern edges to the cloud not seen in this frame. Since it is so narrow, perhaps it should not be classified as a bar cloud; perhaps it is a rotor.*
Photo courtesy of author.

clouds downstream from the Rocky Mountains, they are also seen occasionally in low-elevation, non-mountainous areas (Fig. 2.109, left panels, top and bottom). Most are seen as a periodic series, sometimes as just wisps (not curls), as in Figure 2.111, right panel, but on rare occasions, as just one curl (Figs. 2.114 and 2.115). When this happens, it's possible that the phenomenon is just caused by strong shear along the top edge of an isolated cumulus cloud. K-H waves are seen not only in altocumulus/altostratus clouds, but also in cirrus clouds. Like rotors, they are also aviation hazards. K-H waves are also sometimes seen at the top of fog and low-cloud decks (Fig. 2.115).

Imagine that there are two layers of a fluid, such that the denser fluid below is separated from the less-dense fluid above. This configuration of the lighter fluid (air) above the heavier fluid (air) is stable with respect to vertical displacements. Suppose now that we perturb the line separating the two fluids by alternately pushing up and pushing down, so that the line separating the two fluids becomes wavy (Fig. 2.110), like the surface of the ocean. Imagine now that the winds are from the left and increase in speed with height. If a paddlewheel, whose axis of rotation extending into/out from the page, were placed in the flow, then it would rotate in the clockwise direction. With time (see the distorted waves underneath), in the reference frame of the waves, which will move to the east, the tops of the crests would progress to the east faster than the bottoms of the troughs, so that the line would become distorted asymmetrically and produce what looks like a curl beginning to the right. Eventually the denser air might overlay the less-dense air below

Figure 2.89 *A mixture of an incipient bar cloud parallel to the orientation of the Continental Divide) and cloud bands at an angle to the Continental Divide with iridescence (purple and green), rotor clouds (isolated, scud-like clouds), and a contrail (bottom, left) at sunset on Jan. 18, 2003, over the Foothills in Boulder, Colorado.*
Photo courtesy of author.

and become negatively buoyant, thereby sinking. Also, the air at the top of the cloud could become negatively buoyant if there were a lot of radiational cooling and thereby add to the distortion.

The traditional analysis of K-H waves is for an atmosphere which does not have any buoyant elements (Drazin and Reid 1981), but we do see curls along the edge of clouds having upward-bulging top edges, which may have some buoyancy (Fig. 2.112, bottom panel) or quasi-cumuliform[22] tops (Fig. 2.111, left panel) [and even in negatively buoyant elements such as mamma (Fig. 4.34)]. In any event, the formation of K-H waves is favored by a layer in the atmosphere that is characterized by strong vertical wind shear and relatively low stability with respect to vertical displacements.

Isolated K-H waves may also be seen when there is easterly, cold, moist upslope in the cold low-level air mass below and strong northerly of northwesterly, warm, dry downslope above the cloud deck (Fig. 2.115).

Viewed along, rather than across, they sometimes look like a series of short, parallel bands (Fig. 2.116).

---

[22] In this photograph the edges of the cumulus cloud are ragged-looking, indicative of a lot of mixing.

Figure 2.90 *Isolated rotor clouds. (top) A ragged-looking, stationary, rotor cloud viewed to the west over Bear Peak in Boulder, Colorado, at sunrise on March 23, 2003. (middle) A stationary rotor cloud over Boulder, Colorado just after sunrise on Dec. 26, 2006. (bottom) Ragged-looking stationary rotor cloud with ground fog below in a valley, near Villa Carlos Paz, Argentina, on Nov. 13, 2018.*
Photos courtesy of author.

Figure 2.91 *Rotor clouds with other wave clouds aloft. (top) Stationary rotor cloud with the back edge of an orographic wave cloud above, west of the Foothills of Boulder, Colorado, on Nov. 16, 2020. (bottom) Orographic wave clouds on Dec. 20, 2021 over Boulder, Colorado. These wave clouds were probably associated with trapped lee waves, since there is a periodicity evident. There is evidence of a low-altitude, ragged-looking rotor cloud.*

Photos courtesy of author.

Figure 2.92 *Isolated lenticular clouds with rotor clouds underneath. (top) Isolated lenticular cloud over the Foothills, Boulder, Colorado on Jan. 2, 2014. There are also ragged-looking rotor clouds underneath the lenticular cloud. There are also some higher cirrus clouds (upper left), which may be associated with vertically propagating waves or a hydraulic jump. (middle) Isolated lenticular cloud near sunset, with a stationary rotor cloud (in shadows) below it, on Jan. 18, 2016. (bottom) Isolated, tiny, lenticular cloud forming over the Flatirons in Boulder, Colorado, on June 29, 2014. The rotor cloud (underneath the smooth-looking cloud), which looks ragged, is actually slightly longer than the smooth, lenticular cloud above it.*

Photos courtesy of author.

Figure 2.93 *Isolated, small, flying-saucer lenticular clouds (top, left) Isolated flying-saucer-like lenticular cloud north of Boulder, Colorado, on June 29, 2014. View is to the northwest. (top, right) Isolated lenticular cloud as viewed from above, over the mountain west, somewhere between Denver, Colorado and Seattle, Washington, on Jan. 24, 2011. (bottom, left) Isolated, flying-saucer-like lenticular cloud over Cambridge, Massachusetts, on Dec. 2, 2016. This cloud was downstream from small mountains to the northwest. The cloud takes on a dark purplish hue as not much scattering of the light has traveled back to the viewer. View is to the north. (bottom, right) Isolated lenticular cloud viewed to the north from Caribou Ranch, northwest of Nederland, Colorado, on Aug. 6, 2022.*
Photos courtesy of author.

There are other types of clouds that are probably associated with orographically induced gravity waves. "Whale clouds," described by Richard Scorer in his classic cloud book, are probably just isolated wave clouds (associated with either vertically propagating or trapped waves). They are so-called because they have tails that look like fins (Fig. 2.117). Their tails (seen on the side of their parent clouds) are likely a result of cloud droplet formation on their windward side, growth or conversion to ice crystals, and then streaming out on their lee side. These clouds are probably just lenticular clouds that aren't dominated by the archetypal lens shape. Other small clouds look like upside-down curls, similar to Kelvin–Helmholtz billows (Fig. 2.118); in the spirit of Richard Scorer,

Figure 2.94 *Isolated lenticular orographic wave clouds in the lee of isolated mountains. (top, left panel, viewed from an upper floor of a hotel in downtown Portland, Oregon) on Nov. 10, 2016, in the lee of Mt. Hood; (top, right panel) Airborne view of wave clouds on June 11, 1987, in the lee of Mt. Rainier, Washington. These wave clouds are probably associated with vertically trapped waves owing to the vertical stacking, and, in the latter case but not shown, periodicity. The flow has a component from right to left (west to east) in the photographs in the top panels. (bottom panel) Lenticular clouds in the lee of Long's Peak, seen as a flat-topped mountain at the lower left, on Dec. 23, 2022, viewed to the north from Boulder, Colorado. In this case, unlike in the top panels, the flow has a component from left to right, or from west to east.*
Photos courtesy of author.

I suggest calling them "dolphin clouds," because they look like dolphins breaching the surface of the ocean.

Some other orographic wave clouds which defy easy classification are seen in Figure 2.119.

Figure 2.95 *Isolated lenticular cloud over the Foothills, Boulder, Colorado, on Dec. 9, 2014. There is pronounced iridescence (pink/purple and aqua) on the bottom of the top cloud. There are two more layers of cloud visible underneath the main cloud. The iridescence is a sign of small water droplets or ice crystals of approximately the same size and may indicate that the cloud has formed only recently. Since there is a second cloud layer underneath the larger one, this cloud might be associated with a vertically trapped gravity wave. The third cloud, ragged-looking at the bottom of the stack, might be evidence of a rotor.*
Photo courtesy of author.

2.2.10   Chaotic or unsteady waves

In rare instances, wave clouds do not have a coherent appearance, but rather look chaotic/irregular, and may change their shape rapidly with time. Such "unsteady waves" look like a tangled web of yarn (as seen in Fig. 2.6).

2.3   Boundary-layer rolls and bores

Narrow, parallel bands of low-level clouds are often seen in satellite images and from the ground (Fig. 2.120). These clouds may be found in an either stable or unstable environment. In the latter case, to be discussed in a later chapter, these convective cloud

Figure 2.96 *Isolated, layered, wave clouds. (top, left) Isolated lenticular cloud east of Brainard Lake in the Indian Peak Wilderness, Colorado, on April 5, 2021. View is to the east. (top, right) A lenticular cloud that looks like a flying saucer or space ship landing over Niwot Ridge, Indian Peaks, Colorado, near the Colorado Mountain Research Station (right) of the Institute of Arctic and Alpine Research (INSTAAR) of the University of Colorado, on June 25, 2019. There is still plenty of snow left over from the winter season on Niwot Ridge. (bottom, left) Isolated, stationary, orographic wave cloud, viewed to the north, from Boulder, Colorado, on March 22, 2015. (bottom, right) An isolated stationary orographic wave cloud over the Foothills of Boulder, Colorado, on March 31, 2003. View is to the south.* Photos courtesy of author.

bands are called *cloud streets*. However, in the case when the atmosphere is stable with respect to vertical displacements related to buoyancy forces at the *top* of the boundary layer, they appear to be relatively smooth (Fig. 2.120, top panels; Fig. 2.121). When we chase storms with the objective of finding those that might later produce severe weather and tornadoes, we avoid areas where there are these *smooth* cloud bands, which are particular manifestations of what are called *boundary-layer rolls* or *horizontal convective rolls (HCRs)* (Weckwerth et al. 1997) (Fig. 2.121, lower, right panel). The rolls refer to bands of alternating rising and sinking motion near the surface. Parallel lines of rising motion may produce bands of clouds if there is sufficient moisture for air to be lifted and cooled to saturation. When the bands are found underneath regions of stable air at the top of the convective boundary layer, at cloud top, we would not expect them to grow any further,

Figure 2.97 *Lenticular clouds with a pronounced tail extending off downstream from it, giving it the appearance of a panhandle. If supercooled cloud droplets on the windward side of the cloud grow or turn into ice crystals, they may be blown downstream from the cloud, creating a tail. (top, left) Stationary altocumulus lenticularis over Boulder, Colorado, on Oct. 14, 2020. This cloud has a tail (right) trailing off downstream (lee) and looks like an early twentieth-century Futurist painting that renders an airborne vehicle speeding to the left (west). (top, right) Stationary, but rapidly changing, altocumulus lenticularis with a panhandle over Boulder, Colorado, on Nov. 10, 2020. A larger-scale orographic wave cloud lurks above. The lenticular cloud may have been associated with a vertically trapped gravity wave, owing the periodicity in the clouds undulations. In this case the tail is seen to the left of the cloud. The view is to the southwest. (bottom, left) Stationary lenticular cloud viewed at sunset from Boulder, Colorado, on Dec. 21, 2020. (bottom, right) This photograph was taken slightly later than the photograph in the bottom, left panel. The view is to the north.*
Photos courtesy of author.

particularly not into convective storms (see discussions in Chapters 3 and 4 on how high above the ground convective clouds may extend). Rolls may not necessarily be manifest as clouds at all, however, if it is too dry for condensation to occur. While the adjective "convective" means that the boundary-layer air is unstable (with respect to vertical displacements), if they occur at night or during the day if there is not much heating from the sun, then they probably owe their existence almost entirely to instabilities related to vertical wind shear, which induces vertical mixing and results in dryadiabatic lapse rates. Most HCRs, however, require *both* thermal buoyancy and instabilities related to vertical shear.

Figure 2.98 *Isolated, multi-layered wave clouds. (top, left) Isolated, stationary, pile-of-plates (closely stacked and striated) lenticular cloud viewed to the north, over Boulder, Colorado, Jan. 3, 2018. This cloud was probably associated with vertically trapped waves. (top, right) An isolated, stationary, orographic wave cloud that looks like a slightly tilted pile of plates very closely stacked, on Oct. 9, 1990, near Nederland, Colorado. It is so weird to watch a cloud like this remain stationary, as if it were hovering over the ground, waiting to land. The tilt may be an artifact, created if the viewing angle has a component normal to the mountains. (bottom, left) Isolated, stationary, altocumulus* lenticularis *Boulder, Colorado, viewed to the north, March 12, 2022. (bottom, right) Stacked-plates-looking isolated, stationary orographic wave cloud over the Foothills of Boulder, Colorado, at sunrise on Nov. 2, 2001. View is to the west.*
Photos courtesy of author.

The orientation of the bands may be related either to the wind direction near the surface or the direction of the vertical wind shear in the layer of air between the ground and the clouds: the vertical wind shear in this case is the vector difference between the wind at the cloud level and the wind at the surface. The spacing between the bands over the land is usually around three times the depth of the boundary layer. So, one can infer something about the depth of the boundary layer from the spacing of the rolls.

Sometimes parallel, smooth cloud bands *normal* (rather than parallel to or at a small angle to) to the wind and found behind shallow cold air masses, are associated with *bores*. The discovery of bores in the atmosphere was stimulated by the long-known

Figure 2.99 *Sky with numerous lenticular clouds. (top, left) Fuzzy-looking flying-saucer lenticular clouds invading Norman, Oklahoma, on May 29, 2012, midday, prior to tornadic supercells early in the evening northwest of Oklahoma City (see Fig. 4.110, middle panel). These clouds were probably produced by stationary, vertically trapped gravity waves. The haziness is due to the trapping of pollutants underneath an inversion/stable layer. The haziness may be a sign of a broad spectrum in cloud droplet size, as noted in Chapter 1. (right) Fuzzy-looking "flying saucers" invade Norman, Oklahoma, on May 4, 2013. These clouds are most likely forced by vertically trapped waves owing to their periodicities and some vertical stacking. (bottom, left) Flying-saucer orographic wave clouds, some with fuzzy edges, over Cambridge, Massachusetts, on June 12, 1972.*
Photos courtesy of author.

phenomenon of tidal bores, which are triggered when the leading edge of an incoming relatively dense, salty, ocean tide collides with relatively light, fresh water flowing in a river or other narrow channel down to the sea from higher elevation. The bores propagate upstream against the direction the river is flowing and can produce unexpected waves. Bores in general are high-amplitude, nonlinear gravity waves that occur when very stable air encroaches on a layer of less stable air, but which is still relatively stable (Fig. 2.122). In "solitary" waves there is a leading, high-amplitude wave, followed by a succession of decreasing-amplitude waves (Fig. 2.122d) (Haghi et al. 2017). Bores may be detected from satellite imagery (Fig. 2.123, top panel) and also by radar (Fig. 2.123, bottom panel) (Snyder et al. 2015). At the ground, clouds driven by bores may look like the singular rotor clouds that are associated with orographic gravity waves (Fig. 2.124), but are less ragged looking; they are sometimes called "roll clouds." They may also produce parallel bands of stratus, stratocumulus, or altocumulus *castellanus* (Fig. 2.125).

Figure 2.100 *Pile-of-plate wave clouds at sunset. (top) Isolated stationary orographic wave cloud after sunset over Bear Peak in Boulder, Colorado, on Jan. 14, 2012. The stacked plates have spaces between them. Although they look vertically stacked, they probably actually are since the color and brightness on each cloud plate is nearly identical: If they were horizontally "stacked," they would probably have different colors or appear with different levels of brightness. (bottom) Layered orographic wave cloud over Boulder, Colorado, on Dec. 11, 2021, illuminated by the setting sun. (bottom, right, insert) A real stacked pile of plates, not flying ones!*

Photos courtesy of author.

Figure 2.101 *Tilted mountain-wave clouds. (top, left) A short-stack of plates comprising a stationary orographic wave cloud over the Foothills in Boulder, Colorado around sunrise on Dec. 4, 2016. The cloud is oddly tilted, i.e., not in the upstream or downstream direction. If the cloud were actually oriented at an angle to the mountain range, however, and if the right side were farther away, then it would appear lower and tilted, as seen above. (middle, top) Isolated, apparently tilted, stationary, lenticular cloud from a snowfield on the south slope of Mt. Audubon, Indian Peaks, Colorado, on July 15, 2015. This cloud looks as if it could beach itself against the mountain slope, but it actually was not speeding into it, but instead was stationary. The tilt seems to be upward in the downstream direction (to the left, south). However, if the cloud is actually oriented normal to the point of view, then the right side would be farther away and lower. (top, right) Stationary orographic wave clouds in Boulder, Colorado, July 9, 2019. It is a bit unusual to see clouds associated with lee waves during the summer because the airflow over the Rocky Mountains from the west is usually weak. In addition, during the summer the air is less stable and the boundary layer is relatively deep. The view is to the north. The clouds appear to tilt downward in the lee direction (to the right, east). However, if the clouds are oriented at an angle to the mountains such that right edge is farther away than the left edge, then the right edge would appear lower. bottom, left) An apparently tilted pile-of-plates, stationary, lenticular cloud over the Foothills of Boulder, Colorado, at sunset on Jan. 3, 2021. This cloud appears to be pointed downward in the lee (left) direction. However, the apparent tilt could be due to the cloud being aligned parallel to the mountain range, as the left, more distant side, viewed at an angle to the mountains, is farther away and therefore lower. (bottom, right) Tilted-appearing, orographic wave cloud at sunset, over Boulder, Colorado, on Jan. 7, 2022.*
Photos courtesy of author.

Bores will be revisited in Chapter 4 in relation to cold pools produced by convective storms that "intrude" into a stable, nocturnal boundary layer.

In summary, bands of relatively smooth, shallow clouds could be caused by gravity waves, bores (a special type of gravity wave), or boundary-layer/horizontal convective rolls; the cause may not be obvious from photographs without closer inspection of meteorological data.

Figure 2.102 *Wave clouds with two distinct layers. (top, left) A stack of two plates connected in a stationary orographic wave cloud over Boulder, Colorado, on Jan. 14, 1985. (top, right) Stationary orographic wave clouds consisting of two widely separated plates, over Boulder, Colorado, June 24, 2019. (middle) Two layers of altocumulus lenticularis in the distance northwest of Boulder, Colorado, on Dec. 12, 2021. (bottom) Two separate plates of a stationary orographic wave cloud over Bear Peak, Boulder, Colorado, on April 24, 2021.*

Photos courtesy of author.

Figure 2.103 *Narrow, cylindrically shaped wave clouds. (top, left) Isolated, tilted, stationary orographic wave cloud with striations, Boulder, Colorado, on Jan. 1, 2018. Because there is some leaning of the cloud in the upstream direction, vertically propagating waves are possibly associated with this cloud. (top, right) Isolated vertical column, stationary, orographic wave cloud with striations over Boulder, Colorado, near sunrise, on Jan. 5, 2018. (bottom, left) Vertical columns of a stationary orographic wave cloud, northwest of Boulder, Colorado, on Jan. 3, 2021. Since there is some periodicity and vertical stacking, these clouds are probably due to vertically trapped gravity waves. (bottom, right) A smoothly sculpted, stationary, vertically trapped wave cloud over Boulder, Colorado, Jan. 6, 2018. This cloud looks as if it has been machined-tooled by a lathe. On Dec. 17, 2020 there was an almost identical-looking cloud in the same location (not shown). It is likely that some particular mountain peak (probably Long's Peak) may have played a role in forcing it.*
Photos courtesy of author.

2.4 Isolated cirrus

> "And higher, higher yet the vapors roll:
> Triumph is the noblest impulses of the soul!
> Then like a lamb whose silvery robes are shed,
> The fleecy piles dissolved in dew drops spread,
> Or gently waft to the realms of rest,
> Find a sweet welcome in the Father's breast"
> J. W. von Goethe[23]

---

[23] Although this English translation of Goethe's poem may be found in many places, I first encountered it in a *New York Times* book review by Alfred Corn on Richard Hamblyn's book *The Invention of Clouds*, on July 29, 2001.

Figure 2.104 *Wide, quasi-cylindrically shaped wave clouds. (top) Stationary, orographic wave cloud, with many striations or plates, viewed to the north from Boulder, Colorado, on Nov. 10, 2020. (middle) Back (western) side of an orographic wave cloud with many layers or plates, viewed to the north, from Boulder, Colorado on Dec. 12, 2021. (bottom) As in the top panel, but at sunset on Jan. 3, 2018.*
Photos courtesy of author.

Figure 2.105 *Smooth, fuzzy-looking, isolated wave clouds. (top, left) Isolated stationary wave cloud that looks like an upside-down gum drop, over Boulder, Colorado, on Jan. 4, 2021, late in the day. View is to the north. (bottom, left) Isolated stationary wave cloud viewed to the north of Caribou Ranch, Colorado, northwest of Nederland, on Aug. 6, 2022. There is a small cumulus* fractus, *possibly a rotor cloud, underneath and to the left of the wave cloud. (right) Isolated, stationary, but rapidly changing orographic clouds over Boulder, Colorado, on Feb. 2, 2021, viewed to the northeast, late in the day. The edges of the larger cloud are fuzzy, especially at the top, perhaps indicative of a wide spectrum of droplet sizes. The cloud at the lower left looks like a "whale cloud" (but facing downstream, unlike in Fig. 2.117, where the edges face upstream).*
Photos courtesy of author.

2.4.1 Cirrus with vertical structure

While cirrostratus clouds (featured in Section 2.1.2) may cover much of the sky, individual, feathery, or wispy-appearing, more localized, cirrus clouds are often seen. More isolated cirrus clouds may be characterized by whether or not they have fall streaks, heavier ice particles that fall from the cloud, or not. These fall streaks may be oriented vertically when there is no vertical shear or in a slantwise manner when there is shear (Marshall 1953). The tops of the cirrus above the fall streaks may have a bubbly appearance with small turrets or they may be very diffuse or not have well-defined tops at all. When the cirrus clouds do not have fall streaks, they may appear as flattened patches;

Figure 2.106 *Small, fuzzy, isolated wave clouds. (top) An isolated, stationary, orographic wave cloud over Boulder, Colorado, on Jan. 4, 2021, which looks like an egg. (middle) A field of stationary altocumulus lenticularis over Broomfield, Colorado, on Jan. 20, 2003. There is a fuzzy, egg-like orographic wave cloud in the upper center-right. These clouds were probably associated with vertically trapped gravity waves owing to the periodicities and vertical stacking. A fuzzy-edged, elliptically shaped cloud is seen at the upper right. (bottom) Airborne view of an isolated, egg-shaped, lenticular cloud over a broken stratus or stratocumulus undercast, probably near Labrador, on Sept. 7, 2019.* Photos courtesy of author.

Figure 2.107 *Kelvin–Helmholtz billows. (top, left) Kelvin–Helmholtz (K-H) billows over the Foothills of Boulder, Colorado, on Dec. 15, 2009. (top, right). A series of Kelvin–Helmholtz billows over Boulder, Colorado, on Oct. 13, 2020. In this case it appears as if the vertical shear must have been from the southeast, since the view is to the west southwest. This inference, however, is probably incorrect, because it is not corroborated by a time-lapse video that I took. In this video, one can see overturning to the left, while cloud particles seem to be shed just above in the opposite direction, to the right. I hypothesize that there must have been a jet at the level of the waves, such that the vertical shear in the lower half of the cloud was in the opposite direction to that in the upper half. (middle, left) Kelvin–Helmholtz waves at the top of an orographic wave cloud, viewed to the west, from Boulder, Colorado, on Dec. 11, 2021. (bottom, right) Kelvin–Helmholtz waves over Boulder, Colorado on Jan. 15, 2022. Viewed to the west. Photos courtesy of the author. (bottom, left) Kelvin–Helmholtz wave clouds over Boulder, Colorado on Dec. 19, 2021, popping out from an orographic, stratiform cloud band. They may represent individual, buoyant elements or the tops of waves.*
Photos courtesy of the author.

when the patches are sometimes so thick that sunlight is significantly attenuated and shadows are cast on the ground, they are referred to as *spissatus*.

The portion of the cloud in which the ice crystals grow to a size large enough that they fall out (Figs. 2.126 and 2.127), producing a "fall streak," is called a "generating cell"

Figure 2.108 *Small-scale Kelvin–Helmholtz waves at the top of a cloud. (top, left) Kelvin–Helmholtz billows over the Foothills of Boulder, Colorado, on July 19, 2015. View is to the west. (top, right) A long series of at least nine short-wavelength K-H billows on top of an orographic wave cloud, Boulder, Colorado, near sunrise on Nov. 30, 2012. View is to the northwest. (bottom) A series of at least six short-wavelength K-H billows atop an orographic wave cloud over Boulder, Colorado, on Nov. 22, 2009.*

Photos courtesy of the author.

(Heymsfield and Knollenberg 1972). The generating cell may be in a locally buoyant area comprised of mostly supercooled water droplets. If some ice crystals appear, the water droplets evaporate and are deposited onto the ice crystals (because the saturation vapor pressure over ice is less than that over liquid water) and may grow to the point that they are heavy enough to fall out from the cloud. Alternatively, if too *many* ice crystals were to appear, then the ice crystals would grow much more slowly, never reaching a large enough mass to fall out. The process in which the supercooled water droplets turn into ice crystals is known as "cloud glaciation." The buoyancy of the generating cells is

Figure 2.109 *Thin, isolated Kelvin Helmholtz billows. (top, left) A connected series of three K-H billows over Norman, Oklahoma, on April 16, 2007. The sky is mostly clear, while waves appear in the cloud filament. View is to the north. (top, right) Kelvin–Helmholtz billows (left) and blowing snow over Niwot Ridge in the Indian Peaks Wilderness, viewed to the south from Long Lake, on Jan. 3, 2022. (bottom, left) A series of Kelvin–Helmholtz billows over Norman, Oklahoma, on Nov. 15, 2012. (bottom, right) A series of Kelvin–Helmholtz billows over Boulder, Colorado, looking to the north, on Jan. 19, 2021.*
Photos courtesy of the author.

maintained to some extent by the latent heat released when water droplets freeze. When clouds become glaciated, they tend to become fuzzier looking.

Cirrus *castellanus* have small cumuliform towers extending upward from a common base (Fig. 2.127) and sometimes are seen in lines (Fig. 2.128). The cirrus showing fall streaks and some cloud, though not convective looking at the top, may have evolved from cirrus *castellanus*. Since there must be some buoyancy in these towers, they arguably may not belong in this chapter on non-buoyant clouds. Once again, we see an example of the difficulty in classifying clouds according to physical processes alone, rather than just by appearance alone.

On a few very rare occasions. I have seen mamma (see Chapter 4, Section 4.3.1) at the bottom of fall streaks (Fig. 2.129).

Figure 2.110 *An airborne view of a periodic series of waves (see Fig. 2.49) in cirrus clouds, possibly from vertically trapped waves, over the American intermountain west, above ground fog, on Jan. 27, 2011. Four wave crests are visible in this frame. However, they might also be Kelvin–Helmholtz billows when the waves have not yet broken or they may never have broken.*
Photo courtesy of the author.

When isolated cirrus clouds do have fall streaks (falling ice crystals/snow), the vertical wind shear below the cloud determines their shape. Sometimes fall streaks don't even appear to fall, but instead are seen stretched out horizontally across the sky, perhaps because the vertical shear is extremely strong. If you are watching the sky during the summer at a beach on a hot day and see fall streaks from cirrus, you are, remarkably, watching snowfall, but fortunately not reaching the ground and possibly ruining your day.

Scattered, feathery/wispy-looking cirrus clouds with fall streaks that look like hooks or commas are called *cirrus uncinus* (Fig. 2.130). They are also known colloquially as "mare's tails." Unlike the uniform, stratiform cirrus clouds, these clouds seem to hang in the air, with what appears to be cloud material falling or extending downward. Generating cells may or may not be visible (Figs. 2.131, 2.134, and 2.135). Viewed from an aircraft, they seem to gain three-dimensionality (Fig. 2.135), instead of their two-dimensional appearance as seen from the ground.

Cirrus clouds that have fibrous-looking streaks (Fig. 2.134) are given the name *fibratus* and those that are twisted-looking are given the name *intortus* (Fig. 2.132, middle left). The former clouds are like *uncinus*, but don't have a hook shape and are separate from each other. The latter are often twisted around each other and randomly oriented. Since it is not always easy to tell all these types of isolated cirrus apart, they are not showcased in any figure. Their various appearances are probably related in large part to different vertical wind shear profiles. Sometimes we see a substantial portion of the sky covered by small streaks, close together, almost blending in with each other (Fig. 2.133).

Figure 2.111 *Miscellaneous Kelvin–Helmoltz wave clouds. (left) A series of three K-H billows in ragged-looking clouds, in the Indian Peaks Wilderness of Colorado, on Feb. 23, 2006. (right) Airborne view of K-H billows in cirrus clouds, above a cloud deck below, March 11, 2015, at an unknown location over the US between Florida and Oklahoma, probably over the Gulf of Mexico.*
Photos courtesy of the author.

Cirrus *floccus* are not flat and may have fall streaks pendant from them (Fig. 2.136, top and bottom left panels). The name "floccus" comes from the Latin name for pieces of wool, since these clouds look somewhat like tufts of wool, though I think in terms of tufts of cotton. Altocumulus also exhibit this characteristic sometimes (see Chapter 3).

It obviously helps to have studied Latin to understand cloud terminology more easily. It is, however, as previously noted, not always easy to tell the different types of cirrus apart and there may be a mixture of two or more types. There is some ambiguity in classifying cirrus, just as there is in classifying orographic wave clouds.

2.4.2 Flat cirrus

Scattered cirrus clouds that have flat segments and may be thick enough to block out the sun, arc also called cirrus *spissatus*, as are cirrostratus that are thick enough to block out the sun. When extensive cirrus approach, but are not solid enough in coverage to be considered cirrostratus *(spissatus)* overcast, they may appear to diverge overhead and are called cirrus *radiatus* (Fig. 2.137, lcft panel), while sometimes completely isolated cirrus often appear like cirrostratus broken up into small pieces (Fig. 2.137, right panel).

2.4.3 Cirrocumulus

Cirrocumulus clouds are cirrus that have very small-scale features like wavelike ripples or even tiny granules or speckles (Fig. 2.138). The latter remind one of a pointillist painting, such as those by the late-nineteenth-century French painter Georges Seurat, except that *only* the clouds are represented in the pointillist manner in real life. All clouds really are "pointillist" in the sense that if you looked at them on a fine-enough scale, you would see individual cloud droplets or ice crystals. In the case of cirrocumulus, the cloud ice

Figure 2.112 *Kelvin–Helmholtz waves at the top part of a cloud. (top) K-H billows over Boulder, Colorado, silhouetted against a background of cirrostratus clouds, on Jan. 18, 2003. (middle and bottom) Kelvin–Helmholtz waves embedded in an orographic wave cloud over Boulder, Colorado on Dec. 20, 2021. Each individual billow reminds me of an old cartoon of unknown origin (possibly from* The New Yorker*) depicting a series of lemmings lined up, ready to jump off a cliff.*

Photos courtesy of the author.

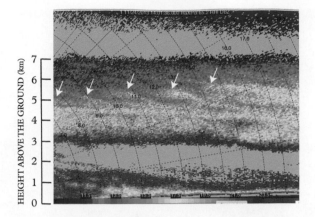

HEIGHT ABOVE THE GROUND (km)

Figure 2.113 *Kelvin–Helmholtz waves noted in a vertical cross-section of radar reflectivity in a winter storm in Oklahoma on Nov. 24, 2013. My graduate students and I collected data during this storm in Norman when there were ice pellets falling. The antenna from the mobile radar RaXPol (described later in Chapter 4) was aimed toward the northeast. Radar reflectivity is color coded (at the bottom) in dBZ (described later; the redder the color, the more intense the precipitation). The top of the precipitation region is at about 7 km and the K-H waves are seen about 5 km above the ground, shown as yellow wisps (~18–19 dBZ) spaced about 1 km apart (white arrows). A layer of strong vertical shear (not shown) was present in a layer just below and within where these waves were detected. An animation of the radar data showed the waves moving from right to left, toward the radar, from the northeast. (See also Fig. 4.61 [bottom panel] for a radar image of K-H waves on top of a gust front emanating from a mesoscale convective system.)*

crystals themselves are clumped into larger entities. The former may belie the presence of short-wavelength atmospheric gravity waves or even Kelvin–Helmholtz billows.

2.4.4 Contrails: Artificial cirrus clouds

When a jet aircraft flies at high altitude, ejecting relatively hot air (hundreds of °C) into the cold environmental air (much colder than ~−30° C), a trail of condensate (cloud), a condensation trail (Figs. 2.139 and 2.140), or *contrail*, is often seen. The mechanism by which the air becomes saturated is the same as that which produces steam fog (Section 2.1.1; Fig. 2.30): the mixing of unsaturated air at two widely different temperatures and different amounts of water vapor may be brought to saturation. According to Richard Scorer, if the cloud droplets do not freeze, they will evaporate because after the temperature of the air reaches the warmer, environmental air temperature, the water droplets should evaporate. If the cloud droplets freeze, however, then it is possible that the water vapor pressure in the air mixture will be above the saturation value for ice, and the contrail will last for a relatively longer time. Contrails may spread out with time and look like natural cirrus clouds. To me, contrails are like air pollution, what I "view" as visual, "cloud pollution," ruining what may have otherwise been a pristine photograph of a natural cloud, with streaks of white.

Figure 2.114 *The most isolated and simplest Kelvin–Helmholtz billow: K-H curl at the top edge of a shallow cumulus* fractus *(broken-looking) cloud, near the Yellow Mountains (Huangshan), China, on July 23, 2012. There might, however, be several others visible to the right, but not as well developed.* Photo courtesy of the author.

There is some evidence that contrails might affect the radiation budget as they decrease the amount of solar radiation reaching the surface, but also might increase the amount of longwave radiation trapped below. So, we would expect that contrails could cause the minimum temperature at night to increase and the maximum temperature during the day to decrease. For several days after the terrorist attacks in the US on Sept. 11, 2001, commercial aircraft were not permitted to fly, thus reducing the amount of artificial cirrus from contrails for a brief period. One study found that the diurnal temperature range in fact was increased during this time in some areas and decreased in other areas.[24] Another study, however, found the higher diurnal temperature range was more due to clearer-than-usual weather in the US during the same period.[25]

Contrails even on occasion have fall streaks or features that look like *mamma*, pouch-like downward protuberances (Fig. 2.141, top panels), to be discussed later in Chapter 4. Contrails like this may be very turbulent and therefore to be avoided by aircraft, though all contrails may be turbulent enough that aircraft should avoid all of them. Unlike mamma underneath the anvils of cumulonimbus clouds, etc., there is often spacing between adjacent pouches that are as wide as the mamma themselves. Richard Scorer proposed that these mamma-like structures are produced by the downward branches of

[24] Travis et al., 2004.
[25] Kalkstein and Balling, Jr., 2004.

Figure 2.115 *Possibly a K-H-like, isolated wave breaking on top of a layer of stratus below, over Boulder, Colorado, on Sept. 25, 2005, from the summit of Green Mountain, viewed to the southeast. This breaking wave looks like that of a ghost from below in the cold low-level air, poking its finger through the cloud layer to escape the gloom below and feel some warmth above. It is also possible that there is a buoyant element breaking through a low-level stable region/inversion and being bent over by the vertical shear, but it is difficult to imagine how any cloud element would become buoyant, owing to the very cold air underneath. The airflow in the cloud deck has a component from the east (left to right), while the airflow above the cloud deck has a component from the west (right to left): Thus, there must have been a westerly component of shear, which would create a wave that breaks to the left, not to the right as seen here. I suspect that there may have been a jet of air with an easterly component just above the cloud deck, so that there is actually an easterly component of shear, not westerly, just above the cloud deck. In looking at time-lapse videos of the top of similar cloud decks, however, one can see the motion of small-scale eddies popping up here and there, which could mean that the direction of the curl in the airflow seen is random and not related to the larger-scale vertical shear, but rather to the shear associated with small-scale eddies.*
Photo courtesy of the author.

a set of counter-rotating, horizontal vortices produced in the wake of the aircraft producing the contrail, when the vortices are periodically closest to each other; the downward air flow produced in between the two vortices and reinforced when they are closest, brings cloud material downward, producing the mamma-like structures. Other contrails exhibit mamma-like pouches that extend upward from the contrails (Fig. 2.141, bottom, left panel), while others exhibit pouches that extend both upward and downward periodically along different segments of the contrail (Fig. 2.141, bottom, right panel).

Figure 2.116 *Kelvin–Helmholtz waves viewed on edge or at an angle, but not broadside. (top, left) K-H billows seen on end, looking to the south, parallel to the Continental Divide, on June 19, 2019, from Mitchell Lake (10, 735 feet MSL), in the Indian Peaks of Colorado. A series of classic K-H curls were observed from Boulder, to the southeast, with a view perpendicular to the undulations. This cloud may have been a lenticular cloud that broke up into billows. (top, right) Orographic wave clouds with billows, viewed to the west, over Boulder, Colorado, on Feb. 2, 2021, late in the day. (bottom) Orographic wave cloud almost overhead in Boulder, Colorado, with possible embedded Kelvin–Helmholtz billows, on Nov. 10, 2020. This cloud looks like an X-ray of the bones of a small, ancient creature.*

Photos courtesy of the author.

An instability called "Crow instability" (named after the author of the paper in which the problem was first analyzed, not after the bird) may be responsible for the wavy inter-action between/among nearby vortices produced behind the aircraft (Fig. 2.142, middle and bottom panels), especially apparent when there are two parallel contrails. This pro-cess and the mamma-like features have been replicated in high-resolution computer

Figure 2.117 *"Whale" clouds. (top) "Whale" cloud (bottom cloud) near Boulder, Colorado on June 18, 2007, in this case appearing to "swim" upstream (to the west, left) against the air flow. (bottom) "Whale" cloud (bottom right) and other orographic wave clouds downstream from the Continental Divide (white-capped mountains on the horizon), on May 5, 2011, as viewed to the north-northwest from the summit of Flagstaff Mountain in Boulder, Colorado.*

Photos courtesy of the author.

Figure 2.118 *"Dolphin clouds" (top) "Dolphin" clouds near Boulder, Colorado, on March 9, 2003. (middle) "Dolphin"-like cloud over Boulder, Colorado, on Feb. 11, 2016. The shape of the Dolphin cloud perhaps suggests the wavelike flow of the air up, over, and down (toward the right, from west to east). (bottom) "Dolphin"-like clouds curled downward (center) amidst other orographic wave clouds, over Boulder, Colorado, on Feb. 23, 2013.*

Photos courtesy of the author.

Figure 2.119 *Odd-shaped orographic wave clouds. (left panel) Unusual orographic, fuzzy-looking wave cloud (upper right), with a tail-like feature, over Boulder, Colorado on June 29, 2014. There are also some lenticular clouds scattered on the horizon to the west and northwest. (right panel) An isolated, sui generis orographic wave cloud, perhaps similar to the "dolphin cloud," almost overhead, Boulder, Colorado, on Dec. 17, 2020. This cloud looks as if it has a frame in the shape of a sidewise-oriented "U" (see also the upside-down "U-shaped" tubular clouds in Chapter 5).*
Photos courtesy of the author.

simulations and attributed to Crow instability, though the details of how the visual mamma-like structures are produced are somewhat complex and not easily described.[26] I think that it is most likely, as originally proposed by Scorer, that the counter-rotating vortices produced in the wake of an aircraft exhaust oscillate toward and away from each other, and that downward motion is enhanced when they are closest to each other and mamma are produced; when the vortices are separated the most, mamma are not produced, leading to a space between the pouches, as is observed in nature. The counter-rotating wake vortices will be discussed later in more detail in Chapter 5, when we look at tornadoes and other clouds that exhibit rotation.

[26] Lewellen and Lewellen, 2001.

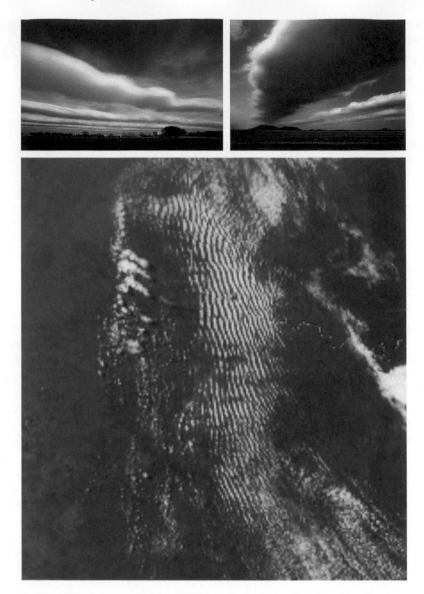

Figure 2.120 *Parallel stratus bands due to horizontal convective rolls when there is a cap preventing any further upward cumulus development. (top, left) Long, parallel, stratus bands in Norman, OK, on Nov. 4, 2006, viewed to the east. (top, right) Long, parallel, stratus bands in southwestern Oklahoma, on May 30, 2012. The Quartz mountains are seen in the lower left. View is to the west-northwest. (bottom) Visible satellite imagery from a NOAA GOES satellite, on May 30, 2012, showing the banded clouds seen from the ground in the top, right panel, in southwestern Oklahoma. These bands look like fingerprints. The spacing between the bands is relatively close on their eastern edge and increases to the west. That spacing between neighboring bands is proportional to the depth of the moist, boundary layer, is consistent with the observation that the boundary-layer depth increased to the west, where the boundary-layer air was drier and deeper. West of the bands it cleared out until the location of the dryline is reached in the Texas Panhandle, where a north-south line of deep convection is beginning. The dryline is a surface boundary separating relatively warm and dry continental air from relatively cool and moist marine air from the Gulf of Mexico.*

Photos courtesy of the author.

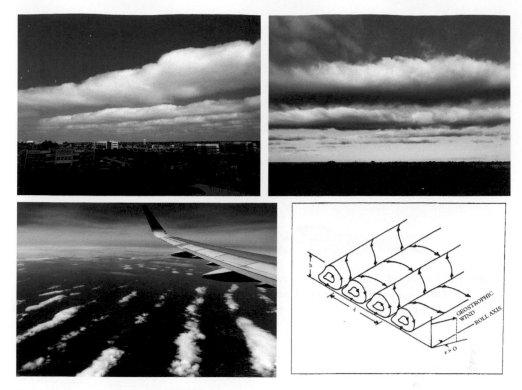

Figure 2.121 *Parallel bands of stratus or stratocumulus due to horizontal convective rolls. (top, left) Bands of stratus or stratocumulus over Norman, Oklahoma, on Sept. 28, 2018, viewed to the north-northeast. (top, right)) Bands of stratus or stratocumulus over Norman, Oklahoma, on Nov. 5, 2018 viewed to the west. (bottom, left) Airborne view of stratus bands over western Massachusetts on May 26, 2022.*

Photos courtesy of the author. (bottom, right) Idealized flow around boundary-layer rolls/horizontal convective rolls (HCRs) (from Weckwerth et al., 1997, their Fig. 1; adapted from Brown, 1980, their Fig. 5a; © American Meteorological Society. Used with permission).

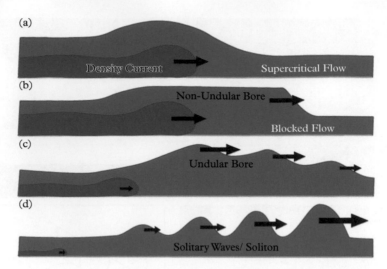

Figure 2.122 *Idealized illustration of the evolution of a bore (from Haghi et al., 2017, their Fig. 4; ©
American Meteorological Society. Used with permission). A dense, heavy, cold air mass (blue) hits a
stable air mass (red) at the leading edge of the cold air mass; the stable air ahead is lifted but does not go
all the way over and past the leading edge (the dynamics will be discussed later in Chapter 4, when we
look at thunderstorm gust fronts), triggering bores, some as singular bumps, others as a series of
ascending (solitary waves) or as a series of bumps descending (undular bores) in amplitude, in the
direction of bore motion.*

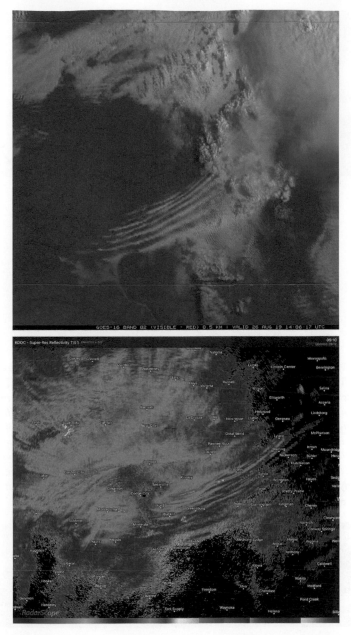

Figure 2.123 *Manifestation of bores by satellite and by radar. (top) Clouds produced by lifting along solitary-wave-type bores (curved, parallel, very narrow, seven cloud bands) over southwestern Kansas, as seen by a NOAA satellite on Aug. 26, 2019 at 9:06 a.m. CDT. In this case the bores were triggered when a southward-progressing cold front made impact on a low-altitude stable layer. (courtesy of College of DuPage Meteorology) (bottom panel) Radar reflectivity (dBZ) image of bores in southwest Kansas on Aug. 26, 2019, at 9:10 a.m. CDT. The green and blue curved, streaky-looking bands probably represent enhanced backscattering of the radar beam from insects that are caught up in narrow zones where air is converging underneath the rising branches of the bore.*

Figure 2.124 *Manifestation of isolated bores. (top. left) A rotor/roll cloud, probably along a bore, at Craigville Beach on the south shore of Cape Cod, near Hyannis, Massachusetts, on July 17, 1975. The mechanism by which this solitary cloud formed is not readily apparent. There is another band of clouds barely visible, on the horizon. (top, right) A band of stratocumulus just offshore of Pacific Grove, California, by the Asilomar conference center, on the morning of Oct. 29, 2015. It may have been triggered by the interaction of the sea breeze with a nocturnal inversion over land. (bottom) This persistent, singular band of cloud just under the low-stratus overcast was seen from Limon, Colorado, on Dec. 9, 2021. I suspect it may have been triggered by a bore at the inversion level in the presence of cool southeasterly "return" flow of moisture, and certainly not a contrail because it is at a low altitude.* Photos courtesy of author.

Figure 2.125 *Panoramic photograph of the leading edge of an approaching cloud probably associated with a bore, in Norman, Oklahoma, on May 3, 2015, viewed to the west (left) through northeast (right). Note the striation in the cloud base near its leading edge. There are other bands partially hidden to the rear of this cloud.* Photo courtesy of author.

Figure 2.126 *Vertically aligned fall streaks. (top, left) Cirrus over the Indian Peaks of Colorado, on Aug. 12, 1984. The relatively large generating cloud is at the top, while ice crystals are falling out and sublimating, literally making a "sublime" image. (top, right) Isolated, very narrow fall streak from an isolated cirrus cloud over Norman, Oklahoma, on April 28, 2020, probably into a layer of weak vertical wind shear, since the fall streak is nearly perpendicular to the ground. (bottom) Cirrus with ice crystals falling out over Norman, Oklahoma on Nov. 12, 2021.*
Photos courtesy of author.

Figure 2.127 *Cirrus with fall streaks and convective tops. (top) Cirrus* castellanus *with fall streaks over Boulder, Colorado, on June 27, 2020. This ghostly looking cloud looks a bit like a generic jellyfish or man-of-war in the ocean, since the fall streaks look like tentacles. This cloud is an example of an elevated convective cell that produces precipitation that sublimates or evaporates in dry air on the way down, producing virga. These look like the cirrus with fall streaks in Fig. 2.126, but with heftier generating cells. (bottom) Cirrus* castellanus *with fall streaks over Norman, Oklahoma, on May 1, 2020.*

Photos courtesy of author.

Figure 2.128 *Lines of cirrus with fall streaks and some evidence of cumuliform tops (cirrus castellanus). (top) An airborne view of a line of cirrus with tiny, cumuliform, convective-looking tops and fall streaks, over an arid portion of the western US, on Dec. 10, 2011. (middle) Fall streaks from small generating cells oriented along a line, over Norman, Oklahoma, on Jan. 22, 2022. The line might have been left over from a contrail. (bottom) Line of cirrus-generating cells with ice crystals falling out, over Norman, Oklahoma, on Jan. 25, 2022. There is an inverted funnel-shaped cloud extending upward from one of the cloud elements (center).*

Photos courtesy of author.

Figure 2.129 *Cirrus band with fall streaks, but not generating cells, but with mamma (pouch-like downward protrusions—see Chapter 4) at the bottom, on Oct. 23, 2018 over Norman, Oklahoma. These are unusual in that fall streaks do not usually exhibit mamma at their bottom edge.*
Photo courtesy of author.

Figure 2.130 *Cirrus* uncinus. *(top, left) Cirrus clouds without bubbly towers, but with slanted fall streaks or "mare's tails," known as cirrus* uncinus, *over Norman, Oklahoma on March 31, 1977. The shape of the fall streaks is probably related to the vertical shear. In this case, ice crystals are falling into much more slowly moving air. (top, right) A repetitive pattern of cirrus* uncinus *over Boulder, Colorado, Jan. 17, 2014: Getting "hooked" on cirrus* uncinus. *In this case the fall streaks take on a slanted and sometimes hooked appearance. (bottom, left) Cirrus* uncinus *strung along at an angle across the Continental Divide, west of Ward, Colorado in the Indian Peaks Wilderness, viewed to the north from Niwot Ridge, on Aug. 13, 2005. The cirrus are partially obscured by fog patches. There are also parallel bands of cirrus above the cirrus* uncinus, *oriented approximately along the Continental Divide. (bottom, right) An isolated cirrus* uncinus *over Boulder, Colorado, Aug. 9, 2019. The brighter, more solid-appearing vertical streak probably is a large generating cell or maybe contains some supercooled water droplets.*

Photos courtesy of author.

Figure 2.131 *Cirrus with fall streaks and some generating cells, but which are nearly horizontally oriented, as a result of strong vertical shear. (top, left) Cirrus in Colorado on Aug. 10, 1986. There are five generating cells, each with fall streaks. These clouds look similar to those that look like sperm cells seen later in the lower-right panel, but the generating cells are larger. (top, right) Cirrus streaks that are nearly horizontal over Shoshone Pk., viewed to the east from the Isabelle Glacier, Indian Peaks Wilderness, west of Ward, Colorado, on Aug. 17, 2006. The generating cells seem to be large and merge with the streaks. (bottom, left) Horizontal cirrus streaks over Woods Hole, Massachusetts, on Aug. 6, 2012. Generating cells are seen on the left. View is approximately to the east. (bottom, right) Cirrus clouds with tails that don't appear to be falling as fall streaks. These clouds look like sperm cells, invading/impregnating the sky. San Francisco, California, on Dec. 28, 1998. There appear to be "generating cells" in most to the left and streaks of ice crystals extend downstream, to the right.*

Photos courtesy of author.

Figure 2.132 *Cirrus with complex patterns of streaks. (top, left) Complex patterns of streaks of cirrus in which some streaks intersect other streaks at right angles, over Boulder, Colorado, on July 20, 2015. (top, right) Cirrus patches, fall streaks, and horizontally oriented cirrus, over Norman, Oklahoma, on Jan. 25, 2022. (middle, left) Cirrus* intortus *(twisted) and a few cirrus* uncinus *swarming over western Oregon, July 17, 1994. (middle, right) A complex pattern of cirrus and fall streaks, over Norman, Oklahoma, on April 20, 2020. (lower, left) Cirrus* fibratus *or* intortus *and possibly some* spissatus *over Norman, Oklahoma, on Oct. 9, 1977. Not many fall streaks are apparent. (bottom, right) Narrow, hook-shaped, cirrus streaks over Half Moon Cay, in the Bahamas, on Dec. 8, 2008. From this vantage point, only one shows evidence of a generating cell. Most seem to have the greatest curvature at the bottom end of the cloud.*
Photos courtesy of author.

Figure 2.133 *Cirrus streaks "invading" and overspreading the sky over (top) Boulder, Colorado, on July 17, 2010 and (bottom) Norman, Oklahoma, on Oct. 22, 2022.*
Photos courtesy of author.

Figure 2.134 *Elongated, fibrous-looking cirrus streaks without well-defined generating cells visible. (top) Ghostlike cirrus streaks from a relatively isolated cloud over Boulder, Colorado, on Dec. 25, 2010. The fall streaks are rather parallel and not hooked shaped. They appear to be aligned at a small angle from the horizontal and probably in an environment of strong vertical shear. (bottom) Long, thin streaks of cirrus, over Albany, New York, on Feb. 24, 2022.*

Photos courtesy of author.

Figure 2.135 *Airborne views of cirrus with "mare's tails." (top) over the western US, on Feb. 26, 1998. While the clouds appear to look two dimensional from the ground, they look three dimensional from above. Generating cells are not apparent in this case. (bottom) Fall streaks over the western US, on July 12, 2019.*

Photos courtesy of author.

Figure 2.136 *Sky covered by patches of cirrus that are not flat or without extensive fall streaks. (top, left) Patches of cirrus* floccus *without much evidence of fall streaks, covering the sky over Boulder, Colorado, on July 16, 2020. (top, right) Flat cirrus patches over Big Bend National Park, Texas, on March 12, 1979, in the absence of any fall streaks. (bottom, left) Three bands of cirrus over Norman, Oklahoma on Feb. 6, 2022. The lower band is a solid line of wispy clouds. In the middle, there are distinct cells, with some indication of buoyancy. On top, there are clumps of wispy cirrus. These clouds on the larger scale were generated along lines parallel to the upper-level flow, which was southwesterly and downstream from an upper-level trough of low pressure. (bottom, right) Flat patches of cirrus over Norman, Oklahoma, on Jan. 22, 2022.*

Photos courtesy of author.

Figure 2.137 *Wide streaks of cirrus. (left panel) Cirrus (radiatus) seemingly streaming from a point over Kennybunkport, Maine, on Oct. 20, 2019. The cirrus clouds appear to be diverging downstream (toward the viewer). (right panel) On Nov. 19, 2022, wide cirrus streaks viewed to the northeast from Norman, Oklahoma.*

Photos courtesy of author.

Figure 2.138 *Cirrocumulus. (top, left) Cirrocumulus with colorful speckles, at sunset, on Feb. 20, 2004 in Norman, Oklahoma: Pointillism in the sky. (top, right) Cirrocumulus with patches, over Boston, Massachusetts, on Aug. 5, 2015. (bottom, left) Cirrocumulus over Norman, Oklahoma, on Jan. 25, 2022. (bottom, right) Cirrocumulus with both pointillist structure (top) and billows, over Salina, Kansas, on Aug. 20, 2019.*

Photos courtesy of author.

Figure 2.139 *Contrails. (top, left) Airborne view of a contrail, viewed sideways, emanating from a jet on March 11, 2019, having the appearance of a comet: The head of the comet is the jet aircraft and the tail is the contrail itself. (top, right) Two contrails emanating from a jet overhead in Boulder, Colorado (not that the location matters!) on Sept. 2, 2020. (bottom, left) "X marks the spot (in the sky):" Crossing contrails and spreading out of one of the contrails over Norman, Oklahoma on Feb. 23, 1989 and appearing like sky writing. The lower-left contrail has some mamma-like protuberances. (bottom, right) A swarm of contrails near sunset viewed to the south from Boulder, Colorado, on Jan. 2, 2022, as a number of jet aircraft had flown along similar flight paths.*

Figure 2.140 *Vertically aligned contrails. (left) Contrail trailing overhead in Norman, Oklahoma, on Dec. 3, 2021, giving the appearance of it sticking upward, just like some orographic wave clouds viewed on end. (right) Contrail piercing an orographic wave cloud over the Foothills of Boulder, Colorado, on Dec. 12, 2021. This contrail looks like a thread going through a piece of material.*
Photos courtesy of author.

Figure 2.141 *Contrails with pouches. (top, left) contrail with a series of pouches or mamma-like protuberances, on Aug. 1, 2021, over Boulder, Colorado. (top, right) Airborne, close-up view of a contrail with a series of mamma-like protuberances, on Sept. 13, 1996, at an unknown location, probably over the Atlantic on the way from London to the US (bottom, left). Contrail over Boulder, Colorado on Dec. 12, 2021, with pouches sticking upward rather than downward, as in the top panels. I suggest that these might be referred to as contrails* castellanus. *(bottom, right) Contrail on Dec. 13, 2021 near Caribou Ranch, northwest of Nederland, Colorado, in which pouch-like protrusions periodically appear both sticking up and sticking down from the contrail, as if the contrail has been twisted about its axis, perhaps due to Crow or another instability.*

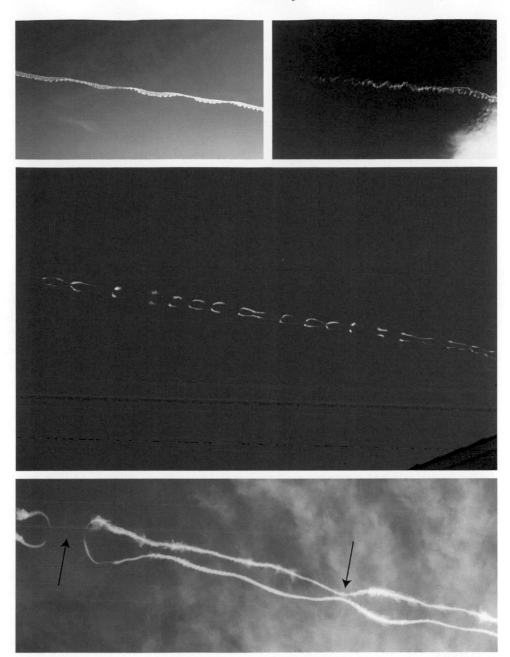

Figure 2.142 *Disturbed contrails. (top, left) Contrail over Norman, Oklahoma, on Nov. 16, 2021 with a wavelike variation appearance. (top, right) Contrail near Caribou Ranch, northwest of Nederland, Colorado, on Dec. 13, 2021, with trailing, wavelike disturbances. (middle) Pair of contrails over Boulder, Colorado, on March 9, 2023, with periodic disturbances typically associated with Crow instability. (bottom) As in the middle panel, but on March 14, 2023. The arrows point to vortices being pulled together (right) and after they have been pulled apart and cut off from the original pair of wake vortices (left). The tubular-shaped cloud lines and segments look similar to those discussed later in Chapter 5 in Section 5.10.2 (high-based funnel clouds and horizontal tubular clouds).*

2.5 Summary

Non-buoyant clouds in a stable atmosphere range from low-level clouds such as fog and stratus to high-level clouds such as cirrus and its many sub-genres. Many non-buoyant clouds at low and middle altitudes have smooth surfaces, owing to the stability of the air with respect to vertical displacements, while those at high altitudes, which are composed of ice crystals, may have feathery and non-uniform surfaces. The organization of non-buoyant clouds is intriguing in that patterns may be repeated periodically as also seen in some buoyant clouds (Chapter 3). The individual patterns may be undulating, banded, chaotic, and otherwise. Some isolated smooth-looking clouds look like gorgeous sculptures in the sky. In some instances, one can guess the atmospheric conditions based on the type of clouds seen and their organization. While understanding what physically is responsible for the appearance of non-buoyant clouds is interesting, the esthetics of the clouds are worthy of contemplation in their own right, as noted for nature in general by Goethe.

Most orographic wave clouds are fascinating to scientists and laypeople alike because they often appear to be stationary in the presence of wind and take on unusual shapes. However, one must recognize that when the air is very dry, orographic gravity waves which are responsible for them may exist but are not evident as any clouds because it is too dry for clouds to form. It can therefore be very difficult, as noted earlier, to identify the nature of the waves responsible for the types of wave clouds we see. For example, an isolated lenticular cloud may be a result of vertically propagating waves, the isolated nature being related to moisture inhomogeneities along the mountain range, or may be a result of trapped waves, but not apparently so because there is insufficient moisture to see any parallel cloud bands farther downstream or because the amplitude of the wave is very high and there is just one big wave. Without enough moisture, wave clouds are "invisible," leaving no visual evidence of the waves. Thus, we come to the conclusion we noted in Chapter 1 that there is a world of invisible, potential clouds out there sculpted by the winds, just waiting for moisture to come out and make them visible. E. O. Wilson, the American naturalist famous for his studies of ants, has noted[27] that "Where scientific observation addresses all phenomena existing in the real world, and scientific theory addresses all conceivable real worlds, the humanities encompass all three of these levels and one more, the infinity of fantasy worlds." We can thus thank the humanities for the capriccios of clouds of artists, the fantasy world of all possible clouds.

In addition, we sometimes see the sky filled with wave clouds of a mixed variety, some from vertically propagating waves and some from trapped waves. Attempts at scientific understanding can be thwarted when different physical mechanisms are at play at the same location, but at different altitudes, as our variation in the Jimmy Buffet metaphor appearing in the title of one of his songs from 1977[28] as: "Changes in attitudes, changes in altitudes."

---

[27] Wilson, 2017, p. 187.
[28] The actual song title is "Changes in Latitudes, Changes in Attitudes."

Although most of the orographic wave-cloud photographs shown here were taken along the Front Range of the Rocky Mountains, many spectacular wave clouds have been photographed elsewhere around the world. A famous photograph by Ansel Adams entitled "Moonrise: Hernandez, New Mexico" depicts a wave cloud over the Sangre de Cristo Mountains of northern New Mexico, as we noted in the first chapter of this book. Although wave clouds are certainly most frequently seen near mountains, they do sometimes occur over much flatter terrain, under certain conditions, well away from any mountains. In these instances, the upward forcing of the air to trigger a wave is due *not* to forcing upward over a land/mountain barrier (orographic), but rather to dynamic forcing, such as is found in some upper-level atmospheric disturbances, which undulate in a snakelike manner, from west to east, in midlatitudes, especially during the winter (as a series of troughs and ridges, as noted earlier in footnotes). I have seen many wave clouds in mountain-starved Oklahoma.[29] Air may also be lifted when less-dense air is forced to flow up and over denser air. Thus, dense air can act like a mountain barrier. Furthermore, a convective updraft in a buoyant cloud can also act like a mountain barrier. These phenomena will be revisited in a later chapter, on convective storms.

Another possibility for generating gravity waves in the absence of mountains is "shear instability," which can happen when the winds abruptly change direction and speed with height. Such a circumstance can occur poleward of fronts, when, for example, winds at the surface have a component of motion from the pole to the equator and the wind direction reverses aloft. In the Northern Hemisphere, this happens when northerly winds at the surface behind a cold front shift abruptly to southerly aloft, above a layer of low-level, relatively cool, dense air.

Yet another possible mechanism for generating gravity waves in the absence of mountains involves boundary layers in which there are buoyant air parcels going upward with compensating downward-moving air parcels. When there is enough moisture for condensation, cumulus clouds (discussed in the next chapter) form. In addition to boundary layers in which surface heating promotes vertical mixing, there must also be vertical wind shear. The gravity waves triggered, known as "convection waves," were first discovered by glider pilots.[30]

Like humans or like nearly everything else, clouds exist in a *context*. The context of the clouds is important from an esthetic standpoint. Wave clouds over mountains suggest to one the importance of the mountains in producing them. The grandeur and vertical sweep of the terrain enhance the feeling of something exciting happening in the air. The context of the Ansel Adams photograph is particularly striking because not only are mountains visible, but there is also a religious mission. Imagine wave clouds over the ocean, which are sometimes observed there: I doubt their visual impact is as great, unless perhaps one were to observe them from ahead of a steep ocean wave! Mountains in a

---

[29] To be fair, Oklahoma does have some beautiful hills in the southwestern, southern, and southeastern part of the state, but their relief is modest compared to that of, for example, the Rockies, the Sierra Nevada, or the Appalachians, in the US.

[30] Kuettner, J. P., et al., 1987.

sense render wave clouds somewhat like a visual analog to onomatopoeia: The clouds visually suggest what they represent, smooth wavelike motions over mountains.

Yet another interesting aspect of orographic wave clouds is that even when there are significant variations in the height of the mountain peaks along the mountain barrier, a stark, long, two-dimensional, lenticular cloud may appear; it thus looks as if the sky's response is a smoothed or filtered version of what one might expect if one were to consider all the variations in height *along* the mountain range. The atmosphere must be doing some filtering of the airflow forced in the vicinity of the mountains. In effect, it might be that gravity waves of relatively short wavelength are not allowed, so that only the very longest wavelengths are responsible for forming clouds. In other instances, however, short-wavelength variations associated with individual mountain peaks do make themselves known visually.

Low clouds such as fog and stratus have the unique characteristic of being able to be observed by most of us on the ground easily from underneath, within, and above. The latter perspective can be most spectacular when there is a sharp contrast in light, as sung by John Denver in "Rocky Mountain High." Optical phenomena such as halos may augment the beauty of the clouds themselves. We will now look for more beauty in clouds, but this time, in clouds that are buoyant or are related to buoyancy, and are much more dynamic.

References

Bluestein, H. B. and J. C. Snyder, 2015: An observational study of the effects of dry air produced in dissipating convective storms on the predictability of severe weather. *Wea. Forecasting*, **30**, 79–114.

Brown, R. A., 1980: Longitudinal instabilities and secondary flows in the planetary boundary layer. *Rev. Geophys. Space Phys.*, **18**, 683–697.

Browning, K. A., 1971: Structure of the atmosphere in the vicinity of large amplitude Kelvin – Helmholtz billows. *Quart. J. Roy. Meteor. Soc.*, **97**, 283–299.

Corn, A., 2001: The Invention of Clouds: How an Amateur Meteorologist Forged the Language of the Skies. *New York Times Book Review*, July 29, 2001. (Available at nytimes.com)

Crow, S. C., 1970: Stability theory for a pair of trailing vortices. *AIAA J.l*, **8**, 2172–2179.

Doyle, J. D., and D. R. Durran, 2002: The dynamics of mountain-wave-induced rotors. *J. Atmos. Sci.*, **59**, 186–201.

Drazin, P. G., and W. H. Reid, 1981: Kelvin–Helmholtz instability. In *Hydrodynamic Stability*, Cambridge, Cambridge University Press, pp. 14–22.

Durran, D., 1990: Mountain waves and downslope winds. *Atmospheric Processes Over Complex Terrain*, Meteor. Monogr. No. 45, Amer. Meteor. Soc., 51–83.

Durran, D., and J. B. Klemp, 1982: The effects of moisture on trapped mountain lee waves. *J. Atmos. Sci.*, **39**, 2490–2506.

Federal Aviation Administration, 1997: *Hazardous Mountain Winds and Their Visual Indicators*, AC (Advisory Circular) 00–57, 90 pp. This documented may be downloaded from https// www.faa.gov under "Advisory Circulars." (The original hardcopy, undated, version I have is only 80 pp. long, contains color photographs, and the following are listed as co-authors: T. Q.

Carney, A. J. Bedard, Jr., J. M. Brown, J. McGinley, T. Lindholm, and M. J. Kraus; the online version lists these people only in an "Acknowledgements" section and the photographs are in black and white.)

Fujita, T. T., 1960: *A Detailed Analysis of the Fargo Tornadoes of June 20, 1957*. Research Paper to the U. S. Weather Bureau, No. 42, University of Chicago, 67 pp.

Greenler, R., 2020: *Rainbows, Halos, and Glories*. SPIE Press, 205 pp.

Haghi, K. R., D. B. Parsons, and A. Shapiro, 2017: Bores observed during IHOP_2002: The relationship of bores to the nocturnal environment. *Mon. Wea. Rev.*, 145, 3929–3946.

Harrison, R. G., G. Pretor-Pinney, G. J. Marlton, G. D. Anderson, D. J. Kirshbaum, and R. J. Hogan, 2017: Asperitas – a newly identified cloud supplementary feature. *Weather*, 72, 132–141.

Heymsfield, A. J., and R. G. Knollenberg, 1972: Properties of cirrus generating cells. *J. Atmos. Sci.*, 29, 1358–1366.

Hills, M. O. G., and D. R. Durran, 2014: Quantifying moisture perturbations leading to stacked lenticular clouds. *Quart. J. Roy. Meteor. Soc.*, 140, 2013–2016.

Houser, J. L., and H. B. Bluestein, 2011: Polarimetric Doppler radar observations of Kelvin-Helmholtz waves in a winter storm. *J. Atmos. Sci.*, 68, 1676–1702.

Kalkstein, A. J., and Balling, R. C., Jr., 2004: Impact of unusually clear weather on United States daily temperature range following 9/11/2001. *Clim. Res.*, 26, 1–4.

Kuettner, J. P., P. A. Hildebrand, and T. L. Clark, 1987: Convection waves: Observations of gravity wave systems over convectively active boundary layers. *Quart. J. Roy. Meteor. Soc.*, 113, 445–467.

Lewellen, D. C., and W. S. Lewellen, 2001: The effects of aircraft wake dynamics on contrail development. *J. Atmos. Sci.*, 58, 390–406.

Marshall, J. S., 1953: Precipitation trajectories and patterns. *J. Meteor.*, 10, 25–29.

Menzel. D. H., 1953: *Flying Saucers*, Cambridge, Harvard Univ. Press, 319 pp.

Reed, R. J., 1958: Flying saucers over Mount Rainier. *Weatherwise*, 11, 43–45, 65–66.

Rotunno, R., and G. H. Bryan, 2018: Numerical simulations of two-layer flow past topography. Part I: The leeside hydraulic jump. *J. Atmos. Sci.*, 75, 1231–1241.

Scorer, R. S., 1949: Theory of waves in the lee of mountains. *Tellus*, 75, 41–56.

Scorer, R. S., and L. J. Davenport, 1970: Contrails and aircraft downwash. *J. Fluid Mech.*, 43, 451–464.

Smith, R. B., and V. Grubišić, 1993: Aerial observations of Hawaii's wake. *J. Atmos. Sci.*, 50, 3728–3750.

Travis, D. J., A. M. Carleton, and R. G. Lauritsen, 2004: Regional variations in U. S. diurnal temperature range for the 11–14 Sept. 2001 aircraft groundings: Evidence of jet contrail influence on climate. *J. Clim.*, 17, 1123–1134.

Weckwerth, T. M., J. W. Wilson, R. M. Wakimoto, and N. A. Crook, 1997: Horizontal convective rolls: Determining the environmental conditions supporting their existence and characteristics. *Mon. Wea. Rev.*, 125, 505–526.

Wikipedia, chinook: https://en.wikipedia.org/wiki/Chinook_wind

Wilson, E. O., 2017: *The Origins of Creativity*. Liveright/Norton, 256 pp.

Voigt, M., and V. Wirth, 2013: Mechanisms of banner cloud formation. *J. Atmos. Sci.*, 70, 3631–3640.

3

Buoyant Clouds

Part I. Convective, Non-precipitating Clouds

> *"Motion in the atmosphere originates almost entirely through the action of gravity on masses of air that are differently heated."*
>
> F. H. Ludlam and R. S. Scorer[1]

3.1 How clouds become buoyant

Different buoyant clouds may appear quite differently. It is therefore not surprising that the physics, microphysical dynamics, and thermodynamics responsible for them are also varied. The way light hits them can significantly alter their appearance; so does the viewing angle of the observer. Since there are so many different types of clouds that are driven by buoyancy, we will describe them in two separate, consecutive chapters. To complicate matters somewhat, we note that some clouds have both buoyant and non-buoyant regions. For example, clouds that harbor convective storms may have non-buoyant regions at low altitudes and also at their very top. Dividing up the chapters as I did is somewhat arbitrary and there will necessarily be some overlap. There are certainly other ways of dividing up our journey through the topic of clouds, but the end of all our possible journeys might arguably be the same: the biggest and the most dynamic, severe thunderstorms and tornadoes (Chapter 5). The least buoyant and shallowest clouds will be considered first, working our way up to the most buoyant and deepest clouds last. In between, the weakly buoyant, deep and strongly buoyant, shallow clouds will be considered.

In Section 3.1, we consider the gross difference in the environment of clouds that are non-buoyant from those that are buoyant: clouds that are smooth and laminar (non-buoyant) from those that are puffy and bubbly looking (buoyant). The former clouds form in a stable environment, while the latter form in an unstable environment. If air in the troposphere is lifted and the air is unsaturated, then as a consequence of the thermodynamics of air, it cools at the fixed rate of 10°C (18°F) per kilometer as it ascends. This "lapse rate" (the rate of decrease of temperature with height) in the atmosphere is called

---

[1] Ludlam and Scorer, 1953, p. 317.

The Architecture of Clouds. Howard B. Bluestein, Oxford University Press. © Howard B. Bluestein (2024).
DOI: 10.1093/oso/9780198870548.003.0003

the "dryadiabatic lapse rate" (no heat is added or subtracted from the air molecules and no condensation is occurring). If the lapse rate in the atmosphere is less than the dryadiabatic lapse rate, air motions are stable (Fig. 3.1); the air that is lifted will be cooler than that of the environment and not buoyant. It turns out, though, that if the lapse rate were greater than the dryadiabatic, the atmosphere is unstable and turbulent motions ensue and warmer air is transported upward and displaced colder air is transported downward

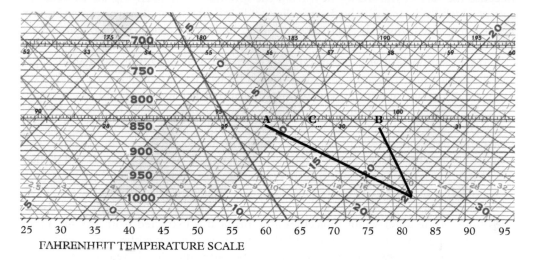

Figure 3.1 *An example of a stable and an unstable lapse rate in a dry (i.e., there is no water vapor at all) atmosphere. In this nomogram, called a "skew-T/log p diagram," lines of constant temperature are indicated by straight, skewed[a] (leaning to the right with height), orange, lines labeled in °C (and at the bottom in °F). The pressure is given at the left in hPa. The vertical coordinate is logarithmic because pressure decreases exponentially with height, so that the vertical distance in the figure approximately represents height. Thin orange lines that slope upward to the left are called "dryadiabats." If an air parcel at the surface, 1, 000 hPa, having a temperature of 25.5°C (78°F) were lifted along the dryadiabat to 850 hPa, then it would have cooled off to a temperature of 12°C (point C). The solid black line (representing the temperature as a function of pressure) ending at point B indicates an environmental lapse rate of less than the dryadiabatic; if the air parcel at 1, 000 hPa were lifted to 850 hPa (C), then it would be about 6°C colder than the environmental temperature at B (~18°C) and be negatively buoyant. The solid black line ending at point A indicates an environmental lapse rate of greater than the dryadiabatic; if the air parcel at 1, 000 hPa were lifted to 850 hPa (C), then it would be about 4°C warmer than the environmental temperature at A (~8°C) and be positively buoyant. It's convenient to use this type of a diagram to test for stability or instability without resorting to actual calculations. This diagram is like a slide rule for atmospheric thermodynamics. These days, however, a slide rule may seem like an ancient artifact, but it helps us easily and quickly visualize the process of the formation of a convective cloud (next figure).*

[a] I have heard from Kerry Emanuel that the reason why the lines of constant temperature are skewed is that during World War II there was an effort to reduce the amount of paper used. I cannot, however, verify the source of this assertion.
Figure courtesy of author.

on very small scales (via "vertical turbulent mixing"); the net effect is to return the atmosphere to one with a dryadiabatic lapse rate. The tiny parcels of air moving upward and downward are called "eddies." On the ground, we may experience eddies as a gustiness in the wind. We typically observe lapse rates greater than the dryadiabatic only right next to the ground during the daytime when there is strong, relentless heating of the ground by the sun; no matter how much vertical mixing goes on, the surface heating maintains a "superadiabatic" lapse rate (i.e., greater than the dryadiabatic lapse rate) because the rate of transfer of heat from the ground to the overlying air exceeds the rate at which the heat gets mixed vertically (cold air is transported downward and warm air is transported upward). Above the ground, however, vertical mixing maintains a dryadiabatic lapse rate.

Now, let's suppose that as unsaturated air is lifted and cooled at the dryadiabatic lapse rate as the pressure decreases, and eventually is cooled to saturation. It then reaches the "dew point" temperature. In this case, after the air has become saturated, latent heat is "released" owing to the change in phase of the water vapor; the heat is absorbed by the air, so that it cools at a rate that is *less* than the dryadiabatic lapse rate. Why does this heat exchange happen? When water vapor makes the transition from vapor to liquid, it is changed from a less-ordered state (molecules are relatively widely scattered) to a higher-order state (molecules are closer together in a more regular fashion) and heat energy is released when the molecules are bonded together; this heat is absorbed by the air. When the opposite happens, i.e., when liquid water droplets evaporate, heat is needed to break the bonds of the water-vapor molecules and this heat is extracted from the air. Think of what happens when you step out of the shower into a dry bathroom or come out of the ocean after a swim on a very dry day. So, after the air has reached its "lifting condensation level" or LCL (Fig. 3.2), if the lifting continues, it now cools at the "moist adiabatic lapse rate," which is less than the dryadiabatic lapse rate. If the lapse rate in the environment of the clouds is less than the moistadiabatic, then the atmosphere is stable because the temperature of the saturated air parcel is less than that of the environment. The air parcel is negatively buoyant and if left alone, it will sink back whence it came.

If, however, the air parcel is forced up higher to an altitude at which the lapse rate becomes greater than the moistadiabatic, then the atmosphere may eventually become unstable if lifted more and a cloud will be buoyant and grow upwards in a turbulent fashion. The level at which the saturated air parcel reaches neutral buoyancy (the temperature of the air is the same as that of the environment) and subsequently upon further lifting becomes warmer than the environment is called the level of free convection (LFC) (Fig. 3.2). When the atmosphere may become unstable only after saturation is reached, but is always stable when unsaturated, it is said to be "conditionally" unstable. The moist adiabatic lapse rate is not a constant, like the dryadiabatic lapse rate, but instead depends upon the altitude: it is least at low levels where the saturation-moisture content is relatively high, increases with height as the saturation-moisture content decreases, and eventually becomes identical to the dryadiabatic lapse rate at high altitudes where the amount of water vapor in the air and its saturation amount is negligible. The buoyant cloud transports heat energy upward and is called a "convective cloud," because convection literally means a "carrying together." In this case it refers to heat transfer (a

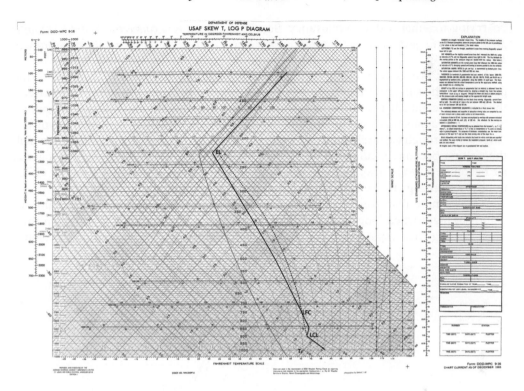

Figure 3.2 *Illustration of how a moist atmosphere may support a buoyant cloud. In this hypothetical atmospheric "sounding" (temperature as a function of pressure) plotted on a skew-T—log p diagram, the air temperature at the surface (1, 000 hPa) is 24°C (79°F) and the dew point (T$_d$) is 19°C (66°F). For this temperature and dew point, the ratio of the mass of water vapor to the mass of dry air (the "mixing ratio") is 14 g kg^{-1} (given by the green dashed line, which represents the saturation water vapor mixing ratio). At 24°C the saturation mixing ratio is ~21.5 g kg^{-1} while the actual mixing ratio is only 14 g kg^{-1}. If an air parcel with a surface temperature of 24°C and a dew point is 19°C is lifted dryadiabatically, along the sloping thin orange line, then the air parcel would become saturated when it reaches 900 hPa, which is called the lifting condensation level (LCL). If the air parcel were lifted more and remains saturated, it would now cool at the slower moist-adiabatic lapse rate, given by the slightly curved solid green lines (the path shown here by a dashed black line). The parcel temperature would remain cooler than that of the environment (solid black line) until it reached the level of free convection (LFC), above which the air parcel would become warmer than that of the environment. If the parcel were left on its own and remain saturated, it would rise buoyantly until it reaches its equilibrium level (EL) near 225 hPa (follow the dashed line along a curve roughly halfway between and parallel to the curved solid green lines labeled 20°C and 22°C. The tropopause here coincidentally happens to be at 225 hPa, above which the lapse rate decreases to nearly zero ("isothermal"). Once the air parcel gets higher than the EL, it becomes negatively buoyant and it decelerates until it comes to a complete halt as it loses all the kinetic energy it was given in its initial push up to the LFC. This level would be somewhere near 13 km (over 40, 000 feet), which is the height of the top of many convective storms over the US during the summer.*

Figure courtesy of author.

"carrying together") due to buoyancy. So, *the likelihood of a convective cloud forming is favored by a large decrease in temperature with height and an adequate supply of moisture at low levels.*

A similar process can occur even if there is no lifting, but heat is absorbed at the ground by the sun and transferred to the air in contact with the ground. In this case, if the lapse rate initially is less than dryadiabatic, then it might, if there is sufficient heating, become dryadiabatic as warm air is mixed upward and colder air is mixed downward (Fig. 3.3) through vertically moving turbulent eddies. In this case, at the altitude at which the air first becomes saturated a cloud will form. If the air is heated enough so that the air becomes buoyant also, then it is called the "convective condensation level" or CCL, in obvious analogy to the LFC if the parcel were forcibly lifted up.

We also note that non-buoyant, stratiform clouds may be converted into buoyant clouds at night when longwave radiation escapes into space as it is radiated from their tops, cooling their tops, while longwave radiation is trapped below, inside the cloud,

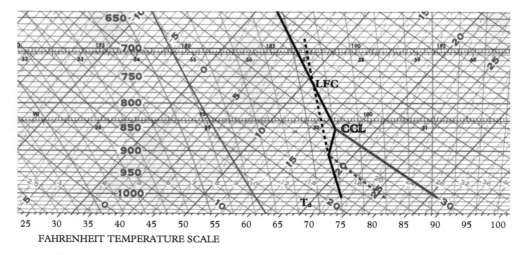

FAHRENHEIT TEMPERATURE SCALE

Figure 3.3 *Illustration of how an air parcel may be heated by the sun to its convective condensation level (CCL). In this sounding, suppose that the temperature at the surface, 1, 000 hPa, is initially 22°C (~72°F) and that the dew point is 19°C (~66°F). If this air parcel were lifted, it would cool off at the dryadiabatic lapse rate and become less than the environmental temperature (located to the left of the solid, black line): The atmosphere here is stable. Now, suppose that the air at the surface is heated up to about 26°C (~79°F); then the air would be "mixed vertically" by turbulent eddies up to ~900 hPa (dashed red line) and become saturated there; a cloud would form but in this case the cloud would not be buoyant because it would be stable if lifted (dashed black line) and would not become buoyant until it reached its LFC. Suppose now that the air is heated even further, to about 30°C (86°F). In this case the air parcel would be saturated and if lifted just slightly (along the thin green line) would become buoyant. The surface temperature required to achieve both saturation and buoyancy is called the "convective temperature" and the height at which this happens is known as the "convective condensation level (CCL)." In this process it is assumed that the water vapor mixing ratio remains "well mixed," i.e., is constant with height. In real life this may not be the case as the water vapor mixing ratio may vary with height.*

Courtesy of author.

warming the cloud interior. It is thus possible that the lapse rate in the upper portion of the stratiform cloud is increased, resulting in instability. This process of destabilization resulting from a vertical change in radiative heating/cooling may be important for some clouds, especially those that appear at night.

In nature, buoyant, "convective" clouds near the surface often begin to form as a result of a contribution of *both* lifting and surface heating (during the day only, of course) and their bases tend to be flat if the LCL or CCL is doesn't vary horizontally and the lift is uniform in the horizontal. The contribution to lift may be from "synoptic-scale" processes (Bluestein 1992, 1993) (on spatial scales of thousands of kilometers and time scales of days) associated with extratropical (by "extratropical" we generally mean in midlatitudes, ~30–50° latitude north or south) cyclones[2] (low-pressure areas) and anticyclones (high-pressure areas) at the ground and north-south undulations in the generally westerly flow of air aloft associated with the jet stream (Fig. 3.4). Synoptic-scale lift, however, is meager, only on the order of centimeters per second, and serves mainly to change the lapse rate of temperature and moisture content in the environment of potential convective clouds.

The contribution to lift may also be from "mesoscale" processes (on spatial scales of tens to hundreds of kilometers and time scales of tens of minutes to hours) along zones of strong horizontals change in temperature such as fronts or gust fronts or outflow boundaries (to be described subsequently) or other boundaries separating dry and moist air and with lift on the order of tens of centimeters per second to meters per second. Convective clouds, by contrast, can have vertical velocities of a number of meters per second up to tens of meters per second. The vertical motions inside convective clouds are usually much stronger than any vertical motions that triggered them.

Lift from slowly rising air currents from synoptic and mesoscale "forcing" may not be sufficient to permit moist air to reach its LCL or CCL. It does, however, make the atmosphere more likely to experience the formation of convective clouds at low altitudes because it cools the atmosphere off more at higher altitudes than it does at lower altitudes, thus reducing the stability of the atmosphere by increasing the lapse rate. Why does this happen? Suppose that there is rising motion. Since air cannot be extracted from the ground (except over holes in the ground such as open missile silos, mines, or subway portals), to conserve mass, the vertical velocity of air at the ground must vanish. So, just above the surface vertical velocity must increase with height. It turns out that from thermodynamic considerations the rate of cooling is proportional to the vertical velocity for a given temperature change with height. So, the rate of cooling increases with height, from zero at the ground to a greater rate aloft. When a *layer* of the atmosphere is stable, but after lift it becomes unstable, it is said that the atmosphere is "potentially" (or "convectively") unstable [3]. The adjective "potential" indicates that the "necessary" conditions for instability are present, but they are not enough or "sufficient" to trigger

---

[2] The word "cyclone" was coined in 1848 by Henry Piddington, an Englishman, on the basis of his experiences at sea.

[3] There is also a special type of instability called "moist absolute instability," which occurs when mesoscale lift of a layer of air leads to saturation of the entire layer and a saturated lapse rate greater than that of the moist-adiabatic (Bryan and Fritsch 2000). It occurs sometimes in the stratiform-precipitation areas of mesoscale

500 mb rawinsonde data 00z Tue 01 Dec 2020

500 mb Heights (dm) / Temperature(°C) / Humidity (%)

0–hour analysis vaild 0000 UTC Tue 01 Dec 2020   RAP (00z 01 Dec)

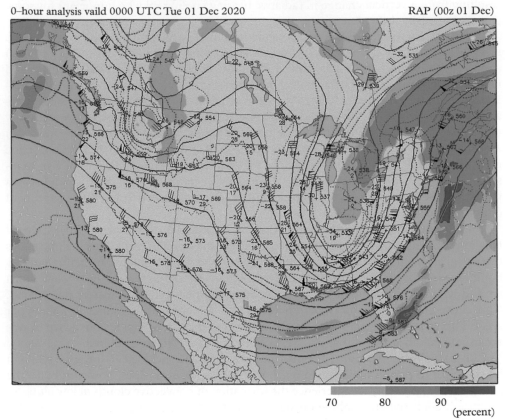

```
        70          80          90
                               (percent)
```

Figure 3.4 *Example of synoptic-scale disturbances in which there are regions of lift (slowly rising air) at 0000 UTC, Dec. 1, 2020. The wind observations at the standard synoptic observing sites are plotted such that the wind direction from which the wind is blowing is indicated by the orientation of the thin lines, attached to which each small barb represents 2.5 m s^{-1}, each whole barb represents 5 m s^{1-}, and each flag represents 25 m s^{-1}; the temperature and temperature—dew point difference is plotted in °C, with the former above the latter. The height of the 500 hPa surface in decameters is plotted to the right. Lines of constant height are analyzed as solid lines. The winds to a first approximation, known as the geostrophic approximation, tend to blow (in the Northern Hemisphere) with the higher heights to the right and lower heights to the left. The isotherms are shown as dotted, purple lines. See any (more advanced) textbook on synoptic meteorology to learn more about how to interpret weather maps such as this. There are two regions of flow that are curved in the counterclockwise direction and that are located in regions of relatively low height. These are known as troughs: one is seen over western Montana and Idaho; the other extends from the Michigan southward into Alabama and Georgia. Rising motion at midlevels such as 500 hPa (~5, 000–6, 000 m) tends to be found just downstream from troughs. The relative humidity is shown color coded in green. Note how the most humid regions are located near and just downstream from the troughs, where there is rising motion. These are the areas where we expect to*

the instability: instability is released only if there is sufficient lift; otherwise, nothing happens, even though there is the potential for something to happen. Upper-level troughs of low pressure are associated with relatively cold air aloft (see Fig. 3.4), so the stability may be lowered also as they approach. Upper-level ridges of high pressure are associated with relatively warm air aloft, so the stability may be increased as they approach. In the case of the upper-level troughs, it turns out that there is rising motion ahead of them, so the stability is decreased both from rising air motion and from the advection (transport of air of a different temperature) of colder air aloft. Synoptic-scale processes are thus said to "precondition" the atmosphere for convection: even though the vertical velocities on the synoptic scale are much weaker than the vertical velocities in a convective cloud (by several orders of magnitude—cm s^{-1} vs. m s^{-1}, or 10 m s^{-1}), they do have an effect on whether or not convective clouds and storms form.

The rate of heating at the surface due to insolation (exposure to radiation from the sun) depends on the type of surface (the nature of the vegetation, if any, whether there are buildings, roads, sand, soil) and if the surface is wet or not. When the surface vegetation is relatively wet, some of the solar insolation goes into evaporating water from the surface, so less heat is actually absorbed by the ground. Bodies of water, such as lakes and oceans, absorb heat less rapidly than land surfaces, unless there is ice or snow, in which case much of the solar radiation is simply reflected or radiated back to space. For example, we often find cumuliform clouds over land areas during the day, but not over the ocean or other water surfaces. However, when relatively cold air flows from snowy and icy land surfaces out onto relatively warm ocean surfaces, the lapse rate can also become relatively high as it is over the land on a sunny day. This situation occurs often during the winter during both the day and night off the east coasts of the US and Asia, as cold air masses produced inland, where there are regions of snow and ice and that receive little if any sunlight, are "advected" (carried by the northwesterly or northerly wind) around the southern portion of an extratropical (in midlatitudes, say from 30° N/S to 50° N/S) cyclone and onto the ocean by offshore winds at the surface (see e.g., Fig. 3.11).

Most of the contribution to the wind by extratropical cyclones and anticyclones is *geostrophic* (Bluestein 1992)[there is an *approximate* balance between the pressure-gradient force, which acts toward lower pressure, and the Coriolis force, which acts to the right (left) of the wind in the Northern (Southern) Hemisphere; such a state is

---

Figure 3.4 (Continued) *find clouds. Clouds may also be found farther downstream, where the winds carry the cloud droplets and ice crystals before they evaporate or sublimate. The regions of clockwise curvature in the flow and relatively warm temperatures are known as ridges. One is found extending from the Dakotas down to Colorado, and others are found just off the west coast and off the northeast coast of the US Ridges tend to be less cloudy or even totally clear, owing to sinking air motion and drying.*
(From NOAA.)

convective systems (see Chapter 4) when the mesoscale lifting that creates it is faster than turbulent vertical mixing can get rid of it.

referred to as *quasi-geostrophic*], parallel to the isobars (lines of constant pressure), while there is some component of motion toward the lower pressure near the ground, owing to surface friction (this effect will be shown to be important in tornado formation in Chapter 5) (Fig. 3.5). The name "geostrophic" refers to the response of the wind to the turning of the Earth through the Coriolis force.

The reason why there is a Coriolis force may be understood in part by considering what would happen if you were to stand at the North Pole and, though possibly shivering from the cold, still have enough control to throw a baseball[4] to the south, which from the North Pole "south" is ambiguous because it actually points in all directions! Imagine, however, that you aimed the ball at someone who is perhaps 20 m away, obviously not at the North Pole. Since the Earth is rotating about its axis in a counterclockwise direction

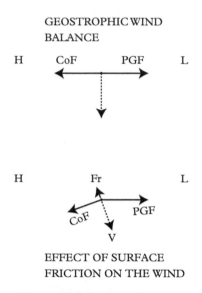

Figure 3.5 *Illustration of geostrophic wind balance (top): The pressure-gradient force (**PGF**) is directed to the right, from relative high pressure to relative low pressure. The Coriolis force (**CoF**) acts equally and in the opposite direction. The wind (**V**) blows from north to south, perpendicular and to the right of the Coriolis force (in the Northern Hemisphere), with the higher pressure to the right and lower pressure to the left. Effect of friction (bottom): Suppose the **PGF** stays the same. The effect of surface friction (**Fr**) is to retard the wind, so it is directed in the direction opposite to that of the wind. Suppose that the **PGF**, **CoF**, and **Fr** are all in balance (add up to zero, so there is no net acceleration): Since the **CoF** is proportional to the wind speed and perpendicular and to the right of the wind, it follows that the wind must blow from north-northwest to south-southeast and be weaker than the geostrophic wind if there were no friction. Thus, there must be a component of the wind directed from high to low pressure near the surface.*

Figure courtesy of author.

---

[4] If the reader is not from the United States or Canada or Japan especially, or many other countries, then he/she is encouraged to imagine using another type of ball.

(in the clockwise direction if you were standing at the South Pole) by the time the ball reaches the intended catcher, it would appear to have been deflected to your right by some invisible force, even though it actually moves along in a straight line (assuming that you did not put some spin on it, and neglecting gravity). This force is due to the rotation of the Earth and therefore as result of your standing on ground that is moving, actually accelerating, since the direction of motion is changing. The Coriolis force at any other latitude and longitude other than the poles is more difficult to understand and we therefore rely on mathematics to compute it precisely. It varies with the latitude, being the greatest at the North Pole and vanishing at the equator. It is not a force in the sense that the pressure-gradient force is since it depends on the motion of the platform on which one is standing.

So much for this diversion on what causes the winds to blow around synoptic-scale cyclones and anticyclones. If there is a stable layer (a layer of air whose lapse rate is less than the dryadiabatic) not far above the LCL or CCL, then the convective clouds will be relatively shallow, as unstable upwardly moving air suddenly finds itself surrounded by warmer air and the clouds consequently lose their buoyancy. This is a common situation during the spring in the Southern Plains of the US as very warm air that has been heated at higher elevation to the west over a mountainous region is advected over relatively cool and moist air that originated over the Gulf of Mexico to the south and southeast. Stable air masses located above unstable air masses below may also be found when there is a broad, synoptic-scale area of gently subsiding air, which warms at the dryadiabatic lapse rate just above the layer of air adjacent to the surface. The stable air mass on top of a potentially unstable air mass is sometimes called a "cap," because it effectively puts a restraining top on air moving upward. When the air temperature simply increases with height the air is very stable. When the air temperature increases with height there is said to be an "inversion." Temperature inversions are notorious for trapping air pollutants, which might otherwise be reduced in concentration if they were vertically mixed and diluted within a deep layer in an unstable atmosphere.

Even if there is no stable layer above the LCL or CCL, if the environmental air is very dry and unsaturated, then the very dry air may be mixed into the buoyant cloud through small-scale, turbulent mixing so much that it dilutes the cloud of water vapor to the extent that the air in the cloud cools as cloud droplets evaporate (the opposite of what happens when water vapor condenses), thereby reducing the thermodynamic buoyancy. The process by which environmental air enters into a convective cloud from its sides or its top is called "entrainment." The opposite is called "detrainment." Entrainment (and detrainment) may be caused by small-scale, turbulent processes (through the mixing of parcels of air from within and without the cloud) just mentioned, or through airflow on the scale of the cloud (i.e., the entrainment is nonturbulent and cloud scale). Clouds that are relatively narrow tend to have their buoyancy more diluted than clouds that are relatively wide, because the mixing tends to occur mainly at the edges of the cloud; thus, the inner region of wide clouds is more protected from the incursion of drier environmental air than the inner region of narrow clouds.

Individual buoyant air parcels or groups of air parcels, which accelerate upwards, whether saturated[5] or unsaturated, are called "thermals." Much of what we have learned about thermals prior to the 1950s came from glider pilots in Germany who soared in their sailplanes and were able to remain in rising, unsaturated updrafts for as long as 20 minutes. I was once given a ride in Arizona by a colleague, Nilton Rennó, in his sailplane. It is an exhilarating experience being suspended in and moving through the air with no engine sounds or vibrations, just the whishing sound of the air flowing around the glider. It is amazing to be able to remain afloat in the air simply by riding thermals. Several conceptual models of thermals were proposed in the mid-twentieth century to describe these thermals:

   a. Bubble: A spherical region of positive buoyancy. As a thermal bubble accelerates upward, the air ahead of it must move out of the way to make room for the rising bubble, and then it curls around the edges of the bubble, returning up into the bubble at its bottom. Imagine that a paddlewheel is placed along the edge of a thermal along a horizontal plane that cuts through the center of the thermal. It will experience an upward acceleration just inside the thermal, but no acceleration at all just outside the thermal, where there is no buoyancy. The paddlewheel will then rotate about an axis that wraps around the thermal bubble like a ring, producing a toroidal-like circulation (Levine 1959).

If the bubble entrains environmental air along its outer edges, it will expand and get distorted as it rises, as shown in Figure 3.6; the outer shell is curled back into the center of the thermal bubble (as seen in Fig. 3.7). In this case air inside the thermal bubble might be detrained along the side edges through turbulent mixing and entrained at the bottom as environmental air is caught up into the bubble.

The bubble was actually not the first conceptual model to be introduced (the plume, to be described next was), but I introduce it here first because it follows more realistically what many convective clouds actually look like. Richard Scorer and Frank Ludlam in England are credited with the introduction of the bubble model in 1953. The bubble has been described as the "'proton' of convection"[6] as it is the basic unit of convective clouds.

   b. Plume: A jet of buoyant air that is continuously being forced by a source of heat at the bottom. A plume spreads with height above the heat source as air from the environment is entrained into it from the sides. Think of a plume of smoke over a smoke stack on a windless day. Forest fires produce buoyant plumes. This conceptual plume model is credited to an Englishman, Bruce Morton, who spent the latter part of his career in Australia, and his collaborators, in 1956 and 1957. This model was the first presented (before the thermal bubble model), owing to its simplicity. Air outside the plume is entrained into the plume, diluting its buoyancy since the air in the environment is cooler than the air inside the buoyant plume. On

---

[5] Thermals are typically associated with unsaturated, buoyant eddies.
[6] Malkus and Scorer, 1955, p. 44.

the other hand, if a plume arises in a moist atmosphere and it becomes saturated, the buoyancy in the plume is enhanced as a result of latent heat released as water vapor condenses, and diminished by the entrainment of colder environmental air or drier air into it.

c. Starting plume: A starting plume is combination of a thermal bubble and a plume. If one were to turn on a heat source and then watch what develops, one would see the leading edge of the plume advance upwards like a bubble, but leave a wake behind like that of a plume because the heat source at the bottom stays turned on. Think of a smokestack on a windless day when something is first burned inside the smokestack. The formulation of this model is credited to an Australian, John Turner, in 1962. A thermal bubble is like a starting plume, but the heat gets turned off quickly.

Convective clouds behave somewhat like a conglomeration of thermal bubbles, but also to some extent like plumes and starting plumes. Sometimes there might actually be a local, constant heat source from the sun heating a relatively local dry surface or an elevated surface like a mountain, accounting for a continuous source of buoyancy.

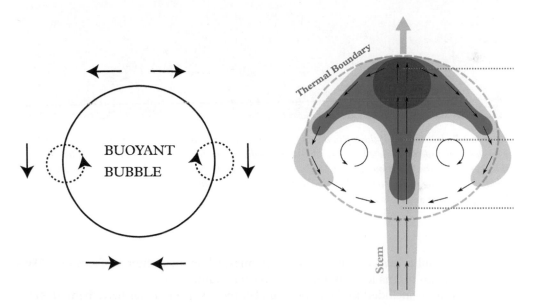

Figure 3.6 *Motion around a buoyant bubble initially let loose. (left) Vertical cross section through a highly idealized spherical buoyant bubble, in the reference frame of an upward accelerating, buoyant thermal bubble. The flow of air is shaped like a torus, with air diverging above and converging below. The circular dashed lines with arrows indicate how a paddlewheel would spin at the edge of the bubble at the locations indicated. Figure courtesy of author. (right) Idealized representation of a vertical cross-section through a rising thermal bubble at some later time in its life. The yellow, orange, and red-shaded regions denote the least, moderate, and the highest buoyancy. The airflow relative to the rising thermal is indicated by the wind vectors. This picture is based on numerical simulations.* (From Morrison and Peters, 2018; their Fig. 2; © American Meteorological Society. Used with permission.)

Figure 3.7 *An example of the top of a group of saturated thermals turning in on themselves high above Boulder, Colorado on June 24, 2014. The arrows indicate the direction of cloud motion. The cloud material in the shadow is moving up, while the cloud material at the edge, above, which is casting the shadow, is curling back down and getting tucked back into the shadow area.*
Photo courtesy of author.

Growing cumulus *congestus* may look like plumes, while their outer surfaces look like a conglomeration of smaller bubbles, like a starting plume.

If moisture is included in the bubble model, then, since bubbles leave behind wakes of enhanced water-vapor content in their environment, subsequent bursts of buoyancy (bubbles) tend to follow the paths of previous bubbles (if there is no vertical wind shear) where it is moister than the environment and therefore entrainment is less inimical, accounting for a "chain" of discrete thermal bubbles. Even if the bubble is dry, the air-flow on the scale of the bubble entrains dry air into the bubble from underneath. If the air in the midlevel environment is very dry, then the bursts of thermals remain discrete, as any thermal is broken up into bursts of narrower buoyant elements. On the other hand, if the environment is relatively moist, then the bursts remain continuous as in a plume.

Since dilution of buoyancy due to entrainment is greatest for narrow thermals, they behave more like bubbles than wide thermals do, for which entrainment of dry environmental air due to mixing is less. Narrow thermals tend to "pinch off"[7] from below as seen in many of the subsequent photographs and are probably also related to the dryness of the atmosphere aloft. These very narrow towers have been called "turkey towers" by storm chasers, while the official terminology is *castellanus*. We don't understand, however, *exactly* what controls the width of thermals. One recent computer-model study[8] demonstrated that dry thermals have a different buoyancy structure than moist thermals: Relatively dry thermals tend to spread laterally more than moist thermals. In the dry thermal rotation is generated at the edges of the thermal, while in the moist case, latent heat release as water vapor condenses inside the thermal leads to circulations that are inside the thermal rather than at the edges.

In this chapter we will focus on convective clouds that either do not grow very high, that is, they are categorized as "shallow," and either do not produce any precipitation that reaches the ground, which is called "virga," or do grow very high, but still do not produce precipitation at all, perhaps because it is too dry in the environment. The depth of shallow clouds tends to be much less than their width and they do not usually produce precipitation. We will continue on and consider "deep," non-precipitating convective clouds, those that are taller than they are wide, but still do not produce any precipitation. We defer any discussion of deep or even shallow, *precipitating* convective clouds until the next chapter, when we will look at convective storms and in particular the ultimate convective storms, severe convective storms. The medieval English poet's Geoffrey Chaucer's words in "Troilus and Criseyde" in 1374, "Tall trees from little acorns grow,"[9] can be transformed for our purposes into "Big, violent, precipitating convective storms from little, non-precipitating convective clouds grow." Thus, we will look at the shallow or non-precipitating clouds first and then gradually build up both the intensity of our discussion and the depth of the phenomena we are describing to monstrous, heavily precipitating convective storms.

Our partitioning of convective clouds into shallow or deep, but non-precipitating clouds from deep or shallow, but precipitating clouds is somewhat arbitrary and not always clear. For example, it may not be apparent from just looking at a cloud whether or not there is precipitation falling underneath it: there could be a curtain of widely spaced raindrops or hailstones which are visually transparent or there could be precipitation growing or suspended, hidden within the cloud. Anvils (to be discussed) are produced

---

[7] The prominent twentieth-century meteorologist Joanne Simpson called these towers "cutoff towers" in Simpson and Woodley, 1971.

[8] Morrison et al., 2021.

[9] I have found this quotation in many places elsewhere, such as from Lewis Duncombe's "The lofty oak from a small acorn grows," and in the eighteenth century, up into the twentieth century, by Henry Ford, but I believe Chaucer's version might be the earliest. It itself might not be original as it apparently is also a commonly used very-old proverb.

when the updraft hits a stable layer such as the tropopause, but their appearance does not mean that there is any precipitation; yet their appearance does indicate a sort of maturity in that a convective tower may have reached the end or is nearing the end of its individual life. In this chapter we will arbitrarily, but for the sake of imposing some order, not consider deep convective clouds with anvils (most of the time), even if they are non-precipitating.

The photographs shown in the subsequent figures reflect my wish to show the wide variety of appearances of the clouds, with regard also for their ability to illustrate specific physical mechanisms, for different times of day (remember Claude Monet's paintings) and for geographic diversity. We will begin in this chapter with the flattest and lowest convective clouds, stratocumulus, then consider the flattest midlevel clouds, altocumulus. We will then work our way up to the more buoyant low level and midlevel clouds, the cumulus and altocumulus, and then on up to the cumulus that tower high up into the troposphere, possibly producing precipitation and therefore becoming cumulonimbi, the object of the next chapter, Chapter 4.

3.2   Stratocumulus

When at low levels there is some buoyancy in the clouds, but convection is quickly suppressed as parcels of buoyant, thermals encounter a strong stable layer or inversion, the tops of the clouds rapidly spread out and become stratiform looking (Figs. 3.8–3.12). The tops are like anvils in deep convective storms, but in the case of the latter, the spreading out is due to the stable tropopause, which is much higher up than the stable layer that inhibits the growth of the lowly stratocumulus.

Stratocumulus clouds are often found over parts of the ocean when it is relatively warm compared to the air aloft but there is a strong subsidence inversion, a stable layer caused by the warming and sinking of air from above. Stratocumulus, since they

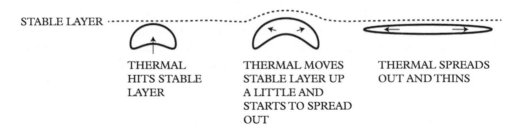

Figure 3.8 *Idealized illustration of how a thermal can hit a stable layer, spread out slightly and perturb the stable layer up slightly, and then spread out just below the stable layer. (inspired by Fig. 1.6.i in Scorer's 1972 book).*

Figure courtesy of the author.

Figure 3.9 *Stratocumulus over Cairns, Australia, on Aug. 8, 2007.*
Photo courtesy of author.

sometimes occupy vast areas, can reflect large amounts of solar radiation, are important components of the global energy budget. It is sometimes difficult to discern stratocumulus from altocumulus when viewed from the ground because they both have similar clumpy structures (e.g., see Fig. 3.13): it may not be apparent how high the cloud is.

3.3   Altocumulus not associated with orography: Patterns of clouds at midlevels not near mountains

Altocumulus are patchy, lumpy, scaly, or wavy bands of horizontally expansive clouds at midlevels. At their altitudes, heating of the air at the surface in reducing the stability of the air (with respect to vertical displacements) cannot play much of a role, if any, in producing convection so high up, as it does in stratocumulus. Instead, the tops of the clouds cool off as longwave radiation escapes above, reducing the vertical stability. The moisture from altocumulus may come from the water vapor from remnants of earlier, deep convection (as is often seen after sunrise over the Plains of the US or over the Rocky Mountains during the monsoon season), or from the gentle ascent of air in synoptic and

Figure 3.10 *Airborne view of stratocumulus and patches of stratus. some looking like pancakes; see also Georgia O'Keefe's painting "Above the Clouds I," described in Chapter 1; she is quoted as having viewed similar-looking clouds from an airplane and said that it looked as if one could walk right out of the airplane, presumably using the cloud tops as stepping stones!) (top, left) over the Southern Pacific Ocean, northeast of Australia, on Aug. 11, 2007. There is mostly cloud, with clear breaks surrounding the patches of cloud. (top, right) off the west coast of Northern Mexico or Southern California, on Sept. 23, 1984, after I had flown into Hurricane Norbert in the eastern Pacific on a NOAA hurricane mission. (middle, left) probably over Illinois (en route from Peoria to the Dallas—Ft. Worth area), Nov. 25, 2005. Some clumps of clouds have a polygonal structure. Clouds that look exactly like this in Scorer's 1972 book (Fig. 7.2.5) are classified as altocumulus, but I think that the clouds in this photograph are at a lower altitude. (middle, right) of stratus turning into stratocumulus downstream, viewed to the north on Dec. 30, 1999, over the Gulf of Mexico, south of Louisiana. Cold air is flowing from the north over the Gulf of Mexico (see a similar situation in Fig. 3.20). (bottom, left) somewhere between Boston and Oklahoma City, but probably nearer the former, as the ground is warmed by the sun underneath relatively cold air, on Sept. 15, 2008. In this case there seems to be mostly clear areas, with smaller patches or patches of cloud of nearly equal area. (bottom, right) the tops of stratocumulus, spreading out at an inversion or stable layer, over Northern Labrador, on Sept. 7, 2019. There is mostly cloud, surrounded by smaller clear areas.*

Photos courtesy of author.

Figure 3.11 *Visible channel satellite imagery over much of Maine, New Hampshire, Vermont, and Massachusetts, on Jan. 2, 2018 during a cold-air outbreak, when cold continental air flowed out over the relatively warm ocean. Bands of stratus appear offshore and then turn into a regular array of cells of cloudiness farther offshore. See Figure 3.10, the middle, right panel. Also see the brief discussion on Rayleigh–Bénard convection a bit later in this chapter.*
(Courtesy of College of DuPage Meteorology.)

mesoscale disturbances such as is found ahead of upper-level troughs, to the east of low-level cyclones where warmer air is being transported poleward, replacing colder air, or over surface warm fronts or stationary fronts.

It is difficult to obtain soundings through altocumulus, especially during the winter months, because in the United States the soundings are released before sunrise and at or after sunset in many locations, when clouds cannot be clearly seen (for confirmation), except perhaps on moonlit evenings. In addition, altocumulus may be patchy and the chances of launching a sounding into one are therefore diminished.

3.3.1   Altocumulus with little vertical development

Some altocumulus, especially those that are rather thin, look like fish scales and contribute to what is called a "mackerel sky" (Fig. 3.13 at top left, top right, and middle right).

On rare occasions, a circular or elliptically shaped clear area called a "hole-punch" or *cavum* is seen in altocumulus, sometimes seen with patches of ice crystals falling out and sublimating way before (melting and) ever reaching the ground as liquid or frozen precipitation, but rather staying up in the air as virga (Fig. 3.14). A very narrow hole-punch

Figure 3.12 *Aerial view just underneath stratocumulus bands, while the author was in an airplane landing at Oklahoma City, Jan. 15, 2011.*
Photo courtesy of author.

cloud is called a "canal cloud." Altocumulus clouds are composed mainly of supercooled water droplets, some of which freeze into ice crystals and grow large enough to bail out of the cloud. In this case, the ice crystal growth mechanism (the Wegener–Bergeron–Findeisen process mentioned in Chapter 1), during which supercooled water droplets evaporate while water vapor is deposited onto existing ice crystals; consequently, latent heat is released as the ice crystals form and the air warms. Local warming induces rising motion in response to its buoyancy and compensating sinking motion occurs outside the ice crystals. The downward motion dries the air and causes the supercooled water droplets to evaporate.

Richard Scorer, in his 1972 book, claimed that the holes begin at local points and then spread outward, as small "splinters of ice" are thrown out from a droplet that grows some ice on its outside surface and is shattered when the inside subsequently freezes and expands because ice is less dense than liquid water (ice floats in water). While there are also a few other explanations proposed that involve cloud microphysics, more recently a dynamical explanation involving gravity waves has been proposed (Muraki et al. 2016): if outside the punch hole the atmosphere is saturated and neutrally buoyant (moist adiabatic), then gravity waves cannot propagate through it. Recall from Chapter 2 that

Figure 3.13 *Altocumulus. (top, left) A "mackerel sky": Altocumulus overcast over Kennybunkport, Maine, on July 4, 1986. (top, right) Altocumulus, Norman, Oklahoma, on Feb. 7, 2019, with the sun partially blocked out. (middle, left) Altocumulus with small elements and some vertical development, Norman, Oklahoma, on Nov. 1, 2010. (middle, right) Altocumulus, Norman, Oklahoma, on Oct. 14, 2019. The National Weather Center is seen on the center, left. (bottom, center) Altocumulus over Norman, Oklahoma, Oct. 12, 2021.*

Photos courtesy of author.

Figure 3.14 *Hole punch clouds. (top, left) Altocumulus from a mackerel sky over Boston/Cambridge with a "hole punch" and ice crystals falling out, on Dec. 17, 1975. (top, right) Altocumulus with a punch hole and ice crystals falling out, on Oct. 20, 2005, near either Martha's Vineyard or Woods Hole, Massachusetts, with a seagull soaring underneath the punch hole. (middle, left) Altocumulus with a hole punch and a patch of ice crystals falling out thick enough to be considered* spissatus, *Norman, Oklahoma, on Feb. 22, 2020. This hole punch cloud is elongated and perhaps may be classified as somewhere between a highly eccentric, elliptical hole punch and a "canal cloud." (middle, right) Airborne view of a hole punch cloud seen from above, probably somewhere near central Illinois, on Jan. 26, 2023. (bottom) Panoramic image of a hole punch or canal punch with a band of ice crystals falling out overhead, from south (left) to north (right), Norman, Oklahoma, on Feb. 21, 2020.*
Photos courtesy of author.

gravity waves require an atmosphere that is stable with respect to vertical displacements. If the clear air in the hole is weakly stable, then gravity waves can occur, but when they hit the edge of the cloud hole, they must stop propagating outward. The collision point of the gravity waves with the inner cloud edge propagates outward as water droplets are evaporated by the drying effect of the sinking air at the outer edge of the gravity wave "front" (Fig. 3.15).

Although it is usually claimed that the origin of the first ice crystals is from aircraft flying through the cloud when supercooled water droplets freeze on contact with the surface of the aircraft or when the air cools quickly adjacent to the top of aircraft wings where the pressure is lowered (see the discussion of how air is forced up and over shallow pools of cold air at the surface in Section 4.3.3), I personally have yet to observe the "smoking gun," direct evidence of an aircraft flying through an altocumulus cloud that subsequently develops a punch hole, but I trust that others have in fact observed this happening. Scorer speculated that fall streaks from cirrus above may also have instigated the hole, but I have always found the sky to be otherwise clear around altocumulus that develop holes, though admittedly, one can't see from the ground everything that is happening above the cloud before the hole appears. Altocumulus clouds sometimes have depth, are not flat appearing, but have some, though very small, vertical development.

Some altocumulus clouds are organized not as a checkerboard of clumps of cloud, but rather as bands of clouds, called "billows" (Fig. 3.16). Their organization may be similar to that of Kelvin–Helmholtz waves in that vertical wind shear may play a major role in their appearance. It is probably noteworthy, though, that K–H billow clouds seem mostly to look two-dimensional, like single, vertical slices through waves, while billows in altocumulus look like the entire three-dimensional loaf of cloud, not just a slice.

3.3.2   Altocumulus *floccus*

Altocumulus *floccus* look like irregular, tufts of cotton suspended in the air (Fig. 3.17).

3.4   Cumulus

3.4.1   Cumulus *humilis*, cumulus *mediocris*, and cumulus *fractus*

Cumulus clouds often occur as a field of clouds at low altitudes organized on a spatial scale much broader than that of the cloud itself, on the "mesoscale." While the nature of the organization is very apparent in satellite imagery (Fig. 3.18), we will focus mostly on what we see from the ground (Fig. 3.19) or from an aircraft (Fig. 3.20). Cumulus clouds often appear as a checkerboard when the surface is heated relative to the air aloft, whether as a result of heating from the sun during the day over land or from a relatively warm water surface, either during the day or night, when relatively cool air flows over it. The history of theories for this checkboard appearance goes back to observational patterns of heated sperm oil (from whales) by Henri Bénard (Bénard 1900), a Frenchman, and from

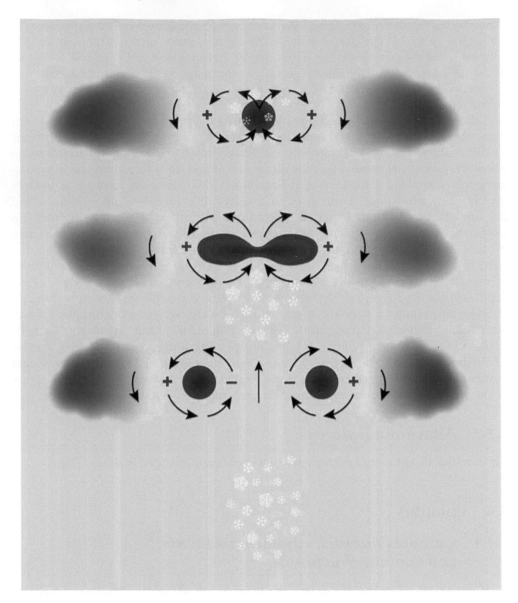

Figure 3.15 *Dynamical explanation of how punch holes grow radially outward from their center. Vertical cross-section is shown for three successive times, increasing downward in each row. The air first locally warms (top row, red circle) and induces a wind field as shown. The cloudy air is depicted by the gray regions. Sinking motion causes the water droplets to evaporate successively more outward. Snowflakes fall out where the latent heating is occurring as ice crystals are deposited.*

(From Muraki et al., 2016; their Fig. 10. © American Meteorological Society. Used with permission.)

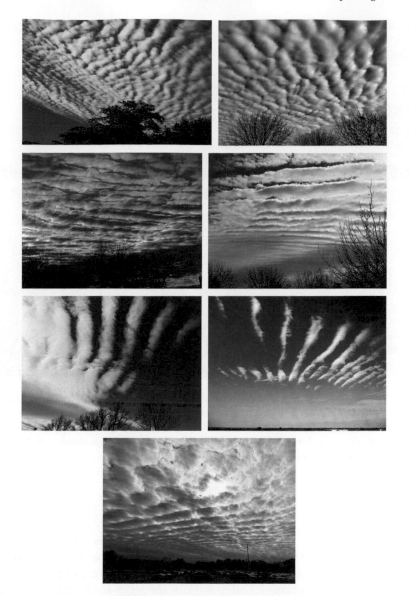

Figure 3.16 *Altocumulus with billows. (top, left) Altocumulus with some billows, Dec. 31, 1997, over Ft. Lauderdale, Florida. (top, right) Altocumulus overcast with closely-spaced billows over Norman, Oklahoma, Feb. 22, 2020. (second row, left) Altocumulus billows overcast, Norman, Oklahoma, on Jan. 27, 2012. (second row, right) Altocumulus overcast, composed of all closely-spaced billows over Norman, Oklahoma, Feb. 7, 2020. (third row, left) Altocumulus with relatively widely spaced billows, transitioning into a solid mass of cloud (at the left), in Norman, Oklahoma, on Feb. 7, 2020. (third row, right) A wide band of altocumulus billows, Norman, Oklahoma, Oct. 26, 2016. (bottom) Altocumulus billows and patches as in a "mackerel sky," over Albany, New York, on Feb. 24, 2022.*
Photos courtesy of author.

Figure 3.17 *Ragged-looking altocumulus. (top, left) Altocumulus floccus, May 20, 2008, Norman, Oklahoma. (top, right) Altocumulus floccus on June 5, 1999, western Nebraska. (bottom, left) Altocumulus floccus, Norman, Oklahoma, May 7, 2021. (bottom, right) Lines of altocumulus castellanus or floccus, on April 10, 2017, in Norman, Oklahoma. These clouds may have been forced by gravity waves because they occur along nearly parallel bands.*
Photos courtesy of author.

mathematical analyses by Lord Rayleigh (Rayleigh 1916), an Englishman, in the very early twentieth century. It turns out, however, that while we use the names Rayleigh and Bénard to describe the cellular/tessellated pattern of cloud, others pre-dating them had also described this phenomenon. I often think of the lyrics to the 2001 Dave Matthews Band song "The Space Between" when I visualize the space between clouds, even though the meaning of the song's title is a metaphor for something else.

When the difference in temperature between the surface and some distance above the surface with an impermeable top is held constant, tiny perturbations (small deviations from the actual temperature) grow, with rising regions (in regions that are slightly warmer) and sinking motions (elsewhere) permeating the region. For those mathematically inclined, this can be shown using a classic type of linear instability

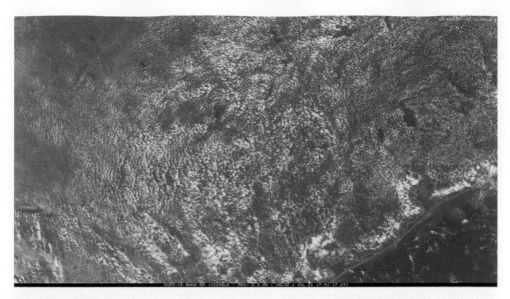

Figure 3.18 *NOAA GOES-East satellite imagery for southeast Texas during the morning of July 1, 2021, in which there is a checkerboard of cumulus clouds. The surface winds were calm or very weak at most locations, except at the Gulf Coast where there was a weak sea breeze from the southeast. The cumulus clouds over land appear to be sprinkled around, except just inland from the coast (lower right), where sea-breeze convergence is enhancing the shallow convection. (See also Fig. 3.20, bottom, left.) The sea breeze occurs as the land is heated more than the ocean and a horizontal gradient of buoyancy develops, which is directed from the ocean to the land and is associated with air coming off the water onto the land and then rising upward; the air sinks just offshore, suppressing convection. This circulation is like the mountain-valley circulation described in Chapter 2.*
(Courtesy of College of DuPage Meteorology.)

analysis, which we will not detail here, lest the reader be led unwittingly into the world of high-order differential equations.

The regions of rising motion that make up the Rayleigh–Bénard cells (like tiles filling the space on a bathroom floor) are associated with the field of cumulus clouds, while the regions of sinking motion are associated with clear air. Over the tropical oceans, when the sea-surface temperature is uniform, a cellular pattern of cumulus clouds can undergo "self-aggregation" (Bretherton et al. 2005)), during which the clouds become broader and deeper, but more widely spaced with time. This phenomenon may be observed over the land also. When there is a background wind and a change in wind speed with height, the organization is likely to be in bands, rather than cells. The meteorologist Ed Zipser[10] has questioned what is actually meant by "organized convection" and "self-aggregation" and whether or not the two are actually words that describe

[10] Heard at a lecture at M. I. T. in June 2022, at a symposium in honor of Kerry Emanuel.

the same phenomenon. I personally think that the latter is a special case of the former, which is discussed in the next chapter, because organized convection can occur especially when the surface conditions are highly heterogeneous, while I think that self-aggregation refers specifically to what can happen only when the surface conditions are homogeneous.

The organization of clouds (which are relatively small in area) into cellular patterns on the larger, mesoscale is *not* a unique example of the organization of a natural feature on a scale much larger than that of the individual feature itself. Alan Turing, the English polymath who devised the basic framework for the first electronic computer (cf. the "Turing machine"), showed how a homogeneous distribution of chemicals can "self-organize" into patterns that are periodic.[11] Another example of tessellated or banded structures is found in vegetation. The feedback in this case results from water transport in a field of vegetation in which there are small variations in vegetation: regions of denser vegetation draw more water and become even denser, while the growth of vegetation in surrounding areas is suppressed because the water supply is depleted there. In the case of clouds, the clouds play the role that vegetation plays, and the transport of water vapor plays the role that the water plays. There are many other examples of this phenomenon in nature, including possibly the formation mechanism of some stars.

Cumulus clouds with the least amount of vertical development are modified by the Latin adjective *humilis*, or humble (Fig. 3.19, top left and right, bottom left panels). Cumuli with more development are referred to as *mediocris*, or having moderate growth (Fig. 3.19, lower right; 3.24). The degree of vertical development may be related to the "degree" of instability or the depth of the unstable layer. In case of cumulus clouds, it is related to the amount of instability in the boundary layer, the layer of air just above the ground, which is affected by the heating of the ground.

Topography sometimes interrupts the checkerboard pattern (Fig. 3.21, left panel), as snow cover and/or an intrusion of a mountain eliminates significant heating of the air at the surface. In Figure 3.21, the right panel, clouds form probably as air is forced upward along an escarpment of the glacier.

Cumulus clouds that exhibit a ragged, "broken" look are categorized as *fractus*, from the Latin word for broken (Fig. 3.22): think of related words such as fragile.

While we often see checkboards or bands of cumulus clouds, we sometimes see just isolated cumulus clouds (Fig. 3.23). These are probably associated with local heating or local lift along mountain ranges or unique types of land surface, which are warmer than surrounding areas (for example, paved surfaces in the midst of vegetation or dry soil amid wet soil or water).

---

[11] Meron, 2019.

Figure 3.19 *Fair-weather cumulus. (top, left) A field of cumulus* humilis (humilissimus,[a] *in this case most-humble looking), which exhibit virtually no vertical development, over Norman, Oklahoma, on April 1, 2005. (top, right) A field of fair-weather cumulus* humilis *or stratocumulus, with very limited vertical development, over Norman, Oklahoma on Oct. 10, 2018, with a view to the west, as cold air flows over a heated land surface. There is a tendency for the clouds to line up in a north-south direction with the wind (bottom, left) Cumulus* humilis, *near Fourth of July Mines, west of Nederland, Colorado, on July 7, 1995. Even though it is July, there is still snow cover at this high elevation, around 11, 000 feet (~3–4 km). (bottom, right) A field of fair-weather cumulus* mediocris *with some vertical development, cumulus* mediocris, *over Norman, Oklahoma, on April 1, 2005.*

[a] This Latin superlative form of *humilis* is my invention and not used officially.
Photos courtesy of author.

3.4.2 Cumulus *congestus*

Cumulus *congestus* exhibit a step up in vertical development from cumulus *mediocris* and I'm not sure where to draw the line between the two types of cloud, especially since we often see a mixture of the two (Fig. 3.24). After all, cumulus *congestus* must have passed through a stage of being cumulus *mediocris* earlier. Some cumulus *congestus* are broad (e.g., Figs. 3.25; 3.26, lower right; 3.28; 3.40; 3.41), while others are narrow (Figs. 3.31; 3.33; 3.34). Some exhibit cauliflower-like (bubbly) growth all around their outer "skin"

Figure 3.20 *Airborne views of checkerboards of mostly cumulus* humilis: *(top, left) en route Ft. Lauderdale to Dallas Ft. Worth, probably over Texas, on March 17, 2012. In this configuration, the cumuli are relatively widely spaced. (top, right) over southeastern Wyoming on July 19, 2013, along with a distant cumulonimbus (a cumulus cloud that has grown up to its greatest potential—see Chapter 4) with an extensive anvil streaming out to the right. Some cumuli are close to their neighbors, while there are also relatively clear regions. Some of cumuli seem to be organized on an even larger scale: For example, note the ring of cumuli at the lower left. The view is to the northeast. (bottom, left) This view on March 11, 2015 of a field of cumulus* mediocris *clouds over southwestern Florida juxtaposed with clear skies over the Gulf of Mexico to the west (left), clearly show the important role of surface heating by the sun in producing the Rayleigh-Bénard field of cumuli. See the satellite photograph of the southeast Texas Gulf coast for a higher-up view of a similar atmospheric setup (Fig. 3.18). There is a strip of clear air over the immediate west coast and a line of cumulus with enhanced vertical development at the western edge of the field of cumuli, most likely as a result of convergence inland underneath the rising branch of a sea-breeze circulation. When there is enough moisture and instability, cumulonimbi form in this location, especially during the summer. View is to the north. (bottom, right) On Sept. 21, 2003, probably somewhere over Michigan. In this configuration, it appears as if there is almost as much cloud material as clear air in between neighboring cumuli.*

Photos courtesy of author.

Figure 3.21 *Cumulus clouds near steep sloping terrain. (left panel) Airborne view showing the disruption of a field of cumulus* humilis *or* fractus *by orography, which in this case is Mt. Fuji, in Japan, on March 4, 2016. It is clear to the left and right behind the volcano, and no cumuli at all around the snow/ice-covered cone or to the northwest. View is to the north or northwest. (right panel) Airborne view of small cumulus lined up along the edge of the top of a glacier and casting distinct shadows to the right (bottom, center), in southern Greenland, on Sept. 10, 2010.*
Photos courtesy of author.

Figure 3.22 *A field of cumulus* fractus, *ragged-looking, but some of which are producing what looks like ice crystals (wispy looking, horizontally stretched out regions of duller, less reflective material), over Boulder, Colorado, on March 22, 2009. It appears as if supercooled water droplets are being converted into ice crystals, which drift with the wind, fall out a small distance, and then sublimate.*
Photo courtesy of author.

Figure 3.23 *Isolated cumulus. (top) Isolated cumulus* mediocris *cloud over North Arapaho Peak, in the Indian Peaks Wilderness, northwest of Nederland, Colorado, on July 31, 2020, viewed to the west from the Arapaho Glacier Trail. This photograph reminds me of Joe Btfsplk, a character invented by Al Capp for his cartoon Li'l Abner, who walked around with a cloud over his head, sometimes spewing lightning, but always implying bad luck. The cloud is situated right over the mountain peak and sticking with it despite the wind trying to blow the cloud downstream. (bottom) An isolated, elongated, cumulus* mediocris, *with small vertical development, over Niwot Ridge, in the Indian Peaks Wilderness, west of Ward, Colorado, on June 8, 2021. View is to the east from a late-spring snowfield on the south-facing slope of Niwot Ridge.*

Photos courtesy of author.

Figure 3.24 *Cumulus clouds with various degrees of vertical development influenced by mountains. (top, left) Cumulus* mediocris *and cumulus* congestus *building during the morning of Aug. 31, 2019 over the Alps, viewed from aboard an aircraft, northwest of Milan, Italy. (top, right) Cumulus* mediocris *and* congestus *growing over Tahiti on Feb. 17, 1984. The island is also acting as a heat island in promoting the growth of the clouds. It is clear over the cooler, ocean. (bottom) Cumulus congestus building on the eastern slope of Niwot Ridge in the Indian Peaks Wilderness of Colorado, west of Ward, on June 26, 2004. Scattered stratocumulus are seen below in the distance, over the Plains, while the base of the cumulus congestus is ragged-looking, indicating the flow of cooler, but moist air up from the east into them. This process is similar to the process by which wall clouds form in supercells, to be discussed later in Chapter 4.*
Photos courtesy of author.

(Figs. 3.27, right panel; 3.28; 3.33, lower right; 3.40; Fig. 4.23, right panel), while others appear so only mainly at the leading edge (Figs. 3.32; 3.34) as in starting plumes, and in rare instances, only at the trailing edge (Fig. 3.31, middle right panel) when an earlier thermal at the top is decaying and breaking away. The presence or absence of bubbly growth in many instances may be a function of how sunlight illuminates the cloud. The "bubbles" are fully highlighted when sunlight strikes them directly, making them look like heads of cauliflower (Fig. 3.26, top, middle panel), but are silhouetted when the sun hits them from behind, producing at their edges what is sometimes referred to, literally, as a "silver lining" (e.g., but not limited to, Figs. 3.27, left panel; 3.28, bottom panel,

Figure 3.25 *Broad areas or bands of cumulus* congestus. *(top, left)* Cumulus congestus *line over Hong Kong, July 4, 1987, with a chiaroscuro effect owing to the sharp contrast between the shadows and the light on the buildings. (top, right) A group of moderate cumulus (*congestus*) on July 31, 2020, viewed to the west from Arapaho Glacier Trail, northwest of Nederland, Colorado. There is great contrast between the sunny and cloudy areas. (bottom, left) Ragged-looking cumulus* congestus *over the Colorado Rockies, viewed from the summit of Mt. Audubon on July 10, 2012. These convective towers appear to have lost most of their buoyancy, perhaps owing to very dry air in the environment. (bottom, right) Cumulus* congestus *building on the west side of the Continental Divide, viewed from the southern slope of S. Arapaho Peak, northwest of Nederland, Colorado, on Aug. 15, 2014. The interplay of light and shadows is like that in the top, left panel.*
Photos courtesy of author.

right most tower; 3.30, right panel; 3.33, top left panel; and 3.54, both bottom, left and right panels; Fig. 4.23, right panel). My vivid first memory of a cumulus *congestus* having a pronounced silver lining was a cloud that moved inland from the ocean and onto the beach I was playing at in Miami Beach when I was very young. As it passed overhead it briefly dumped heavy rain, thus actually being a cumulonimbus, albeit a relatively shallow one without an anvil, not extending up as high as the tropopause.

A potpourri of cumulus *congestus* in different environments is displayed in subsequent figures. Cumulus *congestus* exhibiting various degrees of bubbliness are seen in Figure 3.26. The majesty of some cumulus *congestus* is depicted in Figure 3.29, in which

Figure 3.26 *A potpourri of isolated cumulus* congestus *with varying degrees of bubbliness. (top, left) Cumulus* congestus *with a dark, flat base, over Mt. Audubon, probably taken from the summit of Pawnee Peak (3.8 km MSL, 12, 542 ft MSL), west of Ward, Colorado, on July 16, 1997. (top, middle) A head of organic cauliflower: The top of a cumulus* congestus *(or growing cumulonimbus) looks like a crown of cauliflower. There is fractal structure: There seems to be similar patterns at many scales, as in a fractal. (top, right) Small, narrow, cumulus* congestus *over Brainard Lake in the Indian Peak Wilderness, west of Ward, Colorado, on March 22, 2021. The top of the cloud exhibits a bare minimum of bubbliness. (bottom, left) A cumulus* congestus *building over the Foothills, viewed to the west, on June 27, 2014, in Boulder, Colorado. The Flatirons is seen at the bottom, left and the National Center for Atmospheric Research (NCAR) is seen near the bottom, center. Some prints of the artist Doug West look like this panel.[a] (bottom, right) Cumulus* congestus, *on a beach, on the eastern shore of Cape Cod National Seashore, in southeastern Massachusetts, on Aug. 9, 2012, with a view to the south. (I'm not sure which beach it was, since I spent the day beach-hopping to the north, from one beach to the next,[b] finally ending up at the northern tip of Cape Cod at Race Point and then Herring Cove Beach, in Provincetown. I highly suspect, however, that it was Head-of-the-Meadows Beach.) These clouds were probably forced by heating over the land and then blown out over the ocean.*

[a] https://www.dougwestartist.com
[b] Perhaps in some way as Burt Lancaster hopped from swimming pool to swimming pool in the 1968 movie "The Swimmer."
Photos courtesy of author.

a hiker is seen next to the cloud, giving it a feeling of scale. Figure 3.30 shows silhouettes of cloud over snow-covered mountains. A cumulus *congestus* look-alike, composed of smoke rather than water droplets, is seen in Figure 3.35. A cumulus *congestus* with an

Figure 3.27 *Giant mountains of cumulus* congestus *well on their way to becoming cumulonimbi. (left) A silhouette of a deep cumulus* congestus *close up building into a cumulonimbus, on the summit of Mt. Audubon, near Ward, Colorado, on Aug. 9, 1987. This is the point at which one should be compelled to get down off the summit as quickly as possible[a] for safety. (right) Massive cumulus* congestus *growing into a cumulonimbus on Aug. 7, 1999, directly illuminated by the sun and viewed from Pawnee Peak (3, 945 m MSL, 12, 943 ft MSL), west of Ward, Colorado. The width of this convective tower appeared to become narrower with height.*

[a] Actually, it may be too late to escape danger from lightning in this case.
Photos courtesy of author.

anvil over a tropical atoll is shown in Figure 3.36. Cumulus *congestus* in the tropical South Pacific are shown in Figure 3.37. Miscellaneous cumulus *congestus*, some of whose width varies with height, are displayed in Figure 3.38. Figure 3.39 shows cumulus *congestus* that lean over with height. Figure 3.42, which illustrates how a cloud's environment, in this case mountains and a lake, can enhance the beauty of a cumulus *congestus*. Cumulus *congestus* that look like bombs exploding are seen in Figure 3.43.

Cap clouds, which form over mountains (Chapter 2), can also form over the edges of and the top of growing cumulus and cumulus congestus and even cumulonimbus clouds. Such a cap cloud is called a *pileus* because it looks like the cap a person might wear or the cap of a mushroom (Figs. 3.44 and 3.45), and "pileus" which means head cap in Latin. In a sense, cap clouds are what one might see if a mountain range were to

Figure 3.28 *A contrast in the environment of cumulus* congestus: *Tropical vs. alpine. (top) Cumulus* congestus *building into a cumulonimbus, on July 24, 2017, east of San Diego, California, as monsoon moisture has been advected northward into the eastern portion of southern California from Mexico. (bottom) Cumulus* congestus *building on July 2, 2019, with a view to the east from a lingering snowfield on the eastern edge of Lake Isabelle, west of Ward, Colorado. These clouds built up along a convergence line separating the easterly upslope component of the diurnally forced mountain-valley circulation from the ambient westerly flow aloft and to the west. There is still ample snow for some backcountry skiing or snowboarding on this slope. There is a "lowering" under the cloud base of the cumulus congestus at the right, a clue that moist air from the east, over the Plains, is being ingested into the updraft base.*

Photos courtesy of author.

Figure 3.29 *Towering cumulus in direct sunlight with a hiker shown for scale. (left) Cumulus* congestus *growing just north of Mt. Audubon, in the Indian Peaks Wilderness, west of Ward, Colorado, on July 10, 2010. My friend and colleague John M. Brown is silhouetted against the clouds, leading to a sense of the grand scale of the clouds. (right) A cumulus* congestus *building near the eastern slope of Mt. Audubon, in the Indian Peaks Wilderness of Colorado, west of Ward, on July 29, 2008. The hiker on the slope lends some scale to the grandeur of the cloud, as in the left panel.*
Photos courtesy of author.

Figure 3.30 *Silhouetted, narrow, towering cumuli, with the sunlight coming from the back or side of the cloud. (left) Cumulus* congestus *after an early season snowfall on Sept. 16, 1999, during a hike up to the summit of Castle Peak, a "fourteener" (14, 000 or-so-foot peak) near Aspen, Colorado. (right) Cumulus* congestus *on July 2, 2019, above Niwot Ridge, viewed from a trail just below Lake Isabelle, west of Ward, Colorado. The sun angle allows for a brilliant chiaroscuro "silver-lining" effect of the upper cloud edge. There is still substantial snow cover on the northern slope of the ridge, though it is only a few days before the Fourth of July.*
Photos courtesy of author.

Figure 3.31 *Very narrow cumulus* congestus *clouds. (top, left) A very narrow cumulus* congestus *tower, looking like a "starting" plume, on Aug. 30, 2016, in south-central Kansas. However, unlike a starting plume, the width of the thermal does increase with height. This cloud can be described as having a* castellanus *quality. Only the leading, upper edge of the tower has a cumuliform appearance, while the base is still flat. This is an example of what storm chasers call a "turkey tower." Its "days" (really minutes) are limited. (top, middle) Narrow cumulus* congestus *having a* castellanus *appearance, along the shoreline, Ft. Lauderdale, Florida, on March 9, 2015. The tower is leaning toward the east in response to shear created by a sea breeze (from left to right) near the surface and off-shore winds aloft (from right to left). This tower does not appear to be very buoyant since it is rather diffuse looking and it does not have a flat base. It appears to be a "starting plume" that has ended, as the base is evaporating. (top, right) Narrow, erect, cumulus* congestus castellanus-*like tower during*

grow upwards with height very quickly (on time scales of seconds or tens of seconds, not on geological time scales) and hit relatively humid air. Air is lifted above (where the air is still stable) the cumuliform, buoyant tower. In Chapter 4, however, we will see that under the special circumstance in which the wind speed and/or wind direction vary with height from the bottom of the cloud tower to the top, the cloud actually acts like a mountain obstacle to the wind, even though the cloud is certainly not a solid body. Often the buoyant cloud top punches through the pileus (Fig. 3.44), leaving the pileus looking like a spider web covering the cloud. A number of layers of pileus are often seen, like layers in an altocumulus lenticularis (Fig. 3.46). The layers probably owe their existence to slight vertical variations in moisture in the environment above the cumuliform cloud, as in the case of many orographic wave clouds.

While cumulus *congestus* are composed of liquid water droplets, sometimes fires can produce clouds of smoke that look similar to cumulus *congestus*, as noted earlier (Fig. 3.35). In some instances, however, there may actually be some condensation so that white cumulus *congestus* tops, probably composed of liquid water droplets, are seen (Fig. 3.47).

3.4.3   Altocumulus *castellanus*

Altocumulus *castellanus* (also known by their abbreviation as ACCAS; *castellatus* is an older adjective that was once used) are cumulus clouds (Figs. 3.48 and 3.49) whose bases are relatively high, at midlevels. *Castellanus* comes from the Latin for castle, in the sense that there are turrets in castles, and also refers to buoyant regions in clouds that are taller than they are wide. I would still, however, refer to a high-based cumulus cloud as an altocumulus *castellanus* even if its base were wider than its individual towers. It is ambiguous what is meant by the width of the cloud when there is a line of narrow towers,

---

Figure 3.31 (Continued) *VORTEX-2, on May 23, 2010, in western Kansas. This cloud appears dark against a lighter background of an anvil. Most of this tower appears to be cumuliform, including along its sides. Unlike the narrow towers in the preceding panels, this tower has a distinct flat base and there is therefore still an active updraft throughout. (middle, left) Narrow, erect, cumulus* congestus *tower, like a castellanus, viewed to the south, from below Pawnee Pass, in the Indian Peaks Wilderness, west of Ward, Colorado, on June 15, 2012. (middle, right). The "leaning tower of Ft. Lauderdale": A highly tilted, very narrow, cumulus* congestus *tower having a* castellanus *quality over Ft. Lauderdale, Florida, on March 11, 2015. The upper portion of the tower appears to be dissipated, since it is ragged in appearance, while the bottom portion of the cloud is cumuliform and buoyant. This structure is the opposite of that seen in previous panels, which look like a "starting plumes." Perhaps the structure seen here should be called an "ending plume." (bottom) Very narrow, slightly sheared, cumulus* congestus *or* castellanus *towers over Ft. Lauderdale, Florida, viewed to the west, on March 10, 2015. The tower at the center is spreading out with an anvil-like structure, but at a low-level stable layer rather than the tropopause. The winds at the top of the cloud were probably had a stronger southerly component than the winds at cloud base, thus accounting for the tilt with height.*

Photos courtesy of author.

Figure 3.32 *The author, at the summit of S. Arapaho Peak (4 km MSL, 13, 400 ft MSL), on July 28, 1990, west of Nederland, Colorado, with a narrow cumulus* congestus *whose base is being eroded. Photograph taken by the author's friend and colleague John M. Brown, with the author's camera. The juxtaposition of the author with the cloud was probably not a coincidence. The non-flatness of the base and the narrowness of it along with a diffuse look, suggest that this cloud behaves like a thermal chain in a dry environment or a starting plume that has had its "starting" mechanism disrupted. It is like a puff of cloud material that turns into a "cutoff tower."*
Photo courtesy of author.

Figure 3.33 *Narrow, deep, cumulus* congestus. *(top, left) Narrow cumulus* congestus *on July 26, 2006, from the summit of S. Arapaho Pk. This tower appears to have a cumuliform appearance only at its leading, upper edge, though there may be some cumuliform cloud material hidden on the other side of the cloud. It therefore looks somewhat like a starting plume. (top, right) Airborne view of cumulus* congestus *tower developing into a cumulonimbus over Colorado, on July 26, 2017. This cloud looks like a starting plume, but does not spread with height. (bottom, left) Narrow cumulus congestus from De Palm Island, near Aruba, in the Caribbean, on Dec. 14, 2008, with what might be an old volcanic peak below. (bottom, right) A cumulus* congestus *tower viewed on June 26, 2016 from the Palatine Hill in Rome, Italy. It appears to have active thermals from the bottom all the way up to the top. Clouds such as these certainly must have influenced paintings of classical times by Renaissance and later-period artists.* Photos courtesy of author.

Figure 3.34 *Narrow cumulus* congestus *in a tropical environment over or adjacent to an ocean. (top, left) Narrow cumulus* congestus *building into a cumulonimbus, Dakar, Senegal, on Aug. 27, 1974, during GATE. It is possible this cloud is already a cumulonimbus, but it is not possible to see any rain from this vantage point, and the cloud, as a whole, looks rather small. (top, right) Narrow cumulus* congestus *over Key Biscayne, Florida, on May 27, 1975. This cloud's midsection is diffuse looking, not very buoyant. View is to the north. (bottom) Airborne view of a cumulus* congestus *over the Atlantic, east of Florida, on Sept. 12, 1984. There might be some warm rain falling underneath this cloud, which would make it a cumulonimbus, but we can't see if there is any.*
Photos courtesy of author.

Figure 3.35 *Smoke from a wildfire just east of Boulder, Colorado, the "Marshall" fire, on Dec. 30, 2021, looking like a cumulus congestus, but rather dark looking, not white like a cloud. The buoyancy for the smoke cloud is driven by a fire (see the flames at the right), not latent heat release from condensation.*
Photo courtesy of author.

Figure 3.36 *An airborne view of a cumulus congestus cluster over an atoll in the South Pacific, east or northeast of Tahiti, on Feb. 11, 1984. A small "orphan anvil" is seen jutting out at the upper right, viewed approximately to the northwest, and is not connected to the cloud behind it. Composite photograph. (cumulonimbus anvils are discussed in more detail in Chapter 4; anvils are found mainly in cumulonimbus clouds, but may also be seen in cumulus congestus, even when there is no precipitation)*
Photo courtesy of author.

Figure 3.37 *A cluster of cumulus congestus towers, the largest of which have ragged-looking edges, over the ocean near Moorea, French Polynesia on Feb. 15, 1984. There may be some precipitation falling from these clouds, so arguably they might be considered cumulonimbi. The tops of these clouds look rather ragged or fuzzy, probably owing to weak buoyancy.*
Photo courtesy of author.

as in Figure 3.48 or when there are flat bases all around, with some towers poking up here and there (Fig. 3.49).

There are probably a number of ways in which these clouds can form.[12] "Elevated convection" refers to that in which the air in the cloud originates *above the moist boundary layer* and enters the convective cloud and is lifted via the air currents to its LFC. The nature of the lift is probably associated with gravity waves, bores, or other mesoscale lifting mechanisms, since they often are oriented along and form along lines. It is also possible, however, that there is a very deep boundary layer and that the CCL is simply just very high. These altocumulus *castellanus* may simply be convective clouds over a deep, well-mixed, boundary layer, as we often see during the summer months over elevated terrain, like mountains, with a high CCL. In the former case, the cloud base is elevated because the source of the moisture is elevated, while in the latter, the cloud base is elevated simply because the boundary layer is deep.

---

[12] Corfidi et al., 2008.

Figure 3.38 *Narrow cumulus* congestus, *several of which have an hourglass appearance (middle right, bottom left, and bottom right) or show narrow towers embedded within the top main tower (top left, top right). (top, left) Isolated cumulus* congestus *making landfall as it moves westward, north of Ft. Lauderdale, Florida, on March 9, 2015. It is an otherwise very sunny, pleasant beach day. (top, right) Isolated cumulus* congestus *over the Foothills of Boulder, Colorado, on July 23, 2017. (middle, left) Cumulus* congestus, *June 22, 2001, west of Niwot Ridge, west of Ward, Colorado. The top of the tower is broken up into smaller* castellanus-*like turrets. (middle, right) Cumulus* congestus *viewed to the north from Norman, Oklahoma, on Sept. 16, 2014. The underside of the top of the tower on the right is ragged-looking, indicated a lack of buoyancy there. The turrets at the top of the cloud are narrow and soon dissipated. The profile of this cloud is somewhat like that of an hourglass, broad at the top and bottom, narrow at the center. (bottom, left) Narrow cumulus* congestus *spreading out aloft like an anvil, on May 13, 2009, during VORTEX-2, in west-central Oklahoma. The bottom of the top part of the tower, on its right side, is ragged-looking, indicative of a lack of buoyancy there. The midsection of the cloud is much narrower than the top and the base, assuming an hourglass appearance. (bottom, right) Cumulus* congestus *broadening aloft like an anvil, on July 29, 2008, below James Pk., west of St. Mary's, Colorado. View is to the west.*
Photos courtesy of author.

Figure 3.39 *Cumulus* congestus *that lean over with height. (top, left) Highly sheared cumulus congestus on June 10, 2018, northwestern South Dakota. View is to the east-northeast. This cloud looks as if it is literally being sheared off and not allowed to develop much further. This image reminds me of the name of the avant-garde musician Frank Zappa's group's album title,* Weasels Ripped My Flesh, *which in my mind could be re-phrased as "Wind Shear Ripped My Convective Tower Apart." (bottom, left) Similar to the top panel, but on April 17, 2019, along the dryline in North Central Texas, just south of the Red River. View is to the south-southeast. (middle, right) Cumulus congestus in Boulder, Colorado on Aug. 11, 2022, viewed to the west, leaning over with height, but probably not as a result of vertical wind shear: In this case, the triggering location of the cumuliform tower was stationary, but the wind (uniform with height) carried cloud material off in the downstream direction, resulting in a leaning tower over the Flatirons.*
Photos courtesy of author.

If a few supercooled water droplets in the cloud freeze, they can grow at the expense of the other water droplets because the saturation vapor pressure over ice is less than that over water and may fall out as fall streaks (Fig. 3.50), just as we see in some cirrus clouds.

Figure 3.40 *Cumulus* congestus *building into a cumulonimbus on Nov. 10, 2018, viewed from Villa Carlos Paz, Argentina, during the RELAMPAGO (Remote sensing of Electrification, Lightning, And Mesoscale/microscale Processes with Adaptive Ground Observations) field program. Parallel bands of cloud bases near the leading edge of cool outflow are suggestive of lift due to bores (see Chapter 2).* Photo courtesy of author.

3.4.4   Convective cloud lines

Lines of convective clouds may form for a number of reasons. First, there may be roll circulations associated with vertical shear (Figs. 3.51 and 3.52) in which convective clouds are triggered, owing to the air being lifted to its LFC along the rising branches of the rolls. In other instances, the rising branches of horizontal circulations associated with bores (Fig. 3.53) may trigger convective clouds, which may be elevated if the bore circulations don't make it to the surface. Singular lines of convective clouds may be triggered when the rising branch of a vertical circulation is isolated, for example along a solenoidal mountain–valley circulation (Fig. 3.54, top left), along a cold front (Fig. 3.54, top right, and bottom, both left and right panels), along a sea breeze front (not shown here), along an outflow boundary (not shown here), along a dryline (not shown here), etc.

Figure 3.41 *Cumulus congestus near snow cover at high elevation. (top) Cumulus congestus, June 25, 2019, near Niwot Ridge, viewed from the Arapaho Glacier Trail, northwest of Nederland, Colorado. There is still plenty of snow cover this early in the summer above 11, 000 feet (3.4 km) elevation. (bottom) Cumulus* congestus *building into a cumulonimbus (convective snow shower), on the northeastern slope of Mt. Audubon, in the Indian Peak Wilderness, west of Ward, Colorado, on March 26, 2021, viewed to the north from Brainard Lake.*

Photos courtesy of author.

Figure 3.42 *Cumulus* congestus *near mountains. (top) Cumulus* congestus *building south of Riva del Garda, Italy, on Sept. 1, 2019. (bottom) Cumulus* congestus *building north of Caribou Ranch, northwest of Ward, Colorado, on Aug. 13, 2022. Wildflowers are blooming and a red barn, where artists take residency, is visible. In the past, many musicians recorded now-famous songs on this ranch (but not necessarily in this building).*

Photos courtesy of author.

Figure 3.43 *Isolated cumulus* congestus *that look like explosions from bombs. (top) Isolated cumulus* congestus *tower appearing to explode over the Foothills in Boulder, Colorado, on July 5, 2020. (bottom) Isolated cumulus* congestus *on June 29, 2008, Helsinki, Finland. This cloud is typical of what is seen when the land is heated underneath a relatively cool upper troposphere, north of the polar jet stream.*

Photos courtesy of author.

Figure 3.44 *A sequence of photographs (right to left) just minutes apart of a pileus over a growing cumulus congestus tower, from just below Pawnee Pass (12, 500 feet MSL), in the Indian Peaks, Colorado, on Aug. 7, 2019. Air is forced up above the cloud tower, cooling to saturation if there is a layer of relatively humid, but unsaturated air above, perhaps leftover water vapor from a previous tower that has evaporated. The environment above the cloud is stable. The pileus looks just like a cap cloud over a mountain peak. The convective cloud may push upward past the pileus, leaving the "cap" below, as if you pushed your hand through a hat, puncturing the top, as seen in these images.*
Photos courtesy of author.

Figure 3.45 *Close-up of the pileus shown at the bottom, left, in the previous figure. It looks like a ghostly veil has covered the top of the tower of a convective cloud.*
Photo courtesy of author.

Figure 3.46 *Multi-layered pileus clouds. (top) A cumulus congestus, buoyant tower punching through a three-layered pileus, on Aug. 26, 2020, in the Indian Peaks Wilderness, west of Ward, Colorado, near Niwot Ridge. (bottom) Three layers of stratiform cloud above a growing, but diffuse, cumulus congestus, on March 9, 2015, Ft. Lauderdale, Florida.*

Photos courtesy of author.

Figure 3.47 *A series of cumulus* congestus *tops over the ash plume from a forest fire northwest of Boulder, Colorado, looking to the north from Boulder, on Oct. 14, 2020. These appear to be like a series of intermittent starting plumes.*

Photos courtesy of author.

Figure 3.48 *Lines of altocumulus* castellanus. *(top, left) Parallel bands of altocumulus* castellanus, *Norman, Oklahoma, on May 13, 2019. These clouds were probably triggered by some wave motion. (top, right) Lines of altocumulus* castellanus *over Norman, Oklahoma, on Sept. 21, 2021. Note how the clouds are organized over solid cloud bases. (bottom, left) Altocumulus* castellanus *over Norman, Oklahoma, viewed to the south and silhouetted by the sun, on Nov. 24, 2021. Note how these clouds also emanate from lines of solid cloud bases. The individual convective towers seem to be mimicking the individual treetops below. (bottom, right) Altocumulus* castellanus *cloud over Norman, Oklahoma, on Dec. 5, 2021. Note the very flat cloud base.*

Photos courtesy of author.

Figure 3.49 *Altocumulus* castellanus *above flat bases. (top, left) Altocumulus* castellanus *growing above a flat, elevated cloud base, on May 5, 2020, in Norman, Oklahoma. Note the banded, flat bases. These clouds may eventually grow enough that precipitation forms, in which case they could be considered "altocumulonimbus" clouds, though, as mentioned in the caption for the previous figure, I don't think this a currently accepted name for them. (top, right) Altocumulus* castellanus *rising above a stratiform layer of clouds over Norman, Oklahoma, on Oct. 12, 2021. View is to the west. (middle, left) An area of altocumulus* castellanus *poking above stratus, on May 27, 2015, Norman, Oklahoma. View is to the south. (middle, right) Altocumulus* castellanus, *Norman, Oklahoma, on Oct. 4, 2018. View is to the southwest. (bottom) Altocumulus with flat bases and streamers jutting upwards. These are probably altocumulus* castellanus, *but the towers look different (smoother) than do in the other panels.* Photos courtesy of author.

Figure 3.50 *Altocumulus* castellanus *with fall streaks. (top) Scattered altocumulus with some buoyant tops and with fall streaks, Boulder, Colorado, on Aug. 17, 2013. (middle) A bird approaching a line of altocumulus* castellanus *with fall streaks, Limon, Colorado, on June 1, 2019. (bottom) Line of altocumulus* castellanus *with fall streaks over southeastern Colorado, near Twin Buttes and Lamar, on May 30, 2018.*

Photos courtesy of author.

Figure 3.51 *Airborne view of cloud streets of shallow convective clouds (cumulus* humilis *and* fractus) *over, on July 3, 2003, over the Pacific Ocean, somewhere between Japan and the west coast of the US, probably organized by vertical shear.*
Photo courtesy of author.

Figure 3.52 *Parallel lines of cumulus* congestus *(cloud streets) over the tropical Western Atlantic, about 600 miles west of Dakar, Senegal, during GATE [GARP (Global Atmospheric Research Program) Atlantic Tropical Experiment], aboard the NOAA vessel the R. V. Gilliss, early in Sept. 1974. These convective cloud lines were probably organized by the low-level vertical shear.*
Photo courtesy of author.

Figure 3.53 *Elevated (not rooted in the boundary layer), parallel bands of convective cloud lines (cumulus mediocris/congestus) in Boulder, Colorado on June 8, 2020. These high-based cloud bands may have been triggered by lift from a bore.*
Photo courtesy of author.

Figure 3.54 *Lines of relatively robust cumulus* congestus *that developed into cumulonimbus clouds. (top, left) Line of cumulus* congestus *west of Boulder, Colorado, on Aug. 6, 2002, along the convergence zone separating the easterly upslope wind generated by heating along the eastern slope of the mountains, and ambient westerly winds at higher elevations, as part of the diurnal mountain-valley circulation driven by a horizontal gradient in heating from the sun. (top, right) A line of cumulus* congestus *building into a line of convective storms, viewed to the east from Norman, Oklahoma, on March 21, 2005. This line of clouds was building along the leading edge of a cold front. This photograph is reminiscent of a beautiful painting by John Rogers Cox.* Gray and Gold, *1942, of a line cumulus* congestus *that looks as if the clouds are building into a line of cumulonimbi, may be found at the Cleveland Museum of Art.[a] This painting shows gray clouds building over brilliant golden fields which is probably of wheat. The symbolism in the painting of a crossroads in the future of American democracy during World War II, though, is not mirrored in the photograph, but the clouds depicted are similar. (bottom, left) A line of cumulus* congestus *with distinct, individual towers, growing into cumulonimbus clouds along a cold front near Norman, Oklahoma, on May 4, 2020, viewed to the north. Subsequently, supercells with large hail and some rotation, evolved from this first burst of convective clouds. In some instances, the position of the individual towers can be correlated with the intersection of boundary layer roll circulations that intersect a line of air being forced up along a cold front and enhanced at the points where the upward branches of the roll circulations intersect the line. (bottom, right) Crepuscular rays emanating from behind a line of cumulus* congestus *in Oklahoma, on Oct. 21, 1979, along a cold front. There appears to be a stable cloud produced by a line of lift, with convective elements popping up above. The LFC is apparently above the LCL in this case.*

[a] https://www.clevelandart.org/art/1943.60
Photos courtesy of author.

3.5 Summary

Convective (cumuliform) clouds that are not producing precipitation were considered separately from precipitating convective clouds, cumulonimbus clouds (to be dealt with in the next chapter). I did this in order to keep our discussion as simple as possible, because the non-precipitating[13] convective clouds are pristine in that they are not affected by cold pools and their dynamics, which are associated with precipitation microphysics and their thermodynamic consequences. They vary from clouds having stunted vertical development such as stratocumulus clouds, altocumulus clouds, cumulus *humilis* and cumulus *fractus*, up to cumulus *mediocris*, altocumulus *castellanus* and *floccus*, and finally cumulus *congestus*, the prelude to the cumulonimbus. These clouds may appear isolated, in bands, or in mesoscale checkerboard patterns. In the case of the latter, the Rayleigh–Bénard convection is implicated, while in the case of the former, vertical shear is probably important, or a simple lift along a surface boundary or a bore or gravity waves, or a mesoscale, solenoidal vertical circulation by themselves. In the case of isolated clouds, mountains or inhomogeneities in land use are probably implicated.

References

Bénard, H., 1900: Les tourbillons cellulaires dans une nappe liquide. *Revue générale des Sciences pure et appliqués*, **11**, 1261–1271, 1309–1328.

Bluestein, H. B., 1992: *Synoptic-Dynamic Meteorology in Midlatitudes, Vol. I: Principles of Kinematics and Dynamics*. Oxford University Press, 431 pp.

Bluestein, H. B., 1993: *Synoptic-Dynamic Meteorology in Midlatitudes, Vol. II: Observations and Theory of Weather Systems*. Oxford University Press, 594 pp.

Bretherton, C. S., P. N. Blossey, and M. Khairoutdinov, 2005: An energy-balance analysis of deep convective self-aggregation above uniform SST. *J. Atmos. Sci.*, **62**, 4273–4292.

Bryan, G. H., and J. M. Fritsch, 2000: Moist absolute instability: The sixth static stability state. *Bull. Amer. Meteor. Soc.*, **81**, 1207–1230.

Corfidi, S. F., S. J. Corfidi, and D. M. Schultz, 2008: Forecasters' Forum: Elevated convection and castellanus: Ambiguities, significance, and questions. *Wea. Forecasting*, **23**, 1280–1303.

Levine, J., 1959: Spherical vortex theory of bubble-like motion in cumulus clouds. *J. Meteor.*, **16**, 653–662.

Ludlam, F. H., and R. S. Scorer, 1953: Convection in the atmosphere. *Quart. J. Roy. Meteor. Soc.*, **79**, 317–341.

Malkus, J. S., and R. S. Scorer, 1955: The erosion of cumulus towers. *J. Meteor.*, **12**, 43–57.

Meron, E., 2019: A simple principle relating growth to lateral water transport explains the variety of self-organized vegetation patchiness. *Physics Today*, **72**, 31–36.

Morrison, H., and J. M. Peters, 2018: Theoretical expressions for the ascent rate of moist deep convective thermals. *J. Atmos. Sci.*, **75**, 1699–1719.

Morrison, H., J. M. Peters, and S. C. Sherwood, 2021: Comparing growth rates of simulated moist and dry thermals. *J. Atmos. Sci.*, **78**, 797–816.

---

[13] In some instances, however, there is some precipitation falling out but not reaching the ground.

Muraki, D. J., R. Rotunno, and H. Morrison, 2016: Expansion of a holepunch cloud by a gravity wave front. *J. Atmos. Sci.*, **73**, 693–707.

Rayleigh, J. W. S. (Lord), 1916: On convective currents in a horizontal layer of fluid when the higher temperature is on the underside. *Phil. Mag.*, **32**, 529–546.

Scorer, R., 1972: *Clouds of the World: A Complete Colour Encyclopedia.* David & Charles, London, 176 pp.

Simpson, J., and W. L. Woodley, 1971: Seeding cumulus in Florida: New 1970 results. *Science*, **172**, 117–126.

Turner, J. S., 1962: The "starting plume" in neutral surroundings. *J. Fluid Mech.*, **13**, 356–368.

4

Buoyant Clouds

Part II. "Deep," Convective, Precipitating Clouds

> *"Double, double, toil and trouble;*
> *Fire burn and caldron bubble."*
>
> William Shakespeare, *Macbeth*

> *"Now furious storms tempestuous rage,*
> *Like chaff, by the winds impelled are the clouds,*
> *By sudden fire the sky is inflamed,*
> *And awful thunders are rolling on high,*
> *Now from the floods in steam ascend reviving showers of rain,*
> *The dreary, wasteful hail, the light and flaky snow."*
>
> Franz Joseph Haydn, *The Creation*

4.1 Basic dynamical principles governing the behavior of air motion in convective clouds: Positive and negative buoyancy, and interaction with the environment

> *"Anyone who has been in a thunderstorm has enjoyed it, or has been frightened by it, or at least has had some emotion. And in those places in nature where we get an emotion, we find there is generally a corresponding complexity and mystery about it."*
>
> Richard Feynman, *Feynman Lectures* (vol. 2, p. 9–5)[1]

In this chapter we will "turn up the volume"[2] of the cloud and look at more highly buoyant clouds than we did in the last chapter and also consider clouds that can produce precipitation and severe, violent weather. The convective clouds with stronger buoyancy

---

[1] I am indebted to my colleague Rob Fovell at the University of Albany for calling my attention to this quote.
[2] Not literally, of course! We will not mess with the size of the cloud . . . We will, however, turn up the "volume" in some sense only when the depth of the cloud is considered. We will, however, turn up the level of the volume of the *sound* only in a metaphorical sense.

The Architecture of Clouds. Howard B. Bluestein, Oxford University Press. © Howard B. Bluestein (2024).
DOI: 10.1093/oso/9780198870548.003.0004

tend to be relatively deep; by "deep" I mean the distance between their bottoms and their tops relative to the depth of the troposphere, the layer of air below the tropopause, which separates the stratosphere from the troposphere. Away from cold, upper-level low-pressure areas, where the tropopause may be as low as 5 km, the tropopause is usually around 10–15 km MSL (above mean sea level). In the next chapter, we will look in particular at the most violent weather phenomenon contained within some of these severe convective storms, the tornado, and also other vortices, some of which are found in seemingly benign-looking clouds or are not necessarily visibly connected to any cloud at all. Please notice that I use the term "convective storm" rather than "thunderstorm": Some convective storms, even severe ones, are not necessarily highly electrified and may produce little, if any, lightning and thunder. The electrical nature of storms as manifest by lightning and thunder, while providing clues about what is happening inside a storm and being visually and audibly impressive (and often beautiful), is not typically the prime mover in convective storms. To make a reasonably realistic simulation of a convective storm, you don't need to include the effects of cloud or precipitation electrification.

A convective cloud that produces precipitation is called a *cumulonimbus* (a cumulus cloud producing precipitation—*nimbus*); it behaves according to certain physical laws as we understand them. Strictly speaking, a deep, cumulus congestus cloud, even one that reaches the tropopause, but does *not* produce any precipitation, is *not* a cumulonimbus, even if it looks like a classic cumulonimbus, complete with an anvil. We wonder, why do the clouds look the way they do and what visual clues can we find as to what physical processes are going on inside them and what type of weather are they capable of producing? To consider the dynamics of the more highly buoyant clouds, we will take two approaches to describing how the clouds work, to be explained shortly. Knowing how clouds work will, I hope, enrich your experience of viewing the clouds and cause you to be even more appreciative of them. The basic building blocks for all convective storms are their updrafts and downdrafts, the air currents in them flowing upward and the air currents flowing downward. The amount of buoyancy and the vertical shear of the wind in their environment control much of the behavior of the updrafts and downdrafts, and consequently of the storms themselves. We must therefore understand what physical processes are responsible for producing the updrafts and downdrafts in the presence of their environmental temperature, moisture, and vertical-shear profiles (i.e., how they vary with height). Like their smaller cousins, cumulonimbus clouds can have both buoyant and non-buoyant parts. Some of the buoyancy is negative (acting downward), as well as positive (acting upward). Therefore, their dynamics must include the representation of both positive and negative buoyancy, as well as neutral buoyancy (zero buoyancy).

The most basic method of trying to understand what makes convective storms tick is to consider *Newton's 2nd law of motion applied to a fluid* such as the atmosphere. In basic high school and college classes in physics, one learns that Newton's 2nd law may be applied to the dynamics of rigid bodies, such as apples, with homage to the famous story involving an apple tree and a falling apple (or some variation thereof) and Sir Isaac Newton. So, the familiar equation that states that the force (F) (vector quantities here are denoted in bold italics) applied to a solid object (such as an apple falling from a tree) is given by its mass (m) multiplied by its acceleration (a), or in mathematical terms

($F = $ m a) is altered slightly for a fluid like the atmosphere; instead of considering a solid object, we look at a tiny collection of the same air molecules, an "air parcel,"[3] whose total mass (amount of material) remains the same. For example, if an apple falls off a tree, we can calculate the speed with which the apple is falling to the ground as a function of the time beyond which the apple has broken free from the tree, as it accelerates downward owing to gravity (and ignoring its terminal fall velocity). If a cloud is buoyant, we can calculate the speed that an air parcel in an updraft attains as a function of time and height given its buoyancy or the speed that an air parcel in a downdraft attains.

Since air is a fluid, the shape of the parcel can change and since air is compressible, its volume can also change. We can understand how the air motion in a cloud can change by adding up all the forces acting on air parcels and then deducing how the parcel's motion will change. Doing so can be satisfying intellectually and is one of the bases for numerical weather forecasting (forecasts of meteorological variables such as the wind, temperature, precipitation, etc. using computer models), but also very complicated. Some forces that are significant for clouds are gravity, buoyancy (which contains the effect of gravity), pressure-gradient forces (both vertical and horizontal), the Coriolis force (for the larger, longer-lived convective storms, always acting normal to the wind. . .to the right in the Northern Hemisphere, to the left in the Southern Hemisphere), and friction, described mostly by the small-scale vertical mixing of air parcels moving at different velocities and imparting momentum to one another.

To understand air motion in and around clouds, we take into account that the cloud is not a solid body floating in the air. In actuality, air enters the cloud at some location, such as cloud base, and exits the cloud somewhere, perhaps at the top of the cloud or even around its sides. The cloud is like a living organism, taking in and eating water vapor and dry air and expelling liquid and/or frozen water and water vapor and dry air. Although the boundary between a cloud and the clear air adjacent to it may look unambiguous, there are invisible air currents just outside the visible cloud which are part of the cloud system. We must consider the effects of forces on the motion of *individual* air parcels as they move from the unsaturated area outside the cloud, into the cloud where they get nice and wet, then out from the cloud where they evaporate or sublimate, rather than forces acting on the *entire* cloud itself. A cloud can be changing its appearance and characteristics with time as air parcels move around into it and out from it. The analysis of clouds (and other meteorological phenomena) is therefore much more complicated than that for solid objects.

A more elegant, alternative analysis approach to understanding convective-cloud dynamics involves considering the tendency for air to rotate about some axis, which may be vertical, horizontal, or even tilted with respect the ground. For example, cyclones (both extratropical and tropical) and tornadoes consist mostly (but not completely) of vertical columns of rotating air (there may also be a component of the wind that is converging or diverging from the axis of the center of rotation). In some instances air rotates

---

[3] In fluid dynamics an air parcel is also called a "material volume" (not to be confused with or related to the popular singer Madonna's 1980s hit song "Material Girl"), since it always contains the same amount of material (air).

about a horizontal axis, such as at the leading edge of some convective storms near the ground or in a ring about the top of the updraft in a convective storm. The air itself does not actually need to be rotating, but there is the *tendency* for rotation to be induced by such a wind field if, for example, a paddlewheel were placed in it. Suppose that the wind blows from the west at 1 km above the ground and from the east at the ground, and then imagine that a paddle wheel is placed at a height of 500 m above the ground; then it would begin to rotate as shown in Figure 4.1.

A measure of the rotation that is induced is called "vorticity." In the example shown in Figure 4.1, the vorticity (vector) is horizontal and is directed into the page/screen; in the case of a tornado the vorticity is mainly vertical. By considering vorticity, which is a function only of the wind field and how it varies in space, we can consider how changes in rotation (vorticity) affect the dynamics of highly buoyant clouds *without explicitly considering pressure* (in the case of an atmosphere in which horizontal variations in density are negligible, which is what is in fact the case on the large scale, but not on the small scale; in the latter case, buoyancy forces, which include horizontal variations in density, are significant). Such a simplification is well applied when used to explain what happens in a storm (we did this in Chapter 2, for example, when we considered the diurnal mountain-valley circulation). This is a considerable simplification, not having to consider what the pressure-gradient force is and think *only* about the wind (and buoyancy).

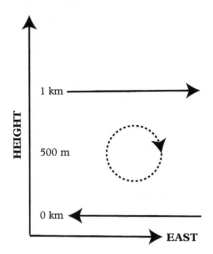

Figure 4.1 *Illustration of how vertical shear induces rotation ("vorticity") about a horizontal axis (in this case directed into the page or screen: the axis is located by curling the fingers of your right hand in the direction of the induced wind, which in this case is in a clockwise direction; the axis is directed in the direction that your thumb points, which in this case is into the page (the vorticity vector is thus directed into the page). The solid lines with arrows denote the wind at the ground (0 km) and at 1 km above the ground. If a paddlewheel were placed in the flow at 500 m above the ground, it would rotate as indicated by the direction of the dotted line.*

Figure courtesy of author.

Why this can be done is too complicated to understand mathematically for readers who have not had courses in vector calculus,[4] but see Appendix 1 even if you are one of the readers who does know calculus for a very simple explanation; otherwise, just have faith that one can combine the different components of the equations of motion (Newton's 2nd law applied to the atmosphere) such that pressure disappears from the equation . . . almost as if by magic. Remember, though, that the science-fiction writer Sir Arthur C. Clarke once said that "Magic is just science that we don't understand yet."[5]

If, however, we consider an atmosphere in which there are large horizontal density gradients, then an analysis of vorticity is still useful, and an additional term, the horizontal buoyancy gradient term appears (as we have used it several times earlier). So, we have the following two approaches: (a) Use the equations of motion acting on an air parcel for each of three, orthogonal directions [up; to the right (usually defined as to the east); and to the left (usually defined as to the north)], including the pressure-gradient force, or (b) use an equation for vorticity, without explicitly including pressure or considering forces acting on air parcels.

However, to get to the real nitty gritty of clouds, we do have to relate the pressure to the wind field in order to be able to consider cloud dynamics, because of the thermodynamic aspects of clouds such as pressure are crucial: for example, we need to know the pressure in order to know if the air is saturated or not or how strong an updraft or downdraft will be. In our subsequent discussions, we will move back and forth between using Newton's 2nd law of motion applied to air parcels in the atmosphere and so-called vorticity dynamics, but will emphasize the latter. Hopefully thinking about vorticity will involve only the air spinning and *not* your head (in confusion).

There is another relationship, which involves the wind field, that is important for cloud dynamics. The different components of the wind field must be related to each other in a way that constrains mass to be conserved; the resulting equation is called the equation of "mass continuity." If air is moved around, none suddenly disappears or appears from nowhere (as postulated can occur in a black hole or worm hole). This common-sense relationship says that if air converges from different directions, most of it must be diverted in the normal direction. For example, if at 5 km above the ground there are northerly winds converging with southerly winds everywhere, then the air must be diverted both upwards and downwards, and also to some very small extent the air parcel must be compressed. When we study clouds, we recognize that the effects of compressibility are relatively small, but do need to be taken it into account when we consider buoyancy, in that air expands when it goes up and contracts when it moves downward. Since sound waves depend upon the compressibility of air, we usually neglect sound (acoustic) waves when analyzing air motion in clouds. Shouting at a cloud will not alter its behavior, even though we may get so emotional that we try to. We cannot hear sound

---

[4] The curl of the equation of motion describes the time evolution of vorticity, The curl of a gradient is always zero: So the curl of the pressure-gradient force is zero and does not contribute to changes in vorticity.

[5] This "quote" is one of several variations on what is known as the third of Clarke's "laws." According to *New Scientist* online, the original was a footnote in his book *Profiles of the Future* (1973) as "Any sufficiently advanced technology is indistinguishable from magic." https://www.newscientist.com/definition/clarkes-three-laws/

from clouds, though there may be some infrasound or ultrasound radiated, which is not audible to our ears, or sound generated at frequencies we can hear, but the amplitude of the sound is too faint to be audible. Sound waves (waves of compressibility) do not affect clouds enough to consider them in understanding cloud behavior. So, we don't really need to consider compressibility when dealing the equation of mass continuity. Perhaps we should call ignoring sound waves in clouds the "anti-Jericho" effect, in recognition of the biblical account of the sound of horns and loud voices bringing down the ancient city's walls. In a similar but visual vein, the founder of Christian Science, Mary Baker Eddy, is said to have "dissipated a thundercloud by simply looking upon it."[6]

We can understand the dynamics of air motion associated with the equation of continuity in the following way: Suppose that air masses collide at the ground. Then the air must be diverted upwards because air cannot go into the ground (well, let's forget about subway entrances and exits, nuclear missile silos, mines, and open underground tornado shelters). How does this happen? Well, we know that the speed of the about-to-collide air masses must decrease to zero at the collision point. So, the air parcels coming from both sides must decelerate. If we consider Newton's 2nd law of motion applied to the atmosphere, then we conclude that in the absence of gravity (and buoyancy), which acts only in the vertical, and friction (which is important at the ground but always acts to decrease the wind speed there), and the Coriolis force (which acts normal to the wind direction), then the only force that is significant is the pressure-gradient force. The only way the air can decelerate is if the pressure-gradient forces acts in the direction opposite to that of the wind. So, the pressure at the collision point must be relatively high. If the pressure at the ground is relatively high, then there must be an upward-directed pressure-gradient force that moves the air upward. This explains the obvious, but is important to a complete understanding. Air rises when airstreams converge near the ground; similarly, air sinks aloft when airstreams diverge at the surface. In the relationship between the convergence or divergence of air and whether air rises or sinks, one does not *cause* the other: rather, they are instantaneously associated with each other.

Understanding the thermodynamics of clouds involves application of the first law of thermodynamics applied to "moist" air, i.e., air which includes a mixture of both dry air and water substance. One must also take into account cloud and precipitation microphysics: how tiny cloud particles form and disappear and how they produce precipitation and how precipitation disappears; these processes involve evaporation, condensation, melting, freezing, sublimation (transformation directly from the ice phase to the water-vapor phase without ever passing through the liquid water phase), deposition (the opposite of sublimation), and fallout. Finally, air motions must be consistent with both the law of motion and the so-called vorticity equation.

---

[6] From a review of *God's Perfect Child*, 1999, C. Fraser, Metropolitan Books/Henry Holt, 561 pp., in the book review section of the *Sunday New York Times*, by P. Zaleksi, p. 8.

4.2 The cumulonimbus: Remembrances of convective bubbles past

> "The world is threatened with peace and quiet.
> I love storm and dread it when the wind abates."
> Christina of Sweden[7]

As noted earlier in Chapter 3, a cumulus cloud that produces precipitation[8] is called a cumulonimbus, the "nimbus" part referring to the precipitation. The cumulus cloud may be deep, extending all the way to the tropopause and possibly a little beyond, and may produce frequent lightning and deafening thunder, or may not extend above the boundary layer, which may be only a kilometer or so deep. Most of the cumulonimbus clouds shown that follow, extend up to the tropopause (e.g., Figs. 4.2–4.4). When the updraft hits the tropopause and loses its buoyancy it produces an anvil, since the cloud material spreads out as the updraft diminishes in intensity and disappears. When the updraft penetrates the stable stratosphere and loses buoyancy is it like hitting a brick wall. In the presence of vertical shear, however, the anvil not only spreads out, but has a component that streams off in the direction of the upper-level winds. The anvil is usually composed of ice crystals, owing to the low temperatures near the tropopause, typically less than −40°C, where homogeneous nucleation (formation of ice crystals without pre-existing nuclei) can occur. Anvils may also contain supercooled water droplets brought up from below, where the air is warmer, by the storm updraft, which gets mixed with the ice crystals. The water droplets tend to evaporate and get deposited onto the ice crystals, which then grow at the expense of the water droplets, according to the Wegener–Bergeron–Findeisen process. Sometimes the updraft completely decays, leaving just the remains of the anvil, which persists for many minutes and possibly even longer. The anvil remains are sometimes referred to colloquially as "orphan anvil,"[9] having lost its parent updraft. Orphan anvils will be mentioned again later on in relation to the life cycle of cumulonimbus clouds/convective storms.

Various other appearances of cumulonimbi, many with anvils, are seen in Figures 4.5–4.9, and 4.11–4.16). A photo that looks like a cumulonimbus anvil, but is really smoke from a forest fire, is seen in Figure 4.10.

---

[7] J. Barzun, 2000, *From Dawn to Decadence*, p. 209, Harper Perennial, 912 pp.

[8] The apparent exact definition of a cumulonimbus cloud seems to vary by source; some sources refer to it as a deep cloud, but without the requirement for it to be producing precipitation, while others note it to be shallow or deep, but with the requirement for it to be producing precipitation. Some note that an anvil aloft is enough to denote it as a cumulonimbus. I prefer to adhere to the "originalist" point of view, going back to the precise meaning of the actual words—cumulus + nimbus, i.e., it must be cumuliform and produce precipitation.

[9] The first one to use similar terminology was probably Walter Hitschfeld, a cloud physicist in the "Stormy Weather Group" at McGill University; *his* original terminology, however, was "orphaned anvil." See Hitschfeld, 1960, p. 278.

Figure 4.2 *Cumulonimbi in a moist, tropical environment and in a dry, high-elevation, mountain environment. (top) Cumulonimbus over the west coast of Florida, as seen from Sanibel Island, on 3 June 1975. View is to the east. The anvil (diverging part of the cloud at the top) seems to be spreading symmetrically, but may in fact have a part that is streaming in the direction into the page/screen. (bottom) Cumulonimbus just northwest of Estes Park, Colorado, viewed to the north-northwest from Boulder, on Aug. 11, 2022. A relatively wide, cumuliform tower is penetrating above an anvil produced from previous convective towers that lost their buoyancy. Storms on this day reminded me of the storms I used to watch to the west, from Miami Beach, when I was a young boy, in that they stayed where they were and* never *moved to where I was. After they formed, to my great frustration, they made no progress at all to the east and remained anchored to a sea-breeze front to my west. In this case the storm remained anchored to the mountains.*

Photos courtesy of author.

Figure 4.3 *Cumulonimbus clouds with sheared anvils. (top) Cumulonimbus over the northern Texas Panhandle, on June 2, 2004, viewed to the south or southeast. A broad penetrating top (penetrating the stable tropopause) is seen on the right above the anvil, which is spreading to the left (east or northeast). (bottom) Cumulonimbus whose anvil is shearing to the south (left), as viewed to the west, while the author was sailing south of Key Biscayne, Florida, on Aug. 28, 1971. This convective storm probably formed along the sea-breeze front.*

Photos courtesy of author.

Figure 4.4 *Cumulonimbus in the Texas Panhandle, northeast of Amarillo, on June 13, 2009 during VORTEX 2. The laminar foot extending out near the bottom to the right is the inflow region of the storm. At the time, unaware to us, the storm we were scanning with the MWR-05XP (Mobile Weather Radar 2005 X-band, Phased array) was located near the Pantex nuclear weapons plant. Having gotten off the main highway onto an unmarked exit, which looked innocent enough, we were shortly greeted by men armed with machine guns. Our deployment was rudely cut short, as was my photography. At least a tornado did not form and inflict any damage over our target, for which perhaps we could have been mistakenly blamed. An anvil cannot be seen from this perspective, but there probably is one streaming in the direction directed into the page. View is to the northeast.*
Photo courtesy of author.

When a cumulonimbus is triggered by the intense heat from a fire, particularly a forest fire, it is called a "pyrocumulonimbus" (Figs. 4.17–4.19) (Fromm et al. 2010; Peterson et al. 2017; Tory and Kepert 2021).

As noted in the previous chapter, it's not always possible to tell whether there is precipitation falling from a cloud because one cannot see through the cloud to see what is inside it and it also might not be possible to see it underneath cloud base if the water drops or hailstones are very widely separated or if it is too dark or hidden by trees, mountains, etc. In the absence of visual confirmation of precipitation, one must "X-ray" a cumulonimbus, that is penetrate through it and detect what possible treasure lies within. Radar is the tool of choice for doing this (e.g., Atlas 1990; Serafin et al. 2003: Wakimoto and Srivastava 2003; Bluestein et al. 2022). Furthermore, a radar can scan

Figure 4.5 *Anvils from weak convective storms. (top, left) Cumulonimbus (producing snow showers) with anvil streaming to the right, over the Indian Peaks, Colorado, from Brainard Lake, on March 26, 2021. The author reached the lake on cross-country skis. Mt. Audubon is seen on the right. Some cumuliform tower tops are seen to the left at the top of the cloud. (top, right) An ultra-narrow, wispy, anvil emanating from a weak convective storm, west of Boulder, Colorado, on June 30, 2021. (middle) Cumulonimbus with an anvil near Guilin, China, on July 9, 1987. Viewed to the west. Stronger winds aloft must be shearing the anvil off to the south or southeast (left). Note the spectacular mountains seen on the left horizon near the Li Jiang (Li River). The cumulonimbus looks rather weak because the anvil is not very crisp and the convective towers below to the left also do not look very crisp. (bottom) Cumulonimbus viewed in the distance to the northwest from Norman, Oklahoma, on Oct. 15, 2022. The anvil narrows downstream, rather than spreading out, because the updraft is probably not very strong. This cumulonimbus with a remarkable long anvil reminds me of a variation on the lyrics by the group Cake in their 2001 song "Short Skirt/Long Jacket": "I want a cumulonimbus with a low base and long anvil."*

Photos courtesy of author.

Figure 4.6 *Very narrow cumulonimbus clouds, probably associated with little precipitation. (left) Narrow cumulonibus tower with a narrow anvil streaming overhead, in western Oklahoma, on April 30, 2003, viewed to the west. The narrowness of the anvil is probably an indication that the updraft in this cloud is rather weak. If the updraft were strong, then there would be strong divergence at the top and anvil would spread out more in all directions, becoming relatively wide downstream. This cloud may be classified as a cumulonimbus, not a cumulus congestus, because it looks as if there is some precipitation fall from it, though some of the apparent precipitation shafts could be shadows cast on the haze in the boundary layer. (right) Narrow, weak cumulonimbus exhibiting no signs of vertical shear at Key Biscayne, Florida, on Sept. 15, 1984. A small precipitation core is seen on the left side of the cloud base. (Compare this photograph to that of the growing cumulus congestus towers seen in Chapter 3; the difference between a cumulus congestus and cumulonimbus in this case is tenuous; the cloud in Figure 3.54, for example, it was classified as a cumulus congestus, not a cumulonimbus, because no rain shafts were visible. It is doubtful this cloud top ever reached the tropopause because it does not appear to be spreading out. Like a chain of bubbles, another burst of convection is seen underneath the cloud top, traveling along a similar path.*

Photos courtesy of author.

an entire storm relatively quickly, providing excellent documentation of a storm's evolution. Flying into it to make in situ measurements is not only dangerous, but one can't be everywhere in a storm at one time to make simultaneous measurements. There are some old accounts from the late 1940s of horrifying flights inside storms, in which tremendous updrafts and downdrafts were encountered. A number of years ago an armored aircraft (T28) made measurements inside convective storms, but it is no longer in use.

Figure 4.7 *Panororamic views of cumulonimbus in midlatitudes. (top) Composite panorama of a cumulonimbus viewed from the Forbidden City (buildings in the foreground), Beijing, China, on July 12, 1987. New cells are forming on the left, while anvil material and older cells and their remains are on the right. While there does not appear to be anything particularly distinctive or unusual about the cloud itself, it is strategically located beyond the roofs of the spectacular buildings of the Forbidden City. Most of the other photographs of cumulonimbi in this book are located in much more pristine settings. Perhaps this is a "forbidden cumulonimbus." (bottom) Panorama of a cumulonimbus over the Rocky Mountains in northeastern Colorado on July 20, 2022, viewed to the north through east from the eastern flank of Mt. Audubon. Long's Peak and Mt. Meeker are visible at the lower right. Rocky Mountain National Park is located to the left of the cumulonimbus.*
Photos courtesy of author.

However, aircraft hardened to penetrate storms safely can provide data along its flight track only and for very brief periods of time. Though not so useful for determining the overall composition of a storm, in situ data are absolutely necessary for confirming that what we infer about the composition of a cloud based on radar observations is valid. Before we look at more of the visual aspects of cumulonimbi, we will look at what we *can't* see because it is hidden from our view.

Figure 4.8 *The beginning stage of convective storms. (top, left) Multicell (to be described later) cumulonimbus, viewed to the southeast, from Boulder, Colorado, on Sept. 9, 2009. Three separate updrafts are evidenced by three discrete anvil bursts. New updrafts are building on the right, while the anvil on the far left is from the first burst, etc. (top, right) Attempt at the formation of a cumulonimbus, in central Nebraska, on June 1, 2018, viewed to the south. An anvil is about to detach from its earlier base and become an orphan anvil, while new convective towers are growing to its west (right). So far, there is no cumulonimbus, only cumulus aspiring to grow into cumulonimbi, a wish that eventually was fulfilled. (bottom, left) An attempt at the formation of a cumulonimbus in central Nebraska on June 1, 2018, viewed to the northeast. An anvil is beginning to form and stream to the right, but the convective tower is rather narrow underneath, probably an indication that this cumulonimbus wannabe won't succeed. (bottom, right) Airborne view of a developing cumulonimbus over Arizona during the Southwest Monsoon, on July 26, 2017. The anvil is streaming off to the northeast (right).*
Photos courtesy of author.

4.2.1   Radar observations of convective storms: Looking inside storms from a safe distance

Radar was originally developed before World War II to detect aircraft. It was noticed serendipitously, however, that radar could also detect precipitation. After the war and some aircraft accidents involving flights through turbulent, convective storms, attempts were made by meteorologists to put radar to use for just identifying regions of precipitation with the goal of making aviation safer. So, the "Thunderstorm Project" was conducted in 1946 and 1947 in Ohio and Florida using radars and in-situ-measurement

Figure 4.9 *Cumulonimbus viewed to the north, at sunset, from Norman, Oklahoma on Aug. 26, 2019. The bottom portion of the anvil and the top portion of cumulus towers are vividly illuminated by the setting sun.*
Photo courtesy of author.

instruments. It was the prototype and gold standard for all observational experiments conducted ever since.

4.2.1.1 *Conventional fixed-site and mobile radars: What's inside a convective storm?*

Conventional radars can detect the "backscattering" of radiation from a collection of raindrops and chunks of ice or ice crystals (and insects, debris, birds, etc.). Most radars transmit periodic pulses of microwave radiation into the atmosphere in the direction of a narrow beam focused by the radar's antenna. The radiation is absorbed and re-radiated by a distribution of precipitation particles (usually raindrops and hailstones) at all angles, but some of it makes it back to the radar (is "backscattered"), though weaker in intensity by many orders of magnitude than the radiation sent out. Since the intensity of the radiation received back at the radar varies inversely with the square of the distance from the radar, the "dynamic range" of the signal backscattered to the radar is very large, many orders of magnitude especially when there are scatterers at many different distances

Figure 4.10 *Sun poking through smoke from a forest fire southwest of Boulder, Colorado, on Aug. 15, 2020. This smoke cloud has a well-defined edge to it and looks like an anvil from a cumulonimbus.*
Photo courtesy of author.

Figure 4.11 *Three cumulonimbus clouds with anvils line up like guards, protecting the southwest coast of Florida, viewed from near Key West, on Aug. 27, 1993. These convective storms probably formed in response to a sea-breeze front. The winds over the intervening waters are nearly calm, as evidenced by the glassy, reflective surface.*
Photo courtesy of author.

Figure 4.12 *Cumulonimbus clouds at night. (top) Cumulonimbus illuminated at night by in-cloud and cloud-to-cloud lightning. View is to the east, in far southwestern Oklahoma, on May 13, 2005. The anvil has several striations, either indications of several earlier discrete bursts of convection or gravity waves near the tropopause forced by the storm's updraft. (bottom) Cumulonimbus illuminated by cloud-to-water lightning (bottom, right) and cloud-to-cloud lightning in the anvil (center, left), Aug. 4, 1971, offshore of North Miami Beach, Florida. The latter is sometimes called "spider" lightning because it branches out like a spider's legs. Viewed to the east.*

Photos courtesy of author.

Figure 4.13 *Cumulonimbus at night and a comet. (top) Cumulonimbus with a penetrating top, illuminated at night by in-cloud lightning, viewed to the northeast over the Plains of northeast Colorado, from Boulder, on July 21, 2020. Stars are visible above the storm. At the time I took this photograph, I was also intermittently photographing the NEOWISE comet, which was off to my left, to the north-northwest (bottom panel). It was quite a two-ring circus, as I switched the object of photography between the cumulonimbus and the comet. Below the comet the top of the Flatirons are visible (bottom, left of bottom panel). The top panel represents the view to the right and the bottom panel represents to the view to the left. Hydrometeors are falling from the storm, to my right, while the comet looks like a hydrometeor falling from the sky, to my left.*
Photos courtesy of author.

Figure 4.14 *Cumulonimbus clouds without visible anvils. (top, left) Cumulonimbus developing near Mt. Evans (as of 2023 re-named Mt. Blue Sky), Colorado, on July 15, 1989. View is to the east or southeast. No anvil is apparent yet. The convection on the right looks very vigorous, with sharp edges and extensive cauliflower-like growth, probably indicative of small, supercooled water droplets. The convective tower on the left, however, looks more mature and less crisp, probably indicative of a weaker updraft and perhaps because there are some ice crystals mixed in. This cloud probably glaciated later and developed an anvil. (top, right) A vigorous cumulonimbus developing over the Indian Peaks, Colorado, on Aug. 7, 1999. Mt. Audubon is seen on the left. View to the northeast from the summit of Pawnee Peak (12, 943 ft; 3, 945 m). No anvil is visible yet. (bottom, left) Massive cumuliform growth in a developing cumulonimbus in eastern Colorado, on June 19, 2011. There is some pileus on the right side of the top of the convective towers. No anvil is evident from this vantage point, but there probably is some anvil material hidden from view, to the east and northeast. (bottom, right) Cumulonimbus growing in the Indian Peaks, Colorado, on July 28, 2004, promising to make an interesting descent later from on a saddle point between Shoshone and Pawnee Peaks, near Pawnee Pass. A silhouette of my friend and colleague John Brown is seen on the right.*

Photos courtesy of author.

from the radar. The "radar reflectivity" is therefore most usefully given as a logarithmic measurement of the intensity of the returned signal. We'll come back to this reflectivity measurement in more detail at the end of this section. It is perhaps appropriate that the word radar is a palindrome, since it involves an instrument that symmetrically sends out something and receives something back, like a boomerang, or a ball bounced off a wall.

Figure 4.15 *Lake-effect, fuzzy-looking, cumulonimbus (snow shower) over Ithaca, New York, on Nov. 10, 2004. View is probably toward the north (Niziol et al., 1995; Kristovich et al., 2003).*
Photo courtesy of author.

In other ways it is like an atom smasher: Hit the storm with a beam of radiation and see what the radiations that flies out looks like . . . an "atom smasher" of sorts for clouds.

To be useful, the return from backscattering in a radar volume (defined by twice[10] the duration of the pulse multiplied by the speed of light, and the width of the radar beam, which increases with distance from the radar) must arrive at the radar before the next in the series of pulses is sent out, or else there will be an ambiguity in how far away the precipitation is: If it arrives soon after the "next" in the series of pulses is transmitted, then one doesn't know if the radiation was backscattered off precipitation close by (because it is detected very soon after the "first" pulse has been sent out) or far away (because it is detected long after the "previous" pulse had been sent out). So, it is advantageous that the pulses are not transmitted too frequently. It will be shown shortly, however, that there is a downside to transmitting pulses too infrequently in determining the wind using radars having Doppler capability.

---

[10] The radiation passes through the precipitation twice: Once on the way out from the radar and once on the way back.

Figure 4.16 *Sheared, weak cumulonimbi during the winter at high elevation. (top) Shallow cumulonimbus (snow shower) near the Continental Divide, northwest of Mt. Audubon (right), west of Ward, Colorado, on March 17, 2021. Viewed from Brainard Lake, which the author reached on cross-country skis. (bottom) Cumulonimbus (snow shower) just east of Brainard Lake, Colorado, viewed from Brainard Lake, on March 22, 2021. The author reached this vantage point also on cross-country skis.*
Photos courtesy of author.

Figure 4.17 *Pyrocumulonimbus viewed to the east-northeast of Nederland, Colorado, on July 9, 2016, while the "Cold Springs fire" was beginning. The famous Nederland Carousel of Happiness is seen in the foreground. I just happened to be in Nederland with my wife at the time the fire began and we were shocked to see the dire situation unfold. The bottom of the cloud is smoke, while the top of cloud, which is white, probably consists of cloud droplets.*
Photo courtesy of author.

In the United States, the network of surveillance radars is comprised of WSR-88D (Weather Surveillance Radar—1988 Doppler) also called NEXRAD (NEXt generation RADar) radars (Crum and Alberty 1993). The WSR-88D radars operate at ~10 cm wavelength (~3 GHz in frequency), which is in the "S-band." The current WSR-88D radars are spaced roughly 150–200 km apart so that much of the country is covered by them, though there are gaps due to some regions of non-overlapping coverage and blockage from mountains; in addition, the radar beams are refracted downward somewhat, owing to index of refraction gradients associated with vertical changes in air density (which are determined from the vertical distribution of temperature and water vapor content); although there is actually some coverage "over the horizon;" at great distances, precipitation is detected only aloft, sometimes no lower than half-way up to the tropopause. In other words, at the outer edge of the coverage of the radar it may be probing a storm at 5 km above the ground, nowhere near the ground where one actually experiences precipitation. There are also transportable S-band radars for research purposes. These radars may be set up at great expense temporarily in desired locations. The

Figure 4.18 *Pyrocumulonimbus and its radar depiction. (left) A pyrocumulonimbus in north-central Colorado, west-northwest of Ft. Collins near Red Feather Lakes, on Sept. 25, 2020, as viewed in the distance to the north from Boulder, Colorado. This pyrocumulonimbus seemed to be producing periodic bursts of cumuliform cloud, which rose to near the tropopause and leaned in the downstream direction of the winds at high altitude. The wildfire that spawned this pyrocumulonimbus was known as the "Cameron Peak" fire. It was not possible to see precipitation owing to the haze, but it was evident on radar (right). There are also a few orographic wave clouds seen at the top and top left. (right) Radar reflectivity (to be explained in subsequent part of this section) pattern at low altitude as seen from the Denver, Colorado WSR-88D, about the same time as the photograph in the top panel. The yellow color-coded core is probably indicative of moderate rain, which is ironic, given that it was triggered by a wildfire, which could certainly have used the rain to put it out. The blue circle south of Boulder was my location for viewing the pyrocumulonimbus seen in the left panel.*

Photo courtesy of author.

advantage of S-band radars is that they are the weather radars that are the least susceptible to attenuation, that is, the decrease in radar signal strength owing to the absorption of the part of the radiation from the radar beam that is *not* backscattered. S-band radars are therefore used when one wants to detect precipitation as far away as possible. They have very large antennas, though in order to have acceptable spatial resolution: the larger the antenna, the narrower the beam.

Other radars, which are primarily used for research, operate at C band (~5 cm wavelength, ~5 GHz frequency) and X band (~3 cm wavelength, ~10 GHz frequency),[11] and to a lesser extent at Ku band (~2 cm wavelength, ~14 GHz frequency), Ka band (~8 mm wavelength, ~35 GHz frequency) and W band (3 mm wavelength, 95 GHz frequency), and have been used to study storms and clouds (Bluestein et al. 2014). The association of letters coded to refer to specific wavelengths and frequency was begun by the military

---

[11] Some surveillance radars, particularly in countries other than the US, operate at C band and even at X band.

Figure 4.19 *Convective clouds from a forest fire and its smoke plume as seen by radar. (left, top) Cumuliform towers (white) poking out high above the smoke plume (brown) from a wildfire in northeast Colorado, near on Oct. 17, 2020, as viewed to the north from Boulder, Colorado. This wildfire, which was northwest of Boulder, near Ward, was known as the Cal-Wood fire. Based on the Denver WSR-88D (right), but despite the ferocious look of the cloud and smoke, it did not appear as if there was any precipitation falling from the clouds and it was therefore probably just a pyrocumulus. Note that the cumuliform clouds were downstream from the source of the fire. They became visible only after the buoyant plume created by the fire had moved far enough downstream to reach a height making it visible above the smoke plume. (left, bottom) As in the top, left panel, but the smoke plume is reflected onto Viele Lake. (right) Radar reflectivity pattern at low altitude in the smoke plume seen in the left panels, several hours later; the smoke plume was streaming out to the east-southeast by the ambient wind, but had relatively low reflectivity, which was not likely indicative of any rain. There is actually a second plume from another fire that originated west of Ft. Collins (the "Cameron Peak" fire), which is not visible in the top panels because it was hidden by the smoke from the Cal-Wood fire.*
Photos courtesy of author.

around World War II. The shorter-wavelength radars are best used to detect small particles such as cloud droplets and ice crystals in clouds owing their higher sensitivity to tiny particles, or to probe tornadoes, which require very high spatial resolution; the longer-wavelength radars are best used to look at precipitation. The shorter-wavelength radars are highly susceptible to attenuation and therefore have relatively limited range, while the longer-wavelength radars are less susceptible to attenuation and are best used when for long-range probing or when there is heavy precipitation.

Since in general the larger the diameter of the antenna,[12] the narrower the radar beam (as previously noted), shorter-wavelength radars tend to have the smallest antennas and are therefore the most portable and easily mounted on trucks or vans or airplanes or satellites. So, S-band radars are usually used at fixed sites or on large ships or transported and then set up, while most, but not all, C, X, Ka, Ku, and W-band radars lead a nomadic life. Ground-based mobile and airborne radars have been used in general to probe convective storms at close range. Airborne radars at X band were used first by The National Oceanic and Atmospheric Administration (NOAA) to probe hurricanes over the ocean in the late 1970s (Marks and Houze 1984), and later to probe convective systems over the tropical oceans, continental rain systems over mountainous areas, convective storms in the Plains of the US (Wakimoto et al. 1996; Dowell et al. 1997), and other rain storms, both convective and non-convective, elsewhere. The National Center for Atmospheric Research (NCAR) in the mid-1990s entered the fray with their own X-band airborne Doppler radar mounted on an Electra aircraft, ELDORA (ELectra DOppler RAdar), which at the time of this writing no longer exists. NASA also operates a high-altitude aircraft, the ER-2, that can carry X-band and higher-frequency radars (Heymsfield et al. 2010; 2013).

Airborne radars are great for getting to a storm quickly, without worrying about road networks and traffic jams, even in the sparsely populated Plains of the US, where many storm chasers are attracted in their vehicles. My most memorable and harrowing encounter with a traffic jam while in pursuit of a tornado with a ground-based mobile Doppler radar was on May 31, 2013 in Oklahoma City, while staying just ahead of a storm that had produced a massive tornado in El Reno, just to the west. We exited the east-west interstate I-40 to get onto another road and found ourselves in gridlock in Friday-afternoon rush-hour traffic, augmented by people in their vehicles trying to get out of the way of the approaching storm. We stopped to collect data, but realized there was no way of leaving, if we had to (sort of like a convective-storm viewing—location version of "Hotel California").[13] I recall mentally trying to decide how to find a place to shelter just in case a tornado came right toward us, and began looking for ditches, etc. Alas, someone in an emergency vehicle with flashing lights realized our predicament and came to our aid, acting as our unofficial escort, as we did a slalom run around cars and trucks on the highway to get farther ahead of the storm. Airborne radars, however, unlike ground-based mobile radars, cannot resolve what is happening right near the ground, owing to ground-clutter contamination, i.e., radiation emitted by the antenna off the axis of the main radar beam (by the "sidelobes") is backscattered off buildings and trees, etc., and effectively masks (overwhelms the weaker signal from the meteorological targets) the radiation actually backscattered off precipitation along the main beam. In addition, it takes a number of minutes to complete a flyby, turn around, and

---

[12] Assuming a parabolic-dish antenna, which focuses the beam.

[13] "Hotel California" was a song by the rock group the Eagles in the late 1970s, in which it was said that once you checked into the hotel, you could never leave.

probe the targeted storm again, a time interval during which significant evolution of the storm may have occurred, which is therefore not documented (like turning on a time machine and finding oneself far in the future without knowing what had happened in the interim).

My research group was the first to use, in the late 1980s, a portable, non-scanning, X-band radar, which was designed and built at the Los Alamos National Laboratory (LANL) (Fig. 4.20), with support from my late, LANL, physicist colleague Wes Unruh, who modified it to make remote windspeed measurements in tornadoes. My research group in collaboration with engineers (such as the late Bob McIntosh, and Andy Pazmany) at the University of Massachusetts at Amherst, was the first to use a *scanning*, mobile, truck-mounted, W-band radar in 1993, though we did not successfully obtain ultra-fine spatial resolution imagery of tornadoes until much later, in 1999. when we collected several excellent datasets of a number of tornadoes (Bluestein and Pazmany 2000). Our early failures reflected (not backscattered) bad luck, a technical error in the field while we were probing a tornado in 1994; poor operational decisions (like choosing the wrong storms); bad luck, not being in the right place at the right time; a number of mediocre storm seasons in the Southern Plains; a fundamental difficulty in being able to replace failing hardware for W-band radars (relatively rare) quickly; and my reluctance to travel very far away from our home base (e.g., it's a long, costly, boring drive to South Dakota from central Oklahoma). We chose the W-band wavelength for specifically targeting tornadoes with very high spatial resolution, since a very fine beam at W-band (only 0.18°, compared with 1° for most X-band radars) is possible with a very small antenna. Josh Wurman (along with Jerry Straka and Erik Rasmussen and radar engineers from NCAR and NSSL) a new colleague at that time at my institution (with an office just a few doors away) pioneered the first scanning, mobile, truck-mounted, X-band radar in 1995 and was the very first to produce stunning radar imagery of tornadoes and their parent storms almost immediately (Wurman et al. 1996). Jerry Straka was also a colleague at OU at the time, while Erik Rasmussen chased with me in the late 1970s when he was an undergraduate at OU but at the time worked for NSSL. Their radar was named the DOW (Doppler On Wheels), a moniker that has stuck so well, that many refer to any truck-mounted Doppler radar as a DOW. The use of this nomenclature is like using the brand name Kleenex to describe any facial tissue. Over the years we found that X-band mobile radars are more versatile and are therefore more used than W-band mobile radars, but I recall Josh Wurman once having remarked at a conference that W-band radars were the "Cadillac" of radars. The wind measurements made with these portable and mobile radars were made possible using Doppler techniques, which we will look at in the next section, but I really want to emphasize first the history of radar development and platforms used to carry them.

The main advantage of using mobile radars to probe convective storms is that the number of storms that can be studied is increased significantly over how many would be studied by just sitting in one place and waiting for storms to come to the radar (but, of course not passing directly over the radar, possibly destroying it). Also, the spatial

Figure 4.20 *My graduate students probing a tornado near Hodges, Texas, on May 13, 1989, using the LANL portable Doppler radar.*
Photo courtesy of author.

resolution is maximized by getting as near to the storm as possible, since the width of the radar beam increases with distance from the radar (through spreading). Of course, one must not get too close, for safety concerns. By getting close to a storm, you can probe very near the ground and also you can actually see cloud features, which can then be correlated with what the radar "sees." This aspect of mobile-radar field operations, to me, is one of the great thrills of conducting field experiments with mobile radars: you get to appreciate the beauty of the beast while also making scientific measurements

("X-rays"[14] of storms) that can be seen in real time. One of my colleagues who is now at UCLA, Roger Wakimoto, is the leading expert on correlating photographs of storm features with radar-observed features, and has accompanied us on many of our storm chases (Wakimoto et al. 2015). He honed his skill working with his advisor, Prof. Ted Fujita at the University of Chicago in the early 1980s. He and I share our excitement of simultaneous radar and visual observations.

Over the years more mobile, X-band radars have been built, as well as C-band radars. The latter, under the direction of my OU colleague Mike Biggerstaff have been most successfully used to probe landfalling hurricanes, though Josh Wurman and his group have also been very successful in using his (X-band) DOWs to study landfalling hurricanes (and of course, also convective storms). My former Ph. D. student Chris Weiss and his students and colleagues, at Texas Tech University, have used truck-mounted Ka-band radars,[15] to probe tornadoes and storms. There are other versions of ground-based mobile X and C-band radars either already out there or under development, especially at the University of Oklahoma in Norman, my home institution, the University of Massachusetts at Amherst, University of Alabama in Huntsville, and in Japan and Italy, among others.

Mobile radars, as noted earlier, have also been used to make observations of precipitation in remote areas. When I was a graduate student, I participated in an international field experiment in 1974 called GATE, a nested acronym standing for the GARP (Global Atmospheric Research Program) Atlantic Tropical Experiment. During GATE, I operated a C-band radar, used at MIT by Dr. Pauline Austin and her radar engineer Speed Geotis, which was mounted on the NOAA research vessel Gilliss, stationed off the west coast of Senegal, for about a month, in the intertropical convergence zone (ITCZ). The ITCZ is a region in this part of the world, typically within several degrees of 10° latitude, where there is a confluence of southeasterly winds with northeasterly winds. Convective storms are triggered there (Fig. 4.21) by rising motion that is created by the convergence of the two airstreams. The ITCZ is also found in the Eastern Pacific, but both north and south of the equator; it is also found in the Indian Ocean. I was interested in the cloud bands that one could see on satellite imagery that were associated with the ITCZ and with other features in the tropics. These cloud bands stimulated me to select a Ph.D. research topic involving their observations and dynamics. When "Mrs." Austin (who had a Ph.D., taught a graduate course, and did research, but was not allowed at that time to be a professor because her husband was already a professor of meteorology at MIT, the same institution at which she was employed) asked me if I were willing to run the radar during part of GATE (Houze and Cheng 1977), my thesis advisor, Prof. Fred Sanders, asked me how it would help me advance my thesis, I naively replied that it probably wouldn't, but it would be fun. It turned out to be more than fun; it actually helped me find my specific research topic. It was a seminal experience for me to run on deck of the ship and see convective storms and clouds and then run inside to see

---

[14] The use of "X" in both "X rays" and "X band" is entirely coincidental.
[15] e.g., Hutson et al., 2019.

Figure 4.21 *Cumulonimbus developing over the tropical Atlantic, in the Intertropical Convergence Zone, west of Senegal, West Africa, on Aug. 29, 1974, from the deck of the NOAA research vessel, R. V. Gilliss.*
Photo courtesy of author.

what they looked like on the radar scope. Shades of things to come for my career . . . Many other meteorological researchers also have gotten their start as graduate students participating in field experiments.

Another radar that has been used to probe convective storms in remote areas of the globe was developed by NASA scientists and engineers. This radar was a dual-frequency, Ka- and Ku-band radar, mounted on a polar-orbiting satellite during TRMM (Tropical Rainfall Measuring Mission) (Kummerow et al. 1998; Nesbitt and Zipser 2003). Remarkably, it ran successfully for almost a decade, and was used to obtain valuable climatological precipitation data in the tropics with a 5 km "footprint" in otherwise radar-data sparse or radar non-existent areas of the globe.

All the conventional radars, whether fixed-site or mobile, or short wavelength or long wavelength, measure the intensity of the radiation backscattered to the radar. But how do we compute/estimate the intensity of the precipitation based on this signal? It depends on the wavelength of the radar and the type of precipitation (liquid water drops, ice particles, ice, ice with a water coating, etc.). Raindrops are more reflective than ice crystal; hailstones covered by water are more reflective than "dry" hailstones, those not covered by water. So, the nature of the surface of the scatterers matters.

When the size of the scatterers (hydrometeors, dust, debris, birds, bats, aircraft, and unidentified flying objects, etc.) is small compared to the wavelength of the radar, scattering is of the "Rayleigh" type and it turns out, based on electromagnetic scattering theory, that the more intense the signal at a given range, the more intense the precipitation is for a given type of precipitation (of course, dust or birds or bats or aircraft...or unidentified flying objects, will corrupt our estimates because they are *not* precipitation, even though they may "rain" down on us). We also need to normalize the measurements for range because the backscattered signal decreases rapidly in intensity with increasing range: a beam of radiation spreads with distance from the radar and the energy in the beam becomes less and less concentrated.

For S-band radars, most raindrops are much smaller (~0.1–0.2 cm) than the wavelength of the radar (~10 cm). Even large raindrops are only ~0.5 cm across. However, giant hailstones, which can be as wide as 10 cm, or bats or birds or aircraft are *not* small compared to the wavelength of the radar. When the scatterers are comparable in size to the wavelength of the radar they produce "Mie" scattering. Based on theory, there is *not* a clear, monotonic relationship for Mie scattering (i.e., if the signal is stronger, the precipitation is *not* necessarily more intense and vice versa) owing to complex interference, both constructive and destructive of the absorbed radiation re-emitted by the raindrops. So, using S-band surveillance radars, one can relate rainfall rate to signal intensity nicely, unless one suspects that there are large hailstones or other large targets out there or if there is just frozen precipitation. For C-band and X-band radars, Rayleigh scattering is also expected for most types of liquid precipitation. When one gets down to the relatively short wavelength of Ku, Ka, and W-band radars (~ mm wavelengths), the sizes of most raindrops can be comparable to the wavelength, so relating precipitation rate to signal strength is complicated and more difficult. It turns out, however, that the shorter-wavelength radars are much more sensitive to radiation from small particles and therefore are useful for looking at the properties of clouds, whose particles are typically much narrower than raindrops. Also, for just measuring the winds, one may not care about the intensity of the precipitation, only that the returned signal is strong enough to detect the precipitation (i.e., is above the "noise level"). Radar reflectivity Z is typically measured in dBZ, a logarithmic measurement of the reflectivity normalized for range and assuming that there is Rayleigh scattering. The actual value of Z is given by the sum of the diameters of all the raindrops in the radar volume raised to the sixth power. Although this relationship may not be intuitive or easily explained with simple mathematics, it is important in relating the radar reflectivity to the distribution of raindrop sizes and densities, which is not trivial, and affects how well we can see through the rain and what cloud features we see. While not perfect, using this unit of measurement is a way of comparing the radar reflectivity to a standard. Heavy rainfall is typically associated with *radar reflectivity factors* of ~50–60 dBZ, moderate rainfall is typically associated with ~35–45 dBZ, and light rainfall is typically associated with 20–30 dBZ or less. Modern operational radar systems usually color-code these intensities, usually with red or purple for the highest reflectivity, yellow for moderate reflectivity, and green and blue for the lowest reflectivity.

A given value of dBZ, however, may be associated with many small raindrops or a fewer number of larger raindrops. When there are just a few large raindrops, the air is relatively translucent, and one may not even see precipitation falling and therefore be able to look right through the region of precipitation. When there are a lot of small raindrops, the region of precipitation may be nearly opaque and one sees that it is relatively dark and that there is a lot of precipitation. In most instances, there is a complex size distribution of raindrops which can vary from location to location and with the type of cloud producing the rain and how long the storm has been around, among other factors. The relationship between rainfall and dBZ for tropical rain is different from what it is for rain over a continent; the nature of the condensation nuclei and vertical profiles of temperature and moisture are different (e.g., there are more salt nuclei over the oceans). The rainfall rate–dBZ relationship is different in a region of convection from what is over a region of stable, stratiform precipitation. The relationship between snowfall rate and dBZ is different from the relationship between rainfall rate and dBZ. An example of this relationship between the translucency/opaqueness of a precipitating cloud to the sizes and numbers of precipitation particles is carried to an extreme when there is very large hail. Hail is highly reflective and when it is large, it does not take many hailstones to cause very high radar reflectivity factors in dBZ. Since very large hailstones tend to be very widely separated when they fall, often one can look right through a region of them even if they are giant sized, but "see" very high values of dBZ on radar. Woe to those who venture into what looks like a clear area inside a convective storm but a high radar reflectivity factor is "seen" on a radar display . . . storm chasers may be in for a broken windshield, or worse. There is not a unique relationship between visibility in a storm and the intensity of the radar echo. When I look at the radar display online using my cell phone or computer of a severe convective storm somewhere far away, especially in another section of the US, I always wonder what the storm actually looks like. Would part of it be mostly hidden by small raindrops, or would it be vividly visible because there are widely spaced large raindrops or hailstones? Is it a mess and not aesthetically pleasing, or is it highly photogenic?

In order to study convective storms and document their evolution, one must scan them with a radar that has adequate temporal resolution. Mechanically scanning radar antennas spin around, driven by motors. The more measurements made at each volume in space (the number of returns from different pulses from the same radar volume), the more accurate the measurement of radar reflectivity (and other radar variables such as Doppler velocity) is. This situation is similar to that of a voting poll, in which the more the participants, the more accurate the results, assuming that there are no voting machine errors, fraud, etc., and assuming that the sample of participants is not biased in any way. There is therefore some practical limit on how fast or slow the antenna should be rotated. Though one may be able to scan an entire storm every 3–5 minutes using a WSR-88D and obtain high-quality data, it might be that the storm shows changes in its appearance every 1–2 minutes, or it takes only 10 seconds or less for a tornado to form or change its intensity significantly. Looking a radar imagery every 5 minutes is obviously not sufficient to document a tornado forming. In this case, a "rapid-scan" radar is required (French et al. 2013).

The most primitive rapid-scan radars still scan mechanically, but some tricks may be employed to enhance their speed of data collection. The more rapidly a radar scans (mechanically), the fewer data points are available, which can result in very noisy, unrepresentative and unusable data. The antenna of our radar called RaXPol (Rapid scan X-band Polarimetric) (Pazmany et al. 2013) can spin at a rate as fast as 180° per second. To increase its data collection efficiency, the frequency of the radar is changed incrementally and quickly eleven times. The amount by which the frequency changes is just enough that each data point can be considered like having a separate, "independent" radar making a measurement, but not so much that the size of the antenna needs to be changed to accommodate the changes in frequency. This technique is called "frequency hopping," and was originally used in airborne radars. With eleven hops in frequency, it is like having eleven different radars, each scanning very rapidly. We have used this radar, originally designed by Andy Pazmany and his colleagues at his company in Massachusetts[16] and supported with funding from the National Science Foundation, for almost a decade now to document tornado formation and dissipation.

Other, more sophisticated radar techniques have also been employed to build radars that scan rapidly and produce accurate data. These radars scan either partially electronically and partially mechanically (hybrid scanning), or completely electronically. There are two main types of electronic scanning techniques we employ. In one, a regularly-spaced series (in one dimension) or an array (in two dimensions) of separate antennas, is fed the radar signal, which is effectively[17] delayed slightly at each antenna, so that the phase of the signal varies systematically with each antenna. The net effect of adding up all the signals with their different phases is that the radar beam can be focused at an angle off the perpendicular from the plane of the antenna (these antennas are usually flat plates rather than parabolic dishes). The changes in phase and the resultant changes in angle can be done almost instantaneously. The radars that use this technique are called "phased-array" radars, and were originally used by the military to track rapidly moving objects such as missiles. The use of phased-array radars by meteorologists is a good example of the application of military hardware and technology for peace-time applications.

Two hybrid-scanning phased-array mobile radars operating at X band have been used to probe tornadoes. Josh Wurman and his group developed the Rapid-DOW. Our group, in collaboration with the Naval Postgraduate School in Monterey, California under the guidance of Bob Bluth and his staff, and ProSensing, Inc. in Massachusetts under the guidance of Ivan PopStefanija and his staff, has used the MWR-05XP (Mobile Weather Radar—2005, X band, phased array) to probe tornadoes (Bluestein et al. 2010). The MWR-05XP is a radar that the military operated on a ship and was adapted for meteorological use on spatially distributed targets such as raindrops in a volume, while the Rapid-DOW was developed from scratch. Both radars scan electronically in elevation

---

[16] The radar has since been modified by engineers at the Advanced Radar Research Center (ARRC) at the University of Oklahoma, Norman.

[17] There are two different techniques for delaying the phase: one involves delaying the signal and the other involves changing the frequency slightly.

angle and mechanically in azimuth. The MWR-05XP, however, also backscans electronically in azimuth slightly to hold the beam fixed on a target in space long enough to obtain more samples at the same point and thus reduce smearing caused by the rapid rotation of the radar (like taking a blurred photograph if the object being photographed moves too quickly).

A second type of electronic scanning, which is similar in some ways to the first, is called the "imaging" technique. Instead of electronically sending out *narrow* beams of radiation at different angles, azimuths, and elevation, by changing the phases of an array of antennas, a single elliptical beam (also called a "fan beam"), which is narrow in one direction (usually the horizontal), but wide in the other direction (usually the vertical) is transmitted, thus "illuminating" a vertical swath of the target storm. It's like taking a photograph (image) with a flash, but only along a strip. The radiation is absorbed and some is scattered back to the radar. But now the backscattered radiation from different points in the illuminated elliptical volume are received by electronically focused antennas. The Atmospheric Imaging Radar (AIR), developed by Bob Palmer and his colleagues at the Advanced Radar Research Center (ARRC) at the University of Oklahoma (Isom et al. 2013), used a "fan beam" ~1° wide in the horizontal and ~20° wide in the vertical and had been used to collect rapid-scan data in tornadoes until just very recently, when a new, more advanced radar is in the works (scheduled to be tested in 2022, within 6 months or so after the time of this writing).

4.2.1.2 *Doppler radar: What is the wind field inside a convective storm?*

Conventional radars had been used by meteorologists for several decades before radar technology advanced to the point at which engineers in the late 1960s began to develop Doppler radars for meteorological use. Doppler radars measure not only the signal intensity of the backscattered radiation, but also the Doppler frequency shift associated with movement of the scatterers probed in a radar volume (e.g., Doviak and Zrnić 2006; Fabry 2015; Rauber and Nesbitt 2018). Like the audible Doppler shift in the frequency (pitch) of sound waves you hear when a train blowing its whistle passes by (you hear the pitch of the train whistle go up as the train approaches and the pitch of the train whistle goes down as the train gets farther away), or the shift in color of the light of a star that is moving with respect to our Earth (it shifts up in frequency to the blue side when it is approaching Earth or down in frequency to the red side when it is receding from Earth; that our universe is expanding as all stars are separating away from each other is evidenced by a red shift [downward shift in frequency] in the light of distant stars), the frequency of the backscattered microwave radiation is shifted upward if the scatterers have a component of motion toward the radar and the frequency of the backscattered radiation is shifted downward if the scatterers have a component of motion away from the radar. Why this happens is easy to understand: When there is motion toward you, the "observer," the crests and troughs of the sound or light waves get shortened, owing to the motion toward you (the back side of the target is closer to you than it had been a very short period of time earlier); when there is motion away from you, the crests and troughs of the sound or light waves get lengthened (because the back side of the target is farther away from you than it had been a very short period of time earlier), owing to the motion away from you. If you assume that the scatterers move along with the wind, then the Doppler shift in frequency is related to the component of the wind field aligned along the line of sight of the radar beam. In some instances, scatterers do not move along with the wind; one of these cases (when bulk of the scatterers are rotating rapidly, such as in a tornado) will be mentioned later. All the WSR-88D radars and most research radars now have the ability to measure the Doppler velocities of moving particles.

There are several techniques we use to compute the frequency shift, but it probably should not concern the reader *exactly* how this is done. In the portable Doppler radar that we first used, which was from the LANL, we listened to the returned signal, which fortuitously was in the audio range, through stereo earphones: The approaching and receding velocities were determined by the pitch of the sound in heard in each ear (Bluestein and Unruh 1989). When we scanned across a cyclonically rotating tornado from right to left in the Northern Hemisphere, the pitch would get higher and higher as we got closer to the center of the tornado and shift back to the other ear as we scanned past the left side of the tornado and get lower and lower. In between the time of maximum frequency shift and the time the radar would be aimed directly at the center of the tornado, the pitch might not change at all or even decrease. The amount of frequency shift is proportional to the Doppler velocity and is up (approaching velocities, usually color coded as purple or blue or green) or down (receding velocities, usually color coded as red or orange or yellow). Zero frequency shift indicates no relative motion and is usually coded as white.

There is a limit to how far positive or negative the Doppler velocities can be measured, in analogy with how one determines the time by looking at an analog clock.[18] If it is 10 o'clock (in the morning) and we add one hour, it is clearly 11 o'clock in the morning. However, if it is 10 o'clock in the morning and we add 3 hours, it is clearly 1 o'clock in the afternoon, even though $3 + 10 = 13$. In the case of a twelve-hour clock, we know the time unambiguously only to within $\pm$ 12 hours; i.e., 4 o'clock could be 4 p.m. or 4 p.m.–12 hours = 4 a.m.

The more frequently pulses are sent out, the less ambiguity there is in the measurement of the Doppler shift and the higher the "maximum unambiguous" Doppler velocity is. If, for example, the maximum unambiguous Doppler velocity is ± 12 m s^{-1} (so that the interval of measurement is from -12 m s^{-1} to $+12$ m s^{-1}) and the actual Doppler velocity is 13 m s^{-1}, the Doppler velocity would be *measured* as -11 m s^{-1} (12 $+1$ "folds" back to -11, just as 12 o'clock, plus one hour, folds back to 1 o'clock). This erroneous value of velocity is caused by "velocity folding" or "aliasing." If pulses are sent out too infrequently, you would never know exactly what the Doppler velocity is because you would get a very poor sample of the signal. Recall though, that if a radar sends out pulses too frequently, then the return from a target far away might get back to the radar before the return from a target by the next pulse from a closer target and it would appear as if a target far away is much closer than it actually is. In other words, one wants to send out pulses from a radar frequently enough that the maximum unambiguous Doppler velocity is not too low, but not so frequently that there is an ambiguity as to how far away

---

[18] The use of "analog" when describing a clock and our use of "analogy" are different.

the radar echo it is. There is a tradeoff in how one selects the rate at which pulses are sent out; this tradeoff is called the "Doppler dilemma." It turns out that the maximum unambiguous range and Doppler velocity are related to the frequency of the radar, so one must be careful in choosing what wavelength is of the radiation that the radar transmits: If you're trying to measure very high wind speeds in a tornado, you might want to use a different wavelength radar than you would if you were trying to measure relatively low wind speeds in a typical rainstorm. However, there are other considerations such as sensitivity of the radar to the scatterers and the amount of attenuation experienced. There are techniques for correcting for velocity folding that involve "editing" the data so that Doppler velocity varies smoothly in space and from scan to scan, or by sending out sequences of pulses at different time intervals so that the folding occurs at different velocities and one can make use of this information as constraints on what the actual velocity is.

Another issue that one must worry about is turbulence, and how much of a spread there is of the motion of scatterers within a radar volume. A measure of this is called the "spectrum width." The spectrum width in a tornado or at the edge of a convective storm may be very high, where winds can vary very rapidly in both space and time, but is very low if the wind field is homogeneous (or simply laminar, smoothly varying in space and time).

Because the winds measured by a Doppler radar are only valid in the line of sight of the radar beam, we don't know the *actual* three-dimensional winds. For example, if the winds are from the south and the radar beam points to the east, the radar will measure zero wind velocity. Only if the beam were pointed to the east or west would it know the actual wind. To really compute the winds, you need to have more than one radar looking at a volume of space from different viewing angles. This has been done with fixed-site radars and mobile radars. But one can still learn important properties of the wind field using just one Doppler radar. For example, a mesocyclone (tornado) on Doppler radar would appear as a region of receding (high) velocities right next to a region of approaching (high) velocities (Fig. 4.22). These vortex features, if they exhibit spatial and temporal continuity (i.e, if they appear at more than one height and for more than one scanning sequence), are known as a "mesocyclone signature" or a "tornadic vortex signature" (TVS) (Donaldson 1970; Brown et al. 1978) depending on their width.

You have to be careful, though, to determine whether or not the TVS is real or an artifact due to velocity folding. The latter can be difficult to correct if the magnitude of the maximum unambiguous Doppler velocity is small compared to the actual wind velocity. For example, if the maximum unambiguous Doppler velocity is ± 34 m s^{-1} (i.e., varying over a range of 68 m s^{-1}), but the wind in a tornado is $+105$ m s^{-1}, then the Doppler measurement might be folded over ("aliased") more than one velocity range (also known as the Nyquist interval): A wind speed of $+105$ m s^{-1} might actually be measured as -31 m s^{-1}, which is $+34$ (one half interval) $+$ 68 (one full interval) $+$ 3 added to -34 m s^{-1}. Editing severely folded Doppler radar data in a tornado can be quite a headache and a good test of a graduate student's fortitude and patience.

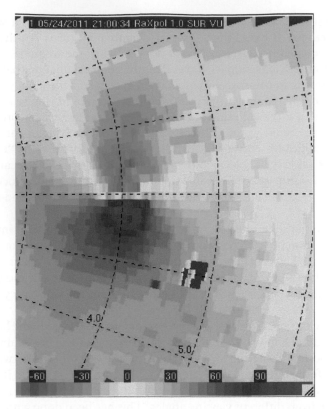

Figure 4.22 *Edited (unfolded, dealiased) Doppler velocities in a large tornado in Oklahoma on May 24, 2011 near El Reno, from RaXPol. The tornadic vortex signature (TVS) is marked by a transition from outbound Doppler velocities as high as 90+ m s^{-1} (coded red) to ~−50 m s^{-1} (coded green) over a distance of just 100–200 m. Jana Houser, a former student of mine, edited this dataset for her Ph.D. thesis. Range rings are spaced 1 km apart. The radar is located to the left, outside the frame of the image.* Figure courtesy of author.

4.2.1.3 Polarimetric radar: What type of hydrometeors and other things are inside a convective storm?

Knowing what physical processes are occurring inside a convective storm involves more than knowing the reflectivity factor and the Doppler wind velocity. We also need to know what kinds of hydrometeors are inside the cloud. This knowledge is important for finding out, for example, how much evaporative cooling goes on when precipitation falls into unsaturated air, how much cooling goes on when ice particles melt, how much latent heat is released when there is condensation, and how much thermally buoyant updrafts are retarded when weighed down by heavy precipitation. We also need to be able to distinguish among targets such rain, drizzle, and hail, etc., and biological targets such as insects and birds, and dust and debris kicked by high winds at the ground. As we noted in the earlier section on conventional radar, the radar reflectivity factor Z gives us only partial information. Radars that send out separate horizontally and vertically polarized beams give us additional information. Radars that send out separate horizontally and vertically polarized beams are called "polarimetric" radars (Zrnić and Ryzhkov 1999). All the WSR-88D radars in the United States are now polarimetric, as are many research radars. Polarimetric radars are like sunglasses with polarized lenses: Light from the sky is polarized, so if a lens that accepts light that is polarized mostly in the direction normal to that of the light from the sky, then the sky will appear to be darker. Some use polarizers over the lenses in cameras to darken the sky and enhance the contrast of bright clouds against a clear blue sky. Or, a polarizer can be used to make a rainbow appear more vivid. Polarized lenses are like polarized antennas. Now, let's go back to polarized radars and precipitation.

When large raindrops fall, they are flattened by drag, while small raindrops or drizzle drops remain mostly spherical because they fall much more slowly or they don't fall at all and experience much less drag. When large raindrops fall, if they don't tumble, then the amount of radiation that is backscattered is greater if the radiation is horizontally polarized than if it is vertically polarized (because the cross-sectional area the radar beam intersects in the horizontal is greater than that in the vertical). We make use of this property in the radar variable called "differential reflectivity" or Z_{DR}, which is a logarithmic measure of the ratio of Z for the horizontally polarized beam to that of the vertically polarized beam. If the scatterer is spherical, then the Z_{DR} is 0; if the scatterer is flattened, then Z_{DR} is greater than one.

There are also other polarimetric variables, which measure how much a particle tumbles around or how irregular its surface is (co-polar cross correlation coefficient, ρ_{hv}), and how much water there is along the path of the beam, resulting in the radar's radiation being slowed down as it passes through the water droplets or ice crystals (differential phase, Φ_{DP}). The latter is not sensitive to the intensity of the backscattered signal, which is useful when the radar is not calibrated or if there is significant attenuation. The important thing to know is that different types of hydrometeors have different polarimetric signatures for different variables (Kumjian and Ryzhkov 2008; Zhang 2016; Snyder et al. 2017). It is quite complicated to model how electromagnetic radiation is affected by the different types of targets, but it has been done successfully. Generally, we know that each type of scatterer can result in ranges of values of each polarimetric variable. Once we have measurements of each polarimetric variable, we can use "fuzzy logic" to guess the properties of the targets. For example, for various sets of values of the polarimetric variables and the conventional measurement of Z, which is typically horizontally polarized, we may infer that there is a high likelihood of large hail, or heavy rain with large raindrops, or debris in a tornado, drizzle, or a flying cow, and a lower probability of other types of hydrometeors or scatterers.

Having briefly looked at the capabilities of radars, which give us an additional sense of what is inside storms, we will now consider the visual appearance and explanations for various types of convective storms, using both cloud photographs and radar observations.

4.3   Ordinary cells and multicells

The life cycle of convective storms was first documented by radar during the "Thunderstorm Project" in the late 1940s mentioned earlier; it was led by Horace Byers, with significant contributions from Roscoe Braham, both at the University of Chicago (Byers and Braham 1949). The Thunderstorm Project provided the foundation for subsequent studies that clarified all the aspects of our understanding of convective storms. Convective storms appear on radar as one or more isolated blobs of precipitation, which look like biological cells as seen under a microscope. For this reason, these radar-echo blobs are called "cells."

In the absence of strong vertical shear (the vector difference in the wind vector associated with wind speed and wind direction at two altitudes), convective storms undergo a regular cycle of behavior. The *first stage* in the development of a convective storm, the *cumulus stage* or the *formation stage*, is characterized by a cumulus congestus, or towering-cumulus cloud (Fig. 4.23, right panel), driven upward by thermal buoyancy (Fig. 4.23, left panel). The upward-flowing air is called an updraft.

CUMULUS STAGE

Figure 4.23 *The cumulus stage of a convective storm. (left) Idealized representation of the cumulus, formative stage of an ordinary-cell convective storm in the absence of vertical wind shear. The cumulus congestus cloud is characterized by a positively buoyant (B > 0) updraft. There are no downdrafts and there is no precipitation. Figure courtesy of author. (right) A deep cumulus* congestus *building into a* cumulonimbus *just southeast of Mt. Audubon, west of Ward, Colorado, viewed from its summit, on Aug. 2, 2019.*

Photo courtesy of author.

(When the vertical wind shear is non-negligible, the updraft interacts with the shear; this process will be described later.) As the blob of buoyant air accelerates upward, air must be pushed out of the way above it, and air must rush in to fill the space below the blob of buoyant air (as noted earlier in Fig. 3.6). Why does the air diverge above the upward-moving air and converge underneath the air? It must do so in response to a region of relatively high pressure above and low pressure below. Thus, there is an outward-directed pressure-gradient force above and inward-directed pressure-gradient force below. A consequence of this high pressure above and low pressure below, is that there is *downward-directed* pressure-gradient force that counters the effect of *upward-directed* buoyancy. This downward-directed force, however, does not completely cancel out buoyancy, but typically simply reduces it. If the buoyant blob is much wider than it is deep, then the buoyancy is reduced more than if the blob were much narrower than it is deep. In the case of the former, a lot of air must get moved out of the way above the cloud and a lot of air must rush in underneath the cloud, so the pressure forces involved are relatively large. In this case we say the pressure is "hydrostatic" and reflects only the *vertical variation of the mass of air above*. In the case of the latter, much less air must get moved out of the way above the cloud and must rush in underneath the cloud, so the pressure forces involved are relatively small and the atmosphere is non-hydrostatic, having a part that is due to mostly to buoyancy.

When the cloud is very narrow, however, unsaturated and cooler air outside the cloud gets "entrained" into the cloud at its edges through turbulent mixing, reducing the buoyancy of the updraft. If the cloud is relatively wide, entrainment affects mainly the outer edges of the cloud, so the buoyancy of the inner portion of the cloud is relatively unaffected. We considered this entrainment effect when we talked about cumulus and in particular, cumulus *congestus*.

If the top of the upward advancing updraft has enough kinetic energy to penetrate the tropopause into the stratosphere, it produces what is known as a "penetrating top" (the updraft penetrates the tropopause, into the stable stratosphere, where the temperature does not change much with height or even gets slightly warmer) or "overshooting top" (the updraft overshoots its equilibrium level) (Fig. 4.24).

When the updraft loses its kinetic energy completely after it decelerates in the presence of negative buoyancy, it stops growing. Gravity waves may be triggered when the updraft hits the stable air above the tropopause just as gravity waves may be triggered when air is forced up and over a mountain when the atmosphere is stable with respect to vertical displacements (Fig. 2.44). Water vapor is injected into the stratosphere as the cloud material evaporates or sublimates, and then may affect chemical reactions with other gases such as ozone. Also, the transfer of gases can go the other way, as ozone and other stratospheric gases or material may get mixed into the cloud below. As the updraft weakens to nothing, it spreads out laterally, producing what is known as an "anvil," because when viewed on end, it looks like the flattened top of an anvil, as used when crafting objects from metals. In the absence of vertical wind shear, the anvil spreads out symmetrically and the cloud top look like the top of a mushroom (Figs. 4.25–4.27) and acts like a canopy.

Figure 4.24 *Penetrating/overshooting tops. (top, left) Penetrating/overshooting top above the anvil of a cumulonimbus in central Nebraska, near Arnold, viewed to the east, on June 1, 2018. View is to the east. (top, right) Penetrating/overshooting top above the anvil of a small cumulonimbus on Oct. 17, 2007, near sunset, in north-central Oklahoma. View is to the southeast. (bottom) Penetrating/overshooting top on a cumulonimbus west of the Continental Divide of the Indian Peaks, Colorado, west of Ward, June 24, 2020. View is to the west.*
Photos courtesy of author.

The appearance of an anvil in a convective cloud does not mean that there is precipitation falling from the cloud, nor does it mean that the cloud is electrified to the point that there is lightning. It simply means that an updraft has reached its equilibrium level and the air spreads out. After an anvil first appears, the sky may eventually be covered by anvil material and the sun blocked out (the anvil being composed of cirrostratus *spissatus*).

Since the air is very cold near the tropopause, anvils, as noted earlier, are usually composed of ice crystals and often have a fibrous appearance, though some supercooled water droplets may also be present, having been brought upward in the storm's updraft. Convective storms, which may have been triggered by daytime heating, produce anvils which then block out or partially block out solar radiation from warming the ground any further. Convective storms also transport the heat energy produced at the surface upward to where it is cooler and reduce the amount of instability, thus lessening possibility of the triggering of new convective clouds. Anvils thus can act as natural temperature

Figure 4.25 *Mushroom-shaped anvils from cumulonimbi. (top) Mushroom-shaped anvil in a convective storm on June 13, 2001, viewed to the southwest from Hays, Kansas. (middle) Cumulonimbus with a symmetrical anvil, viewed to the southeast from Norman, Oklahoma, on May 21, 2011. The National Weather Center at the University of Oklahoma is seen to the right. (bottom) Mushroom-shaped anvil on top of a cumulonimbus viewed to the northeast, from Boulder, Colorado, on July 24, 2016. The active updraft towers are seen on the right, the south side of the storm.*
Photos courtesy of author.

Figure 4.26 *Small anvils. (top, left) Mushroom-shaped anvil in a decaying convective snow shower near Steamboat Springs, Colorado, on Jan. 15, 2003. (top, right) Mushroom-shaped anvil from a cumulonimbus on July 16, 2007, viewed to the east from the summit of James Peak, just to the west of St. Mary's, Colorado. (bottom, left) Mushroom-shaped anvil on a cumulonimbus over the Foothills, viewed to the southwest, from Boulder, Colorado, on June 27, 2015. (bottom, right) Two wild mushrooms in Norman, Oklahoma, Sept. 4, 2022, which look like the tops of anvils from cumulonimbus clouds in an environment of weak vertical shear.*
Photos courtesy of author.

regulators. In a sense the response by convective storms is to reduce the forcing that caused them in the first place. Updrafts stabilize the lapse rate by warming the air aloft through the release of latent heat of condensation, and evaporative cooling of rain as it falls out from the cloud stabilizes the lapse rate at low levels, and anvils diminish the daytime heating at the ground, which by itself acts to destabilize the lapse rate. This process of overall stabilizing the atmosphere is an example of Le Chatelier's principle in chemistry, in which a system responds to a forcing by acting to mitigate the forcing and to force the system back to equilibrium.

Anvils are "forerunners" like the cirrus clouds that flow far in advance of synoptic-scale systems, but in this case, they forewarn us of a convective storm or storms. Without access to radar, storm chasers may see cirrus from an anvil emanating from a distant, not yet visible, cumulonimbus, and head toward the ostensible storm, taking into account that the storm itself may not be propagating in the same direction as the anvil is moving.

Figure 4.27 *Cumulonimbus anvil spreading out over eastern Colorado on the Palmer Divide, on June 3, 2004. There is still some active convection on the left-hand side (southern side).*
Photo courtesy of author.

Of course, one may be disappointed to find that the active portion of the storm has dissipated by the time one reaches the source of the anvil and that the anvil has been "orphaned" or that the active convective clouds occupy only a very small area.

4.3.1   Mamma and reticular clouds: Upside-down convection

The underside of an anvil from a convective storm is sometimes populated by "mamma,"[19] downward-extending pouches of cloud, which look like upside-down convective, cauliflower-like elements, but much smoother (Figs. 4.28–4.32) and have sometimes been described as looking a bit like "stalactites" in caves. Mamma have been seen in paintings as far back as the 1500s. The name comes from William Clement Ley in the late nineteenth century in his book, *Cloudland*. While Scorer's 1972 book contains a nice discussion on what makes mamma, there is a much more recent, comprehensive summary in the scientific literature about what we know now about them.[20]

Mamma, however, unlike cumulus clouds, usually look much smoother than upright, positively buoyant elements in convective clouds, especially at their edges (contrast, e.g., Figs. 4.28 and 3.27). They are associated with negative buoyancy as ice crystals in the anvil fall out and sublimate and cool the air at a rate that is greater than that with which the

---

[19] Also known as "mammatus."
[20] Schultz and Coauthors, 2006.

Figure 4.28 *Mamma. (top) Cumulonimbus mamma, illuminated overhead by a late-day sun, on June 1, 2018, underneath the rear anvil of a mesoscale convective system in north-central Nebraska, near Taylor. I always try to include ground features in cloud photographs, except when impressive features are nearly overhead, as in this case, when it is difficult or almost impossible to do so. (bottom) Mamma underneath a cumulonimbus anvil in the Nebraska Panhandle, near Scottsbluff, on June 7, 2018. The anvil is very thin in between the mamma, exhibiting a bluish tint in between the mamma from the sky above.*

Photos courtesy of author.

air is warming as a result of subsidence (encountering higher pressure, which compresses the air and heats it). Time-lapse videos I have taken show that they grow downward much more slowly than the positively buoyant elements on the top of many cumuliform clouds grow upward.

My graduate students and I have collected radar data slicing through mamma and can associate their shape with what we see and also note that air ascends in between them and descends with them (Fig. 4.29), as had been suggested by Ludwig Prandtl, a prominent German fluid dynamicist. We were not the first to determine this, as there are some nice earlier observations with a radar looking upward from research aircraft that also show this. A number of measurements of the vertical velocity in mamma have been made by Doppler radar and instruments aboard aircraft: Mamma have downward wind speeds of around 3 m s^{-1} and are surrounded by upward motions in between them of around 1 m s^{-1}. Their average width is around 1–3 km, they hang down around 500 m from cloud base, and each individual mamma pouch lasts around 10 minutes.

Mamma have been simulated using a computer model in which only ice particles are allowed (no supercooled water droplets), to test various hypotheses about why they form.[21] Since anvils from which mamma are pendant are usually relatively high up where it is cold enough to expect to find mainly ice particles and since observations have found high concentrations of ice in anvils, it is not thought that supercooled water droplets are essential to their formation. Scorer originally suggested that the reason mamma do not look like the mirror image of right-side-up convection (i.e., although they protrude downward, in the opposite direction, they are much smoother and larger) is that the cloud particles (ice crystals and snow crystals) fall into stable air and small particles evaporate or sublimate and cool to maintain negative buoyancy for a relatively short distance, but when all the particles have evaporated or sublimated, the warming of the air as it encounters higher pressure and is compressed rapidly decreases the negative buoyancy and the mamma do not extend too far below the cloud base (e.g., like an upside-down anvil) and spread out and even curl back up (Fig. 4.34). Since the smallest ice particles sublimate first, the larger ice crystals inside the mamma pouches remain and are opaque, and the pouches cannot be narrower than the width of the larger, ice crystal pouches. In upwardly buoyant cumulus clouds, however, there are small supercooled water droplets, which can evaporate quickly on contact with unsaturated air, and result in sharp edges to the cloud, whose turrets can be very small. In addition, clear, unsaturated air is drawn up in between the negatively buoyant pouches in mamma, thus further giving the appearance of smoother, larger elements.

The numerical simulation experiments showed that when mamma form, negative buoyancy is maintained, which requires the air below the parent cloud (e.g., anvil) which is mixed with the air in the cloud not be too cool or humid, because then the rate of cooling due to evaporation or sublimation is less than the rate of warming due to compression and the air loses its negative buoyancy too soon; if the air is too warm or dry then all the particles disappear too quickly and mamma also do not form. However, even if the negative buoyancy is maintained below cloud base, mamma still do not necessarily form.

[21] Kanak et al., 2008.

Figure 4.29 *Mamma underneath an anvil and their depiction on radar. (top) RaXPol probing cumulonimbus mamma just before sunset on June 10, 2018, in northwestern South Dakota, southwest of Mud Butte, which is north-northeast of Rapid City. (bottom, left) Vertical cross-section of radar reflectivity through the mamma seen around the time the photograph seen in the top panel was taken, but with the radar beam aimed to the southwest. The pouches of mamma are clearly evident as pouches of moderate radar reflectivity (color-coded as orange to red, or ~25–35 dBZ). Many mamma extend downward ~1 km from the anvil. (bottom, right) As in the panel on the bottom, left, but for Doppler velocity. The purple at 11 km range, and extending outward, represents a jet of airflow from the southwest, toward the radar. Individual splotches of purple, which indicate locally enhanced approaching velocities, are adjacent to green areas, which indicate weaker approaching velocities. This pattern is consistent with sinking motion at the center of each mamma pouch and rising motion around the edges (as evidenced by divergence underneath each pouch, with more rapidly approaching Doppler velocities on the near side than on the far side).*

Photo courtesy of author.

The most important condition favoring mamma formation, in addition to the requirement there be negative buoyancy when air is mixed up with cloud air at cloud base,
seems to be that ice crystals sublimate underneath it.

Mamma are especially striking when illuminated from underneath at sunset. In the
most spectacular views of them, they are spread out over an extensive portion of the
anvil and are colored bright red or orange (Fig. 4.30). It is convenient that anvils from
mesoscale convective systems (MCSs) tend to be most massive late in the day, when the
systems are mature, and I have frequently tried to position myself on the western edge
of the system, beyond the region of precipitation, just to have the chance to see mamma
come out at sunset. Mamma also look beautiful as gray/white blobs under a blue-tinged
anvil (Fig. 4.31).

Figure 4.30 *Mamma with an orange glow, at sunset. (top, left) Cumulonimbus mamma southwest of
Lubbock, Texas, on May 17, 2021. Mamma at sunset on the western side of convective storms can be
spectacular. View is to the south. (top, right) Cumulonimbus mamma underneath rear edge of an anvil
of a mesosocale convective system in north-central Nebraska, near Taylor, at sunset on June 1, 2018.
Note the linear striations along the western edge of the anvil. These striations might be manifestations of
gravity waves in the anvil. Striations like these are seen in some satellite images. (bottom)
Cumulonimbus mamma under the trailing anvil of convective storms over northwestern Kansas on
May 24, 2021, after a tornadic supercell had moved on to the northeast. View is to the south.*
Photos courtesy of author.

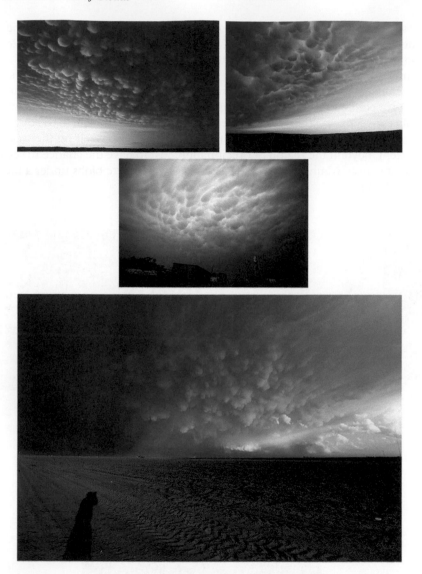

Figure 4.31 *Mamma with only a little or without any direct sunlight. (top, left) Cumulonimbus mamma after sunset near Taylor, Nebraska, on June 1, 2018. (top, right) Cumulonimbus mamma over southeastern Wyoming/the Nebraska Panhandle on June 12, 2017. View is to the east. (middle) Cumulonimbus mamma during the late afternoon of May 29, 2008 over Cawker City, in north-central Kansas. (bottom) Cumulonimbus mamma, viewed to the east, near sunset, on May 17, 2021, in West Texas, southwest of Lubbock. The contrast between the texture of the mamma at the top of the frame between the texture of the furrowed soil at the bottom of the frame is striking. There are a few anticrepuscular rays seen just above the horizon. A shadow, unintended "selfie" of the author is seen at the bottom left.*
Photos courtesy of author.

Even during the day, when there is not much contrast in color, their texture and striations can be striking (Fig. 4.31, middle panel). In some instances mamma are much larger and irregular in appearance and spacing (Fig. 4.32).

I have flown by mamma in research aircraft and have seen that when viewed as cross sections, one sometimes can actually see right through them and also see ice crystals falling out (Fig. 4.33).

At one time, it was thought that mamma are associated with severe weather. In fact, when I was very young, I was inspired by a small "Golden" book in the 1950s called *Weather*, in which it was noted that tornadoes can form from mamma. Figure 4.30, bottom panel, for example, indeed shows mamma from a storm complex that earlier had produced tornadoes. We now know, however, that mamma occur in all types of convective storms and that they are pendant not only from cumulonimbus anvils, but also from ordinary cirrus or cirrostratus or altostratus clouds associated with midlatitude synoptic-scale systems (Figs. 4.34–4.35).

Mamma may be seen pendant not only from stratiform ice clouds, but also from ash clouds. They have been seen underneath ash clouds from forest fires (Fig. 2.40, bottom panel) and also from ash clouds from volcanic eruptions. It is likely that ice particles are nucleated by the ash and fall out and sublimate.

Related to mamma are "reticular clouds,"[22] so-called because they resemble a network of cloud filaments (Fig. 4.36). They are, in a sense, the opposite of mamma, because instead of pouches of cloud hanging down, with cloud-free material in between the pouches, they appear to be cloud-free pouches upside down, separated by a network of clouds. The network may assume a polygonal appearance, but the cells of the network are not necessarily regular. The vertical-motion field associated with reticular clouds is not well known.

4.3.2 Convective cloud tops as mountains

On satellite imagery, penetrating/overshooting tops can be seen in visible imagery most easily by the shadows they cast (Fujita 1982), which are best seen late or early in the day when the sun angle is low (e.g., Fig. 4.37, top left). On infrared (IR) imagery, which provides an estimate of the temperature of the cloud top, one can also detect penetrating/overshooting tops, day or night (Fig. 4.37, bottom left). The radar reflectivity image (Fig. 4.37, right panel) shows where the precipitation is in relationship to the cloud top features (see also Bluestein et al. 2019). In this instance a hook echo is located in the vicinity of the coldest IR (highest) cloud top.

Infrared (IR) satellite imagery, however, must be viewed with a lot of caution. If the temperature decreases with height, then a penetrating top, the highest cloud top in a convective storm, should be detected as a local region of relatively cool temperature. However, sometimes cloud material can form in the stratosphere, where the temperature may actually be warmer than it is at the tropopause. If this is the case, then a region of

---

[22] Kanak and Straka, 2002.

Figure 4.32 *Irregular mamma. (top) Cumulonimbus mamma over Boise City, in the far western Oklahoma Panhandle, on June 2, 2004. The mamma seen in the top panel look like dark pouches surrounded by brighter clouds. The bright regions probably mark thin portions of the anvil. These apocalyptic clouds are reminiscent of the parting-of-the-Red-Sea scene in the epic 1956 movie* The Ten Commandments. *(bottom) Mamma underneath an anvil on June 10, 2018 in western South Dakota. The mamma in the bottom panel are very irregular and many are not smooth and bulbous as in other images. Bluish-tinted sunlight appears to be filtering in where the anvil deck is relatively thin.*
Photos courtesy of author.

Figure 4.33 *Airborne view, from a NOAA aircraft, of three mamma, close up, at flight level, over the Texas Panhandle, on May 26, 1991, during the COPS (Cooperative Oklahoma[a] Profiler Studies) experiment. View is approximately to the east.*

[a] Despite the name of the field program, we sometimes crossed into neighboring states such as Texas. Photo courtesy of author.

warming may be seen. On satellite IR imagery one sees U- or V-shaped cold areas (Fig. 4.38, bottom left) extending downstream from the penetrating (or "overshooting") top. One is then left with a quandary: Is the local region of low IR temperature a depression in the anvil cloud top, or is it due to cloud material that is actually above the top of the storm, which is called an "above anvil cirrus plume" (ACCP)? There is a lot of evidence that these regions are not depressions in the anvil, but rather separate cloud material aloft.[23] In visible imagery, shadows can be seen cast by the ACCP on the top of the anvil below (Fig. 4.38, top left). In Figure 4.38, bottom left, there are two ACCPs, but the easternmost one (IR top, coded green), is the more prominent. Each ACCP in this case is associated with a separate convective storm, as seen in radar imagery (Fig. 4.38, right panel).

It is thought that gravity waves form as air is lifted over the penetrating top as air is lifted over a mountain and that these waves "break" like ocean waves do, producing what the prominent mesoscale and storm-scale meteorologist Ted Fujita called "jumping cirrus." There is some recent evidence that hydraulic-jump-like waves (see Chapter

[23] Bedka et al., 2018

Figure 4.34 *A line of mamma underneath a thick layer of cirrostratus, over Norman, Oklahoma, on Oct. 28, 2014. Curls, like upside-down Kelvin–Helmholtz waves/billows, on the left-hand side of the down-hanging pouches of cloud are evident in two of the four most distinct mamma: the one on the extreme left and the third one from the left.*
Photo courtesy of author.

2) form like those in the lee of mountains.[24] How can a cumulonimbus top act like a mountain?

Suppose that a cumulus congestus or cumulonimbus cloud, which by definition has a buoyant updraft, propagates more slowly than the speed of the wind at the top of the cloud. This might be the case when the cloud is triggered by a stationary surface feature such as a mountain, while there is wind at the top of the cloud. Or, more generally, suppose that a deep convective cloud is embedded in vertical shear, such that wind is from the same direction, but the wind speed increases with height. Then in the reference frame of the cloud, air approaching it must slow down and be forced up and over its top, as a region of relatively high pressure forms on the upwind side (with respect to

---

[24] O'Neill et al., 2021

Figure 4.35 *Mamma pendant from isolated cirrus and cirrostratus/altostratus. (left) Mamma pendant from a patch of cirrus* spissatus *over Norman, Oklahoma, on May 1, 2020. (top, right) Mamma pendant from an isolated cirrus cloud over Norman, Oklahoma, on April 21, 2019. This cloud looks like a jellyfish, with mamma at the bottom of its tentacles. (bottom, right) Mamma underneath cirrostratus/altostratus in Cuba, Illinois on Nov. 23, 2018.*
Photos courtesy of author.

the cloud). It is assumed that the air inside the cloud, which is a large, buoyant bubble or composed of a bunch of smaller, buoyant bubbles, originated at some altitude below where air moves more slowly. The only way this can happen is if there is a pressure-gradient force acting opposite to the wind, so there must be relatively high pressure at the outer edge of the cloud (on the left). If this is so, then there must be a pressure-gradient force acting upward outside of the buoyant cloud. So, air that approaches the cloud top acts as if there were a mountain top ahead of it, and get lifted upward and over it, as air does over an air foil or density current, to be discussed shortly. If the relative humidity is high enough and the air outside the cloud is stable, then a cap cloud (pileus) will form, as mentioned in Chapter 2, but this time the lift is a result of the vertical shear, not simply air that is being lifted above the cloud in the absence of vertical shear.

Figure 4.36 *Reticular clouds. (top) Reticular clouds over Norman, Oklahoma, on Jan. 21, 2020, which are marked by the filaments of darker cloud. The network of clouds appears under a cirrostratus overcast, not associated with any convective storm. (bottom) Reticular clouds underneath a cirrus deck, over Norman, Oklahoma, on March 4, 2020. The network of clouds in this image are composed of wider, more diffuse elements than those seen in the top panel. View is to the north.*

Photos courtesy of author.

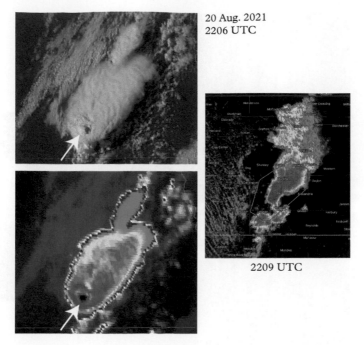

20 Aug. 2021
2206 UTC

2209 UTC

Figure 4.37 *Locating penetrating/overshooting tops in cumulonimbus clouds vis satellite imagery. (top, left) Visible NOAA/GOES-16 satellite imagery on Aug. 20, 2021 at 22:06 UTC in south-central Nebraska. The arrow points to a supercell with the penetrating top with a shadow cast just to its right (east). (bottom, left)) As in top, left panel), but for the longwave infrared channel; in this channel the penetrating top appears as having the coldest IR temperature, which is coded as black. (right panel) WSR-88D reflectivity imagery from Grand Island, Nebraska. Red and purple denote the highest reflectivity. The storm was severe warned at this time, but was tornado warned shortly thereafter; there is a hook echo near Hebron, Nebraska. The horizontal scale of the radar image is not the same as that in the satellite images.*
(Courtesy of College of DuPage Meteorology.)

4.3.3 Downdrafts, gust fronts/density currents and outflow boundaries (arcus/shelf clouds): Jumping over cold pools

In the *second stage* of the life cycle of an "ordinary-cell" convective storm, the *mature stage* (Fig. 4.39), precipitation develops in the upper portion of the cloud, where there are supercooled water droplets (liquid water in the presence of sub-freezing temperatures) and ice crystals (especially very high up in the cloud where the temperature is much below 0°C and either ice nuclei or water nuclei are present). If there is a mixture of supercooled water droplets and ice crystals, the ice crystals grow at the expense of the water droplets via the Wegener–Bergeron–Findeisen "cold rain" process. Ice crystals melt on the way down and become raindrops. On radar, the first echoes appear high up

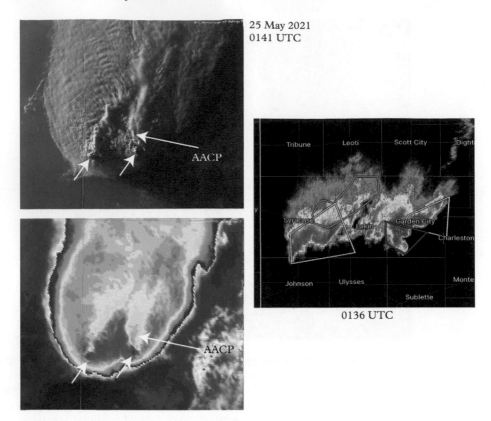

Figure 4.38 *As in Figure 4.37, at a time when above anvil cirrus plumes (AACPs) are evident. (top, left) Visible NOAA/GOES satellite imagery at 01:41 UTC on May 25, 2021 in southwest Kansas. The two short arrows point to two separate penetrating/overshooting tops in two supercells and have deep shadows to the right of the tops. The long arrow points to an AACP, which appears as a diffuse swath of clouds streaming off to the north-northeast. It may also represent lower cloud tops from relatively new convective towers. (bottom, left)) As in top, left panel, but for the longwave infrared channel; in this channel the AACP/new convective towers appear coded by the warmer, green color. The two penetrating tops appear as having the coldest IR temperatures, which are coded as dark red. (right panel) WSR-88D reflectivity imagery from Dodge City, Kansas, at 01:36 UTC, just several minutes prior to the satellite imagery. Red and purple denote the highest reflectivity. The easternmost storm was tornado warned at this time and there is a hook echo southwest of Garden City.*
(Courtesy of College of DuPage Meteorology.)

in the cloud: It takes time for rain to form and air must ascend to a higher altitude before the rain appears. When we storm chase, we check the appropriate National Weather Service surveillance radar at elevation angles large enough to see high up in the storm, for evidence that a storm is about to matures and dump precipitation. Or, we check a radar at low elevation angles, but far from the storm.

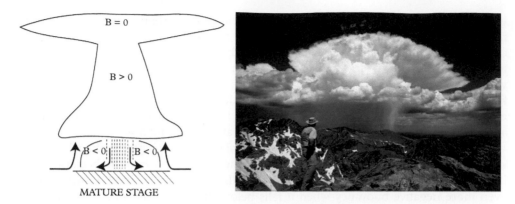

MATURE STAGE

Figure 4.39 *The mature stage of a convective storm. (left) Idealized illustration of the mature stage of a convective storm in the absence of vertical wind shear. The main convective tower is positively (thermally) buoyant (B > 0), while underneath the cloud base precipitation is falling into unsaturated air and producing negative buoyancy (B < 0). A cold pool is produced, which moves outward and produces upward motion near and under cloud base along the outer edges of the cloud (for reasons to be discussed soon). The anvil at the top is characterized by neutral buoyancy (B = 0). Although the main convective tower is thermally buoyant, water loading renders areas actually negatively buoyant, so precipitation falls out despite the positive thermal buoyancy. In the mature stage, both updrafts and downdrafts are found. Figure courtesy of author. (right) An example of the mature stage of an ordinary convective storm. Cumulonimbus with a mushrooming anvil and a probable hail shaft (white precipitation streak) viewed to north from the summit of South Arapaho Peak (13, 326 feet MSL) in the Indian Peaks Wilderness, northwest of Nederland, Colorado, on 3 Aug. 2009. The person on the left is an unidentified hiker obviously admiring the view along with me.*
Photo courtesy of author.

If the temperatures even at moderate altitudes are above freezing, as they are in the tropics, then precipitation may form via the aggregation "warm rain" process, but in this case the clouds are not too deep and do not produce anvils composed of ice crystals. If there is a wide spectrum of droplet sizes, the larger ones fall relative to the smaller ones, and overtake the smaller ones, resulting in the growth of the water droplets. I have seen many tropical showers form in relatively shallow clouds.

Precipitation (rain, snow, hail, and other hybrid forms of liquid and ice) then grows and its mass contributes to negative buoyancy (it is much heavier than dry air and weighs down the updraft). When its total mass becomes so great that the positive thermal buoyancy of the air is completely counteracted and then some, downward buoyancy accelerates the air downward. The falling precipitation's descent itself is countered by frictional drag, which acts upward. If the downward-directed negative buoyancy is exactly counteracted by the upward-directed frictional drag, then the air continues moving downward as a downdraft at its "terminal fall velocity." If you drop a rock off a cliff, it will accelerate downward owing to gravity, but eventually reach its terminal fall velocity owing also to the drag of air on it. Do not try jumping off a cliff to find out when you will reach *your* terminal fall speed without having a parachute handy just in case terminal fall velocity is not reached before you reach the ground and you don't have a soft spot on which to land! Terminal fall velocities of raindrops vary from ~5–10 m s^{-1}, while the terminal fall velocity of large hail can be as high as 35 m s^{-1}. It obviously takes only a weak updraft to keep raindrops suspended aloft, while it takes a very strong updraft to keep a large hailstone suspended. It is easy to understand why large hail breaks windows and vehicle windshields. The formation of large hail must require very strong updrafts and large hail means that its parent storm's updraft must be intense.

When precipitation falls (i.e., when the terminal fall velocity is greater than the upward speed of the buoyant updraft), it becomes even more negatively buoyant because it cools if it is frozen and melts on the way down as it falls below the 0°C level. It then acquires even more negative buoyancy when it falls below the cloud base into unsaturated air and evaporates and cools. The air underneath a convective storm therefore develops a "pool" of cold air underneath it, near the ground. This "cold pool" may be as shallow as just a few hundred meters or as deep as several kilometers, depending upon how high the cloud base is, how dry the air is underneath it, and what the composition and concentration of hydrometeors is that is falling out from the cloud. Some hydrometeors are more or less efficient in evaporating (or sublimating). The coldness of the cold pool also depends upon the same factors that determine the depth of the cold pool.

In the tropics, especially over the ocean where it may be very humid, the cold pool is very weak, if there is one at all, because the potential for evaporation or melting is only slight, if any. The density of the cold pool depends mostly on the temperature, but if the environment is humid then the water loading above also may play a significant role. In arid regions, where the cloud base is high, the cold pool below may be very strong. In the instances when the precipitation evaporates completely before reaching the ground, the "fall streaks" of precipitation that never make it to the ground are called *virga* (Figs. 4.40–4.41). Figure 4.42, on the other hand, shows precipitation that is making it all the way to the ground.

Other views of virga and precipitation from convective storms or convective towers are seen in Figures 4.43, 4.44, top panel, 4.45, 4.46, and 4.47. If the raindrops are spaced very far apart it is difficult or even impossible to tell visually if precipitation is actually reaching the ground. In this case, a radar can detect precipitation that the naked eye may not. Or, one could drive underneath the cloud and experience the rain, if there is any, but this brute-force method is obviously the more difficult way to do this.

When a downdraft hits the ground, it spreads out laterally and the resulting horizontal wind may be strong enough to cause some damage or even blow structures completely away. Why does the downdraft spread out? You know that it cannot dive into the ground (unless it is absurdly strong, like the exhaust from a jet engine or a rocket), so it must come to a halt. An upward-directed pressure-gradient force created by relatively high pressure (not the hydrostatic part, which is *associated* with the coldness of the air, but the dynamic, non-hydrostatic part associated with air motion) must be responsible for slowing the downdraft down. An outward-directed pressure-gradient force then drives the air to diverge at the surface.

Figure 4.40 *Virga. (top, left) Virga falling from a shallow cumulonimbus over the Indian Peaks Wilderness, west of Ward, Colorado, viewed from the south side of Arapaho Peak, on Aug. 20, 2008. The body of water at the lower right is Lefthand Reservoir. There may actually be a few large raindrops making it to the ground, but are not visible because they are so widely spaced. (top, right) Virga falling from a developing cumulonimbus on June 29, 2007, viewed from Niwot Ridge in the Indian Peaks Wilderness of Colorado, (middle, left) An elevated cumulonimbus producing virga over Norman, Oklahoma, on April 24, 2020. This storm was rooted above the boundary layer. (middle, right)*

The strong, damaging winds at the surface underneath downdrafts are called *microbursts*[25] (Fig. 4.48) (Roberts and Wilson 1989). When the negative buoyancy that drives them is mainly due to heavy liquid-water loading in an environment that is too humid for evaporative cooling to be significant, they are called *wet microbursts* (Wakimoto and Bringi 1988; Atkins and Wakimoto 1991). On the other hand, when the negative buoyancy that drives them is mainly due to evaporative cooling in a dry environment where the cloud base is relatively high up, perhaps augmented by cooling due to melting above of frozen particles on the way down, they are called *dry microbursts* (Wakimoto 1985; Srivastava 1987; Parsons and Kropfli 1990). In the United States, wet microbursts tend to occur in the southeastern US or elsewhere east of the Rockies, while dry microbursts tend to occur over the High Plains and Plains areas, which are more arid (Hjelmfelt 1988). Microbursts are dangers to aviation because they cause rapid changes in the lift of an airplane wing, as an aircraft flies through them.

When rain falls out from high-based convective storms or from an anvil, spectacular rainbows or full double rainbows may be seen, particularly when the sun angle is low, late in the afternoon or before sunset, or early in the morning or after sunrise (Fig. 4.49). Those appearing later in the day are much more common than in the morning, because convective storms are more likely to form late in the day. The rainbow seen in Figure 4.49, bottom panel, is a rare, morning exception.

When the cold pool forms as precipitation melts and/or falls through unsaturated air, it constitutes the bottom portion of a convective storm and is characterized by a hydrostatic pressure at the surface that is *higher* than that of its surroundings, outside the storm, in the ambient atmosphere (the "environment"), because according to the ideal-gas law (pressure = density X gas constant X temperature) cold air is denser than warm air at the same pressure, so the air column under the storm is relatively heavy. This hydrostatic dome of high pressure was first analyzed in great detail by Ted Fujita and others in the 1950s using a meso-network of surface observations and called a "meso-high" (mesoscale high; a high-pressure area on a horizontal scale of tens to hundreds of meters) (Fujita 1963).

Figure 4.40 (Continued) *A high-based cumulonimbus coming off the Foothills, over Boulder, Colorado, near sunset on Aug. 6, 2020. Virga is seen hanging down from the anvil. A few large raindrops might actually reach the ground, but it is not possible to tell visually if they do. (bottom, left) Rain from a high-based cumulonimbus falling near the Foothills of Boulder, Colorado, late in the day on July 8, 2020. It appears as if the storm is dumping its precipitation into a vessel composed of mountains. It appears as most of the precipitation has not evaporated before reaching the ground. (bottom, right) Virga falling from a dissipating, high-based cumulonimbus, over Boulder, Colorado, on July 8, 2020, along with crepuscular rays (lower left). The crepuscular rays seem to be radiating outward to meet the falling precipitation.*

Photos courtesy of author.

[25] Microbursts are small-scale "downbursts" (Fujita 1981; Wilson and Wakimoto 2001).

Figure 4.41 *A very high-based cumulonimbus with virga falling from the anvil in eastern Wyoming, viewed to the north, from a location north of Cheyenne, on June 7, 2018. This convective cloud is so high-based that the bottom of the anvil is amazingly around the same altitude as the cloud base under the updraft.*
Photo courtesy of author.

The cold pool that forms in an environment of very weak or non-existent vertical shear is relatively symmetric about the center of the updraft way above and the downdraft below. So, at the edges of the cold pool there is an outward-directed hydrostatic pressure-gradient force that moves the edges of the cold pool outward (Fig. 4.50, top panel). As the edge of the cold pool passes you by, the wind shifts direction to that coming from the cold pool, directed outward from the storm. The wind speed picks up and becomes gusty, the temperature drops, and the pressure rises. This leading edge of the cold pool behind which there is colder air is called a "gust front" (Charba 1974). The gust front moves beyond the region where the precipitation is falling, so one usually first experiences the cold air and the onset of the precipitation typically lags the passage of the gust front. In the 1943 musical *Oklahoma!*, which I first saw as a young kid in the 1955 movie version, "the wind comes right behind the rain." In fact, it is the reverse in a convective storm . . . the wind comes ahead of the rain. Also, owing to the dynamic high-pressure area just ahead of the cold pool, the pressure jumps just before the cold pool passes by, before the hydrostatic contribution to the pressure jump commences.

Suppose that there is no wind at all outside the storm and the gust front underneath the storm moves outward. Imagine that you keep yourself situated right at the leading edge of the cold pool, along the edge of the gust front; perhaps you are doing so in a car, keeping up with the gust front (or surfing along the edge of the gust front). Then in your reference frame you feel air coming toward you, but it stops right at the leading edge of the gust front (Fig. 4.50, bottom panel). You feel air rushing toward your hand at the same speed as the car is moving. In your reference frame, suppose there is no ground-relative wind *behind* the gust front (which is sometimes referred to as a "resting" state). So, air coming toward you must decelerate and come to an abrupt halt at the edge of the gust front. How can this be? There must be a dynamic pressure-gradient force just

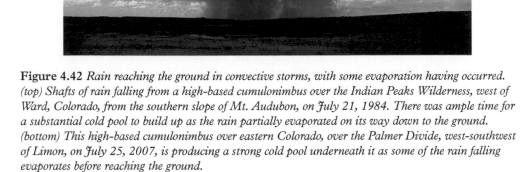

Figure 4.42 *Rain reaching the ground in convective storms, with some evaporation having occurred. (top) Shafts of rain falling from a high-based cumulonimbus over the Indian Peaks Wilderness, west of Ward, Colorado, from the southern slope of Mt. Audubon, on July 21, 1984. There was ample time for a substantial cold pool to build up as the rain partially evaporated on its way down to the ground. (bottom) This high-based cumulonimbus over eastern Colorado, over the Palmer Divide, west-southwest of Limon, on July 25, 2007, is producing a strong cold pool underneath it as some of the rain falling evaporates before reaching the ground.*

Photos courtesy of author.

Figure 4.43 *Anvils with virga. (top, left) Snow shower with a small anvil coming off the Foothills into Boulder, Colorado, on Feb. 20, 2021. This cumulonimbus is very diffuse looking and some cooling must be taking place as some of the snow crystals sublimate on the way down. The National Center for Atmospheric Research (NCAR) is visible on the lower left. (top, right) An anvil from a dissipating snow shower approaching Boulder, Colorado, on Feb. 20, 2021, with virga underneath the anvil. The shape of the cloud reminds me of the scene in the Disney animated movie* Fantasia *from 1940, set to the music of the* Sorcerer's Apprentice. *(bottom) Anvil from snow showers over the Continental Divide streaming northeastward over Boulder, Colorado at sunrise, on Dec. 14, 2021.*
Photos courtesy of author.

outside and ahead of the gust front which is directed outward, so that the air decelerates as it approaches it. How far ahead of the gust front does the air "feel" the effects of the gust front? This seems like a mysterious, action-at-a-distance-like phenomenon. It turns out that from a consideration of mass conservation, the approaching air feels the effect of the gust front at a distance of approximately the same as the depth of the cold pool. The change in pressure acts as a sort of radar or feeler in informing the air that there is an obstacle ahead. The "information" about what's ahead is actually carried by sound waves, which in the atmosphere propagate at a speed of around 300 m s^{-1}. Because we are ignoring sound waves, we are in effect assuming that the information is obtained instantaneously.

Figure 4.44 *Virga and precipitation from convective storms moving off the mountains in winter at sunrise. (top) Virga and snow falling from convective snow showers over Boulder, Colorado, on Jan. 3, 2006. At sunrise the chiaroscuro effect of the sunlit features to the adjacent dark areas is striking. The Mesa Lab of the National Center for Atmospheric Research (NCAR) is illuminated by the sun. (bottom) Edge of rain from a squall moving off the mountains west of Boulder, Colorado, on Dec. 15, 2021. Winds blew 70–90 mph later in the morning. The leading edge of the precipitation is sharply contrasted by early-morning illumination by the sun.*

Photos courtesy of author.

Figure 4.45 *Elevated, narrow convective tower shearing off to the south over south-central Kansas, on Aug. 25, 2015. Viewed to the west during the morning, well before the ground has heated up to convective temperature. An orphan anvil with virga from a previous tower is seen immediately to its left (south). The bottom of the haze layer in the distance is probably at the top of an inversion where the vertical mixing of particles is inhibited. Such convection can result in precipitation over night and early in the morning well before convective temperature at the surface is reached. There is some virga evident in the decaying tower on the right.*
Photo courtesy of author.

As the air decelerates, it is then forced upward by an upward-directed dynamic pressure-gradient force. This explanation is the first of three we will look at and it depends on our assumption that there is no mixing of air outside the density current with air behind it: Oncoming air does not penetrate the cold pool. Underneath the narrow region of upward motion there is convergence to provide air into the upward current of air (Fig. 4.50, bottom). The most important aspect of gust fronts is that air can be lifted up at its leading edge, while the cooling and windiness are usually a bit less significant, unless, however, the winds are strong enough to cause damage. If there are insects flying around, they tend to get trapped along the leading edge of the gust front near the ground in the region of convergence where the wind speed normal to the cold pool decreases and appear as a "fine line" of enhanced reflectivity on radar (Fig. 4.51).

Figure 4.46 *Very narrow, highly sheared convective towers with just a hint of virga. (top) Very narrow, cumulonimbus with a highly sheared, ultra-narrow anvil, shedding a shallow layer of virga (hazy patch to the center, left), along the dryline in central Nebraska on June 13, 2017. (bottom) Highly sheared, narrow, orphan anvil from a decaying updraft and a new cumulus congestus tower to its rear, with crepuscular rays, on April 15, 2017, viewed to the west, in north-central Oklahoma. The crepuscular rays appear to diverge from the horizon. There may be some virga falling from the western edge of the anvil. Anticrepuscular rays are seen in the next figure.*

Photos courtesy of author.

Figure 4.47 *As in Figure 4.55, bottom panel, on April 15, 2017, but anticrepuscular rays in the opposite direction, to the east. The anticrepuscular rays appear to converge at the horizon.* Photo courtesy of author.

Gust fronts can therefore be detected easily by a radar, even if it does not have Doppler capability to provide actual wind information. Some insects to some people are pests; to others they act to pollinate flowers and keep the ecosystem in balance; but to radar meteorologists they act to render gust fronts "visible" when the gust fronts would otherwise not necessarily be detectable. Scattering by insects if there is no precipitation nearby is part of what is referred to as "clear air" radar return (there is no precipitation). In addition to scattering by trapped insects (Achtemeier, 1991) some of the radar return may also be due to dust or pollen or other particles suspended in the air.

The edge of the gust front behaves like a mountain obstacle to the air into which its moving, like a penetrating top, but unlike a mountain obstacle the gust front is *not* a solid wall, but is rather a fluid boundary. It is also similar to airflow over an "airfoil," like an airplane wing (or the airflow over the top of a convective cloud that produces a pileus). Gust fronts associated with convective storms (Figs. 4.52 and 4.53) are examples of what fluid dynamicists call a "density current" or "gravity current" (Benjamin 1968; Simpson 1997), a block of relatively dense fluid that moves through or intrudes into, but not mixing with, a region of less-dense fluid. Density currents are also found at the leading edge of some strong cold fronts (Fig. 4.54) and the leading edge of cold pools in mature MCSs (Fig. 4.55).

Figure 4.48 *Heavy precipitation forcing strong winds at the surface outward from the precipitation core (microburst), resulting in a flared-out edge to the precipitation (seen at the right), on May 16, 2021, in West Texas, southwest of Amarillo.*
Photo courtesy of author.

One can also look at the dynamics of gust fronts using vorticity, without any explicit regard to pressure, just as we noted in the first section of this chapter. Ahead of the gust front there is no buoyancy, while the air within the cold pool, behind the gust front, the air is negatively buoyant (it's cold with respect to the ambient air ahead of it). Imagine a paddle wheel placed right at the leading edge of the cold pool, along the gust front: Behind the cold pool, air is being forced downward due to the negative buoyancy, while ahead of it there is no vertical force at all. So, the paddle wheel will rotate about a horizontal axis, as shown in Figure 4.56, top panel. This represents vorticity about a horizontal axis, in this case pointing out from the page (curl your right hand in the direction of the curved line and your thumb sticks out, toward you). Air thus begins to rise right along the leading edge, just as we had concluded a bit earlier, but in this analysis, we did not have to invoke any pressure perturbations (even though they are there) to explain what forces the air to flow upward.

Finally, there is a third way of explaining the circulation produced at the leading edge of cool outflow from a convective storm, which essentially combines the first two. Consider in Figure 4.56, bottom panel, another vertical cross-section through the leading edge of a cold pool. At the very top of the cold pool, the air ahead of the cold pool has the same temperature as the air above it. Consequently, the hydrostatic pressure is the

Figure 4.49 *Double rainbows. (top) Double rainbow looming over RaXPol, in the Texas Panhandle, on May 17, 2018, on the back side of a convective storm, as rain falls out from the anvil as the sun is setting. One of my graduate students at the time, Dylan Reif, is seen running toward the radar. (middle) Double rainbow in Boulder, Colorado, on Aug. 21, 2021, as rain falls from the high base of a dissipating convective storm that has moved off the mountains, late in the afternoon when the sun angle is relatively low. (bottom) Double rainbow at sunrise over Boulder, Colorado, as snow from snow showers over higher terrain to the west is blown eastward and melts, falling as rain, on Nov. 9, 2002. View is to the west. Unlike most rainbows late in the day when dark clouds are in the background, these early rainbows appear against a bright, cloudless sky.*
Photos courtesy of author.

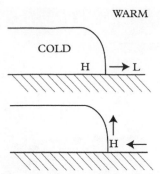

Figure 4.50 *Illustration of air motion and the distribution of pressure behind and just ahead of a cold pool of air. (top) The hydrostatic pressure is relatively high (H) underneath the dense, cold pool and relatively low (L) ahead of the cold pool, where the weight of the warmer, less-dense air is less. The left-to-right hydrostatic pressure-gradient force (directed from high to low pressure, represented by the arrow) moves the cold air to the right. (bottom) In the reference frame of the moving cold pool, air approaches from the right, but comes to a grinding halt at the leading edge of the cold air if there is no turbulent mixing there. There must therefore be a dynamic high-pressure area right at the leading edge, so that a dynamic pressure-gradient force acting to the right slows the air down as it approaches the cold air. The air is also forced up by an upward-directed dynamic pressure-gradient force above the region of high pressure.*
Figure courtesy of author.

same ahead of the cold pool as it is behind it. There is therefore *no* horizontal hydrostatic pressure-gradient force across it at the height of the depth of the cold pool. On the other, there is a hydrostatic pressure-gradient force at the surface, as shown in Figure 4.56, bottom panel. In fact, the hydrostatic pressure-gradient force directed from behind the cold pool to ahead of it decreases in intensity with height. If a paddlewheel were placed along the leading edge, it would rotate as shown and inferred in Figure 4.56.

If the air is stable and relatively humid, a smooth, laminar-looking cloud called an "arcus" or "shelf cloud" forms when the air is humid enough for condensation; the air cools to saturation when it is lifted over the cold pool behind the gust front (Fig. 4.57). These clouds look like the clouds associated with mountain waves and cap clouds. In some instances when there are extensive cold pools the leading arcus cloud may be huge, with many striations, like some mountain-wave clouds (e.g., Fig. 1.1, bottom two panels), again, probably as a result of small variations in water vapor in the vertical. In a vertical-cross-section view, the leading edge of the gust front assumes a wedge shape (e.g., Figs. 4.50 and 4.57). The reader is encouraged to view the magnificent painting by John Steuart Curry "The Line Storm," from 1935[26] to see an artist's rendition of an arcus cloud at the leading edge of a line of thunderstorms in a representation of the quintessential farming region of the Midwest region of the US. In my opinion, this is a great example of mimesis of both a natural phenomenon and an old way of life in the

[26] https://commons.wikimedia.org/wiki/File:John_Steuart_Curry_-_The_Line_Storm.jpg

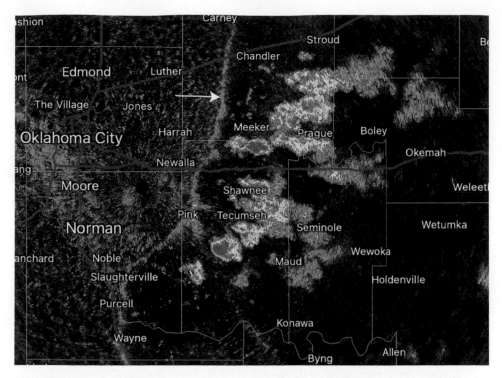

Figure 4.51 *Radar reflectivity from the WSR-88D just southeast of Oklahoma City (located at the black dot just east of Moore) during the afternoon of Aug. 30, 2021. A fine line of enhanced radar reflectivity is seen (white arrow) extending from the north-northeast to the south-southwest, just to the west of a group of rain showers (heavy rain is color-coded red/orange). The fine line marked the boundary of convergence between cooler air produced as rain has evaporated, and ambient warmer air to the west. The fine line/outflow boundary was moving westward.*
Figure courtesy of author.

US by capturing the feeling of the gust front and leading edge of a line of convective storms rolling over scenic farming land. The poem later in this chapter by Robert Frost also relays the feeling of experiencing a line of convective storms.

When a gust front has spread out far beyond the precipitation region of a storm or the precipitation has completely evaporated, the wedge shape of the leading cold outflow may take the form of a wedge of dust, called a "haboob" (Fig. 4.58). Haboobs are common in Arizona during monsoon season (Idso et al. 1972) and recently some particularly spectacular ones have been photographed near Phoenix. In Figure 4.58, top panel, however, a haboob-like feature is seen along the leading edge of a gust front even though there is still plenty of precipitation falling a short distance behind it.

The region underneath many arcus clouds looks very granular (Fig. 4.59, top panel) and has therefore been called "the whale's mouth," after the appearance of the inside of the mouth of the giant whale that swallowed Pinocchio in Walt Disney's animated

Figure 4.52 *Gust fronts in convective storms. (top, left) The gust front and leading edge of precipitation from convective storms propagating off the Foothills of the Rockies in Boulder, Colorado, on July 10, 2011. Note the small-scale arcs in the cloud base above, just ahead of the precipitation curtains. (top, right) The surreal look of a smoothly-sculpted cloud along and above the gust front of thunderstorms, viewed to the north and northwest, in north-central Oklahoma on May 15, 2009, during VORTEX 2. (bottom, left) A cloud driven by a gust front passing over Boulder, Colorado, on June 20, 2004. The cold pool driving the gust front is produced by the evaporation of rain seen to the right as virga. The edge of the gust front is seen as cloud band that curves back toward the west at its southern end. View is to the west.*

Photos courtesy of author.

musical feature from 1940. The granularity is probably an indication of turbulence along the zone of strong wind shear at the top of the cold pool, behind a gust front. Just behind other gust fronts, the cloud base appears smoother (Fig. 4.59, bottom panel). Since the leading edge of the gust front is usually wedge-shaped and the arcus cloud is above and to the rear of the leading edge of the gust front, an observer will note that the wind shifts and temperature falls *before* the bulk of the arcus cloud actually passes by overhead.

As air is lifted over a cold pool, there is strong vertical wind shear (an abrupt change in wind direction with height) at the upper interface between the cold, dense air below and the ambient, warmer air from ahead of the gust front. At this interface there are Kelvin–Helmholtz rolls (and turbulence), as mentioned earlier in Chapter 2 (but those shown were not in convective storms) (Figs. 4.60 and 4.61, bottom panel) (Droegemeier

Figure 4.53 *Striated clouds along the leading edge of gust fronts in severe convective storms. (top) A horizontally striated shelf cloud approaches the University of Massachusetts mobile, X-band, polarimetric Doppler radar, which is probing its parent storm, in the northern Texas Panhandle, on May 21, 2007. The contrast between the cloudy, bright sky outside the storm and the dark region behind the gust front is stark and foreboding, as if a spade were being driven into the atmosphere and scooping up air. View is to the north-northeast. (bottom, left) A gust front from a supercell, approaching the University of Massachusetts mm-wave Doppler radar from the northwest, in northern Kansas, on May 29, 2008. Note the complex layers of striations. View is to the west. (bottom, right) As in the bottom, left panel, but with a view to the north. This base is long, laminar, and striated like some altocumulus lenticularis are (see Chapter 2).*

Photos courtesy of author.

and Wilhelmson 1987). However, I have never seen cloud features that I could definitely associate with K-H billows to the rear of a gust front when viewed from underneath it. The leading edge of the gust front tends to have an elevated "head" and a "nose" that protrudes above the ground in the direction of motion of the gust front (Fig. 4.60), which is also seen in laboratory models of density/gravity currents.

Figure 4.54 *Panoramic view of the leading edge of a cold front passing through Norman, Oklahoma, on Nov. 21, 2013, viewed to the northwest. The density current in this case was not from a cold pool in a convective storm, but rather from the cold air brought in from Canada.*
Photo courtesy of author.

Figure 4.55 *Panoramic view of a highly striated shelf cloud associated with a mesoscale convective system, viewed to the north through northeast, during the morning of Sept. 18, 2014, from Norman, Oklahoma. This cold pool of air backed up from convective storms that were to the northeast and then moved southwestward.*
Photo courtesy of author.

4.3.4　Bores: Gravity waves triggered by density currents

Bores were introduced in Chapter 2 when we discussed gravity waves. We now briefly review what causes bores in order to refresh your memory and show a few examples of clouds most likely formed by bores generated by gust fronts from convective storms (Figs. 4.62 and 4.63) (see also Doviak and Ge 1984). When a density current hits a layer of air that is "neutrally stratified," that is, the lapse rate is dryadiabatic as one would find during the day time when the sun is strongly heating the ground, the temperature in the environment above the gust front is not changed by the air flowing up and over it: The rate of cooling of air as it is lifted matches the rate of decrease of temperature with height in the environment. However, if the air is very stable (much cooler at low levels), as happens especially at night, then air has difficulty making it up and over the gust front, just as air has difficulty making it over a mountain if it does not have enough kinetic energy (to make up for the increase in potential energy the air acquires as it is lifted up and over it). It cools at a rate which makes it cooler than that of the environment.

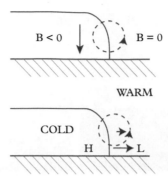

Figure 4.56 *Illustration of how a tendency for rotation (vorticity) as indicated by the circular closed, circular dashed line and arrow to develop along the leading edge of a cold pool. (top) Behind which there is negative buoyancy (B) (solid arrow) and ahead of the cold pool there is no buoyancy. (bottom) An alternative explanation: Schematic showing the induced circulation (dashed circular streamline) at the leading edge of a cold pool owing to the decrease with height of a horizontally directed hydrostatic pressure-gradient force (PGF, arrows) associated with the cold pool. The magnitude of the PGF decreases with height because the weight of the air mass above decreases with height.*
Figures courtesy of author.

If the air is thus blocked, it may simply rise just a bit and then get turned back from the direction it had come. However, under some conditions, the air rises upward at the leading edge of the gust front, but does not continue to pass rearward over the cold pool or get turned back completely from the cold pool. In this case, the depth of the relatively stable air into which its moving may rise enough to trigger what is known as a *bore* (as also described in Chapter 2) (Rottman and Simpson 1989). Bores can take a number of forms, ranging from a series of gravity waves to singular drops in pressure followed by sudden rises, whose form then propagates away from the leading edge of the gust front (Fig. 2.122) (Haghi et al. 2017). Bores are a type of gravity wave, in which air just bobs up and down and the top edge of the low-level stable layer (into which the density current is moving) *propagates* away from the cold pool, whereas density currents actually *move masses* of air. It is thought that there is a continuum of phenomena that range from density currents to gravity waves, depending on various environmental characteristics and other details about their structure, and that one should not consider density currents and gravity waves to be *completely* different phenomena (Simpson 1997). It is the zones of rising motion created in each member of the continuum that may cause lines of clouds to form as parallel, narrow bands.

Cloud features thought to be associated with bores are lines of altocumulus *castellanus* (see Chapter 3) or just altocumulus or altostratus that are triggered ahead of them. Sometimes there are many parallel lines of clouds. (See Chapter 2 for more on bores.) These tend to be seen in the morning, when the boundary layer is still relatively stable. [On satellite imagery, one sometimes sees a series of parallel cloud bands (Fig. 2.123,

Figure 4.57 *Shelf clouds. (top, left) "Arcus," also known as a "shelf" cloud, along the leading edge of a gust front from a complex of nocturnal thunderstorms, passing over Norman, Oklahoma, early in the morning of May 27, 1977, viewed approaching from the northwest. The airflow that produces a shelf cloud appears to behave like a giant shovel diverting warm, environmental air up and over the cold pool below until it reaches saturation and a cloud appears. (top, right) Arcus cloud along the leading edge of a gust front from convective storms in northeastern Colorado, on June 10, 1994, as viewed from a NOAA P-3 aircraft. A tiny portion of a rainbow is also visible. (bottom) A panoramic, composite view of an arcus/shelf cloud rapidly approaching from the west, in northwest Texas, on June 4, 2002.* Figures courtesy of author.

top panel), and on radar one can see parallel fine lines (Fig. 2.123, bottom panel) along with banded signatures in Doppler velocity (not shown).]

Some bores occur with some regularity if they are produced by convective storms tied to some topographic feature and if the low-level environment is relatively moist and stable. Some may not be triggered by convective storms at all, but by other features related to topography. In the southern portion of the Gulf of Carpentaria of northeastern Australia, a cloud feature associated with a bore (Christie 1992) is known as a "morning glory"[27] because it is seen during the morning when the air near the ground is coolest

[27] I have no photographs of my own of this phenomenon and refer the reader to do an online search; there are many available to be viewed.

Figure 4.58 *Dust storms. (top) A haboob-like feature (dust storm created by the strong winds at the leading edge of a gust front) at the leading edge of a gust front in a damaging mesoscale convective*

and most stable (Clarke et al. 1981). It is thought to be triggered by colliding sea-breeze fronts (Wakimoto and Atkins 1994; Atkins et al. 1995) or the interaction of the sea-breeze front and a nocturnal stable layer over land. The morning glory looks a bit like an arcus cloud ahead of a gust front and some orographic wave clouds. Coming out of otherwise clear skies, it can be rather foreboding looking because it has no context, coming right out of the blue.

The *third* and final *stage* of the life cycle of a convective storm is called the *dissipation stage*. During the dissipation stage, the updraft is gone completely and the storm is characterized by downdrafts only (Fig. 4.64, left panel). If air lifted up along the edges of the outward advancing gust fronts does not reach its level of free convection, then no subsequent convective clouds/storms are triggered and nothing is left but anvil material and light, stratiform precipitation. Leftover anvil material after a cell has dissipated, as noted earlier, is called an "orphan anvil" (Fig. 4.64, right panels).

The full evolution from the first through the third stages of the convective storm just described is that of an "ordinary cell." It is difficult to take photographs that illustrate *all* three stages of a *particular*, ordinary-cell convective storm. To do so requires continuous visibility of the cloud base and keeping up with the storm, if it is moving; in addition, the photographs must clearly show precipitation or the absence of it, and the top of the storm must also be visible. Owing to these difficulties, we synthesize pictures from a number of different storms, just as paleontologists combine dinosaur bones from a number of different animals to make a composite skeleton of "a" dinosaur. Some meteorologists refer to an ordinary cell as a "Byers-Braham" cell, named after its "discoverers."

An ordinary cell is a "one-hit wonder": The complete life cycle takes around 30–50 minutes or so, which is approximately the length of time it takes an air parcel to enter cloud base, climb to the tropopause, and then fall to the ground with precipitation. For example, an updraft of ~10 m s^{-1} takes ~1,200 seconds to make it up to the tropopause at 12 km, and then another 1,200 seconds to fall to the ground at the same speed. Since there are 3,600 seconds in an hour, this time span of an ordinary cell is approximately 2,400 seconds, or 40 minutes. Some of the cloud/precipitation material may remain with

---

Figure 4.58 (Continued) *system in the northern Texas Panhandle on May 27, 2001. We tried to stay ahead of the wedge-shaped wall of blowing dust while we were probing the predecessor storm with a mobile Doppler radar from the University of Massachusetts, seen, in resting mode (with the antenna pointing upward) along with one of my graduate students, who is photographing it. Note the distinctly different-looking layers of clouds above it, the lower of which is cumuliform in appearance while the upper ones look like shelves. (bottom) As in the top panel, but later on as the gust front outran the clouds above it and a wedge of blowing dust (haboob) crossed the road to the east ahead of us. Needless to say, we did not make it by without being blasted by the wind and dust. Strictly speaking this photograph shows a true haboob because it is far ahead of any precipitation, while the feature shown in the previous figure is just haboob-like because it is adjacent to copious amounts of precipitation and its parent storm is still active. View is to the east.*

Photos courtesy of author.

Figure 4.59 *Behind the leading edge of gust fronts. (top) The "whale's mouth" behind a gust front of the storm is shown in Figure 4.63 on May 15, 2009, during VORTEX 2., along with the University of Massachusetts W-band Doppler radar. The cloud base arches overhead and there are scattered areas of thinner cloud, where more light is getting through the cloud. (bottom) Behind an arcus cloud (at the right) at the leading edge of outflow in central Kansas on May 8, 2021, with a view to the east; the cloud base here does not look like that of a "whale's mouth," but instead is rather smooth. A precipitation core is seen at the far left. The most unusual aspect of this image is the hole in the stratiform cloud base marking the outflow, above which air is being lifted. The origin of the hole is unknown, but it is vaguely reminiscent of a science fiction movie I once saw in which aliens abducted Earthlings from an apparent hole in the sky, like the real one seen here.*
Photo scourtesy of author.

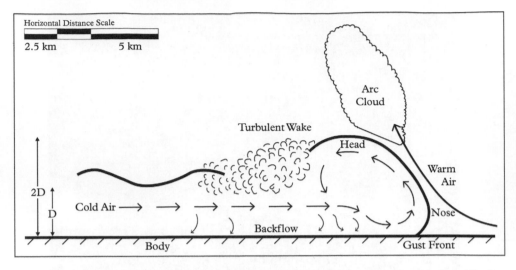

Figure 4.60 *"Curses, foiled again"*[a]: *Airflow over a density current in a gust front that looks like flow over an airfoil. Schematic of the vertical cross-section through a gust front, which is moving from left to right. From Reif et al., 2020, their Fig. 1.* © *American Meteorological Society.*

[a] Uttered frequently by the villain Snidely Whiplash in the late 1950s and early 1960s popular children's cartoon show on television in the US called *The Adventures of Rocky and Bullwinkle.*
Used with permission; based on a number of earlier figures from other journal articles.

the anvil (as orphan anvil) and not fall out to the ground, but eventually sublimate or evaporate back into the atmosphere as water vapor.

The term "cell" these days has been expanded from referring not only to the radar echo, but also the updraft associated with a convective storm. Since it takes an updraft to produce precipitation and the radar-observed "cell," there is a one-to-one correspondence between the two. One typically tracks convective storms by the radar-echo "core," but in terms of the basic physical mechanism responsible for producing the storm, it is the updraft that is tracked, though this is done mainly from computer simulations, not from observed radar data; it is not usually possible to pinpoint the updraft based on the radar-echo core alone.

4.3.5   Regeneration of cells along gust fronts

In the ordinary-cell classification, the two basic physical processes that explain the behavior of most of these convective storms and which make up their building blocks are recognized as: (1) the buoyancy in an updraft and (2) precipitation falling into unsaturated air, producing a cold pool, which is characterized by negative buoyancy. It was recognized early on, though, that convective storms usually persist for much longer than the span of one ordinary cell. Convective storm systems frequently take the form of a sequence of ordinary cells, one, or more than one, right after another. These complexes of cells are known as *multicells* (Figs. 4.65–4.68).

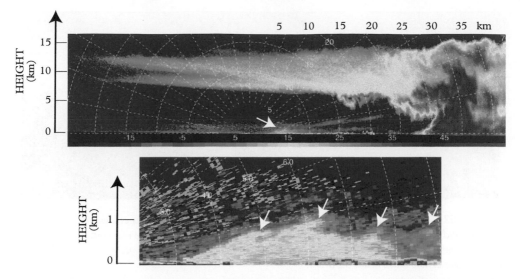

Figure 4.61 *Vertical cross sections of radar reflectivity from RaXPol, looking toward the west-northwest, (top) of a mesoscale convective system in western Kansas at night on July 15, 2015 at 06:11 UTC (01:11 a.m. CDT), during the PECAN (Plains Elevated Convection At Night) field experiment. See Reif et al. (2020, Mon. Wea. Rev. 77, 3683–3700) for more details on the storm and data collection, although the figure here did not appear in the publication. The color scale at the bottom of the top panel is given in dBZ; the red regions are indicative of heavy precipitation, while the white and green areas are indicative of ice crystals or small water droplets or a mixture of both, particularly in the anvil streaming off to the left (toward the east-southeast). The bottom panel is an expanded view of a portion of the upper panel. The white arrow points to the area expanded below (bottom), which shows Kelvin-Helmholtz rolls/billows (white arrows) at the top of the gust front, way out ahead (to the left, east-southeast), of the main part of the precipitation, which was about 25–30 km away. Some virga is seen (top) falling from the anvil at about 15–20 km range. The gust front was moving from right to left, toward the east-southeast.*

Figure courtesy of author.

Multicell convective storms tend to form when there is low-level vertical wind shear over the depth of the cold pool such that the horizontal vorticity that travels from the environment into it (Fig. 4.69) acts in the *opposite* direction to that produced at the leading edge of the gust front by a horizontal gradient in buoyancy (as described earlier). When the rate at which horizontal vorticity associated with the low-level vertical shear in the environment approaches the gust front is exactly counteracted by the rate at which horizontal vorticity is being generated by the buoyancy gradient at the leading edge of the gust front, the air at the leading edge of the gust front tends to move straight up and is therefore most likely that air parcels will be lifted to their level of free convection, thereby triggering new convective clouds and storm cells. Many years ago, radar meteorologists noted these new cells in multicell storms and some named the clouds associated with them "daughter" clouds, the offspring of the original parent cell. Rich Rotunno, Joe Klemp, and Morris Weisman, scientists at NCAR, in the late 1980s first proposed the

Figure 4.62 *Cumulonimbus associated with bores. A series of bores (visible as evidenced by parallel, smooth cloud bands at the left) generated by the gust front on the southern side of a convective storm, viewed to the north, reflected off Lake Thunderbird, in Norman, Oklahoma, on April 25, 2017. Mamma are visible underneath the anvil.*
Photos courtesy of author.

importance of vertical shear and the cold pool (Fig. 4.69). Their theory is known as RKW theory, in honor of the authors of their seminal paper published in 1988.[28] When the vertical wind shear is too weak, the air flows less vertically, so it is less likely that air will be lifted all the way to its level of free convection and trigger new convective clouds and storm cells. If the vertical shear is too strong (or from the "wrong" direction, in that the horizontal vorticity that is imported from the environment is of the *same* sign as that of the horizontal vorticity generated along the gust front, not the opposite), then the air flows backward with respect to the advancing gust front and it also less likely that air will be lifted all the way to its level of free convection and trigger new convective clouds and storm cells.

Rob Fovell and his graduate students at UCLA in the early-to-mid-1990s used numerical simulations to show how the timing of new cells in an idealized, multicell convective storm depends on the strength low-level vertical shear in the environment and amount of negative buoyancy in the cold pool (Fovell and Tan 1998). When a new cell is triggered along a gust front, it moves rearward with respect to the gust front (because the gust front moves from the cold pool toward the warmer, environmental air), precipitation forms, and the cold pool is reinforced as more air is evaporatively cooled (Fig. 4.70). The colder the cold pool, the faster the gust front advances. If it is too cold, the gust front outruns the new cell, which does not have as much access to warm, humid surface air. If the cold pool is too weak, there may not be enough vertical motion produced at the leading to trigger new convective cells. The new convective cells produce compensating sinking motion around their edges, which acts to suppress any new convective cells from being triggered until the older cell has gotten out of the way and moved to the rear

[28] Rotunno, et al., 1988.

Figure 4.63 *A complex set of parallel bands of clouds approaching Boston, Massachusetts from the northwest, with a view of Cambridge, from atop the Prudential Center tower in the Back Bay section of Boston, on Aug. 4, 2015. I was attending a scientific conference at the time on a floor below and rushed up to the top of the building to get this unobstructed view. (I had an inkling of what might be coming because I had been sneaking peaks at the local National Weather Service radar display on my cell phone during talks by colleagues). Convective storms are visible in the far distance, to the northwest. Ahead of the storms near the ground, is a separate cloud, a roll cloud, formed along the edge of a surge of cold air. Closest to me (in the upper half of the photograph) is a line of altocumulus* castellanus, *probably driven by a gravity wave/bore above the low-level cool air from the thunderstorm outflow cold pool.*
Photo courtesy of author.

of the gust front. Timing is everything. Because new cells are triggered discretely and periodically (or quasi-periodically), the storm complex, as a whole, lasts much longer than any ordinary cell by itself.

When you look at a multicell storm, you can see convective clouds in varying stages of their lives. Looking at a multicell storm is like looking at a series of snapshots of ordinary convective storms at different stages in their lives, from the cumulus stage through the mature stage to the dissipation stage (Fig. 4.65, from left to right, respectively). In a way it is like looking at tree rings to look at the changes in the amount of annual growth and relating it to annual variations in temperature and or moisture. The new and old clouds

Figure 4.64 *The dissipating stage of a convective storm. (left) Idealized depiction of the dissipating stage of an ordinary-cell convective storm in an environment of no vertical wind shear. The top of the cloud, the remnants of an anvil, is neutrally buoyant (B = 0), while the area underneath is characterized by negative thermal buoyancy (B < 0), a downdraft, which spreads out at the ground, and by precipitation. Figure courtesy of author. (middle) An "orphan anvil" over the Foothills, Boulder, Colorado, Aug. 14, 2021. Some ice crystals are falling out from this anvil and melting or sublimating. The anvil used to be connected to a positively (in the upward direcction) buoyant convective tower. The edges of the anvil are patchy rather than solid, probably indicative of the weakness of the former updraft. (far right) As in the top panel, but for three orphan anvils in south-central Kansas on May 30, 2022.* Photos courtesy of author.

tend to line up along what is known as the "flanking line" (Lemon 1976) (Fig. 4.65 and 4.68, left side of convective storm).

4.3.6   Lowered cloud bases: Wall clouds and scud

When new cells form along a gust front in a multicell storm, colder, more humid air from neighboring outflow may get ingested into the nearby, new updraft (Fig. 4.71). When this happens, one often sees a scud cloud, a ragged-looking piece of cloud, that rises up to a cloud base and where it connects itself so that it appears as if the cloud base has lowered. Storm chasers often refer to these features by the colloquial name "lowerings," as more humid air is forced upward and the cloud base is effectively lowered. The lowerings are visual evidence that there is an updraft above. We will see subsequently that in supercells these lowerings, especially when they rotate, are called "wall clouds." Lowerings are ubiquitous in multicell convective storms and many storm chasers get a bit overly excited when they see them. Enthusiasm should probably be curbed, unless a lowering is rotating about a vertical axis.

4.3.7   Cloud electrification

In ordinary cells and multicells (and also in supercells, to be described soon), after precipitation has begun to fall, electrically charged precipitation particles may separate enough that the electrical potentials (gradients of electrical charge) increase to the point

ANATOMY OF A MULTICELL CONVECTIVE STORM

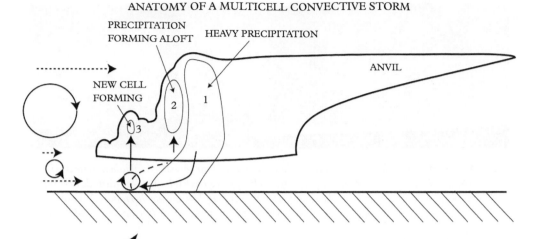

Figure courtesy of author.

Figure 4.65 *Idealized vertical cross-section through a multicell convective storm with a view, for example, from the southwest (left) to the northwest (right). The environmental winds are given by the dashed vectors (arrows) at the left. If the winds are strongest from the southwest aloft, weakest from the southwest near cloud base, and from the southwest, but intermediate in speed between that aloft and that near cloud base, then rotation is induced in the clockwise direction above cloud base and in the counterclockwise direction underneath cloud base. The outlines of the thin, solid lines surround regions of high amounts of raindrops, possibly mixed with ice crystals high up, and which can be seen as moderate to high radar reflectivity. Cell #1 forms from an initial updraft as rain and snow form aloft, the snow melts as it falls, while the entire liquid mess falls as heavy rain. Some of the lighter/smaller rain and snow and cloud material are transported to the right (perhaps out of the page, to the northeast) to produce an anvil because their terminal fall speeds are less and they remain lofted. On the way down, the precipitation becomes even more negatively buoyant as some of it evaporates and cools underneath cloud base, producing a cold pool, the leading edge of which is shown by the long-dashed line, and moves to the left (southwest). If the counterclockwise rotation induced the vertical winds shear in the environment underneath cloud base, shown at the left, is exactly counterbalanced by the clockwise rotation induced by the gradient in buoyancy at the leading edge of the cold pool, then air will be lifted most efficiently over and at the leading edge of the cold pool, triggering a new cell, cell #2 defined by precipitation aloft. Once the precipitation from cell #2 begins falling, a new cell #3 may then form ahead of cell #2, and so on. Thus, new cells are periodically produced. The "flanking line" consists of the cumulus and cumulus congestus whose tops increase in height to the right, culminating in the tallest (and oldest) to the right, above the heaviest precipitation. The "flanking" line is the section of growing cumulus and cumulus congestus towers on the left-hand side (southwestern edge).*

(the intensity of the electrical field is proportional to the magnitude of the gradient in charge) at which there is a breakdown of the air as an insulator and electrical currents flow along channels, resulting in lightning flashes (see Rust and MacGorman 1998 for a general review). In a sense, the buildup of electrical potential and its breakdown is like the buildup of potential instability as the difference in temperature between the air

Figure 4.66 *Panoramic view of a multicell convective storm in northwestern Oklahoma, near Aline, on May 18, 2019, with a view to the west. An arcus cloud is seen extending at low altitude over most of the eastern portion of the storm, while heavy precipitation is found on the northern half. New convective towers are growing on the southern edge of the storm, as warm, moist air is lifted over the cold pool.* Photo courtesy of author.

near the ground and the air aloft increases, resulting in a buoyant updraft, though in a more dramatic fashion. The channels heat up rapidly, producing sound waves that are detected as the rumbling of thunder.

Lightning flashes that are seen to travel from the cloud to the ground (Fig. 4.72) are called "cloud-to-ground lightning," and from one portion of the cloud to another, without hitting the ground (Fig. 4.73), "cloud-to-cloud lightning." When lighting occurs inside a cloud, but you can't see the branching channels, the lightning is called "in-cloud lightning." Sometimes you can see part of the channel, but neither end (Fig. 4.74). Lighting is often seen branching out, creeping along the underside of an anvil (Fig. 4.12, bottom panel); this type of lightning has been referred to by some storm chasers as "spider" lightning or, when seemingly creep along slowly, "anvil crawlers." One sometimes sees cloud-to-ground flashes followed by anvil lightning branching outward from the storm. Lightning makes it possible to see tornadoes and funnel clouds at night that would otherwise not be visible (Fig. 5.25). Lightning also illuminates convective towers in storms at night (Figs. 4.12; 4.13, top; 4.74) that would otherwise be difficult to see except by moonlight.

The lightning channel emits electromagnetic radiation (which perhaps is nature's ironic response to all the radiation that is sent through convective storms by radars) and is detectable by receivers tuned to certain relatively low radio frequencies. Using receivers at more than one location, the location of the lightning discharge can be pinpointed using the differences in the time the radiation from the lightning is detected by different receivers at different locations, and from geometrical considerations. There are operational lightning detection networks in various countries that make use of this technique.

Most cloud-to-ground lightning strokes carry negative charge to the ground, but a small minority of them carry positive charge to the ground. The most impressive and scary lightning flashes are those that seem to travel from the anvil to the ground (Fig.

Figure 4.67 *Radar depiction of an east-west oriented severe multicell convective storm propagating southward, on Aug. 26, 2019 in central Oklahoma. 0.5° elevation angle scan showing the gust front as a fine line of enhanced reflectivity (color-coded blue) deformed into an arc shape. New cells kept popping up behind the gust front, attaining intense radar reflectivity (red).*
Figure courtesy of author.

4.75), away from the storm's precipitation and updrafts. These seem to strike "out of the blue." Just because you are not near the storm and its precipitation does not mean you are completely protected from getting struck by lightning. These anvil-to-ground lightning strokes often carry positive charge to the ground. Many storm chasers have been jolted (not literally, just emotionally) by anvil-to-ground strikes and dealt a wake-up call while watching a storm (supposedly safely) from a distance.

The mechanisms by which charge is separated seem to require a mixture of liquid water droplets and ice particles such as graupel (soft hail), and the presence of down-drafts. For this reason, cumulonimbus clouds in the tropics that are very shallow do not

Figure 4.68 *Example of a multicellular convective storm evolving over central Nebraska on June 13, 2017, with a view to the west. The new cells are being triggered on the left (south) end of the complex of cells. The anvil gets fuzzier and fuzzier off to the right (north), indicating that as the anvil gets older and older, it becomes more and more diffuse. The flanking line is located here at the left of the complex, on its southernmost side.*
Photo courtesy of author.

produce lightning because their tops are so low that ice particles cannot be produced and also downdrafts may not be very strong. To get lightning in the tropics, the storm tops need to be high enough that some ice particles are produced. Lightning also seems to be more prevalent in general over continental areas than over the oceans, perhaps because the relatively cool temperatures over the ocean compared with those over the land (at least during the summer) are associated with weaker buoyancy, weaker updrafts, and downdrafts. Often, we experience the sudden onset of a brief period of enhanced lightning activity just before a convective storm unloads its precipitation as a gush of rain and hail.

Sprites are lightning flashes that occur at very high altitudes and are much rarer than ordinary lighting. I have never seen a sprite and refer the reader to an online search to find good examples, some of which are quite spectacular. They can be found above electrically charged cumulonimbus clouds.

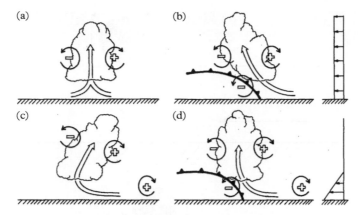

Figure 4.69 *Idealized schematic of how the intensity of the cold pool and the low-level, vertical shear (below cloud base), both affect the tilt of the circulation in a convective storm. For (a) and (b), there is no vertical wind shear (see the vertical profile of the winds on the right). For (c) and (d), there is vertical wind shear (see the vertical profile of the winds at the right). In (a), we begin with a buoyant updraft, but no precipitation or cold pool have formed yet. Rotation induced around the sides of the cumulus congestus as a result of the positive buoyancy in the cloud is as shown and the updraft induced in between is erect. However, in (b), precipitation has formed, and a cold pool has developed. The winds transport the cumulus congestus to the left over the cold pool while the cold pool itself moves to the right, which induces rotation at the edge of the cold pool in the same sense as that on the left side of the buoyant updraft aloft. The result is that the cumulus congestus leans to the left with height and eventually the cloud stops growing as the updraft in the buoyant cloud gets tilted over more and more and the LFC is never reached. In (c), however, we also begin with a buoyant updraft, but owing to the vertical shear in the environment below cloud base, the updraft gets tilted to the right, in the direction of the low-level vertical shear.* But as precipitation develops and falls out, a cold pool develops and induces rotation along its leading edge which is in the counterclockwise direction. In this case, however, the counterclockwise rotation induced at the leading edge of the cold pool is counteracted by the clockwise rotation induced by the vertical wind shear in the environment and the updraft becomes more erect and more likely to produce a new cell, and so on. Thus, multicell growth is encouraged when there is low-level wind shear of an appropriate magnitude and direction.

(From Rotunno et al. 1988, their Fig. 18. © American Meteorological Society. Used with permission.)

4.3.8 Summary of ordinary and multicell convective storms

Convective storms in the absence of vertical shear are called "ordinary" cells and owe their existence to positive buoyancy and a cold pool produced in large part by the evaporation of precipitation into unsaturated air, but also by water loading and melting. When there is some vertical shear at low levels, it is possible that lift along the gust front produces a long-lived period of a series of ordinary cells. In some midlatitude continental areas during the summer ordinary and multicell convective storms are prevalent when the deep—(surface to the tropopause) layer shear is weak; these are sometimes referred to as "garden-variety" storms or "air-mass" storms.

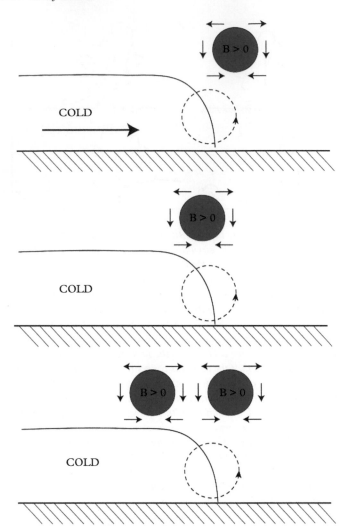

Figure 4.70 *Idealized illustration of how cells are triggered periodically along the leading edge of a gust from a convective storm, in the absence of vertical wind shear (for the sake of simplicity). In a reference frame moving with the gust front. In (top), a buoyant bubble (red circle) is triggered along the leading edge a cold pool as air is lifted to its LFC by the rising branch of the circulation induced at the leading edge of the cold pool. In (middle) the buoyant bubble moves rearward (to the left) with respect to the leading edge of the cold pool because the cold pool moves to the right. Upward motion at the leading edge of the cold pool is suppressed by the sinking branch of the vertical circulation generated at the leading edge of the buoyant bubble. In (bottom), a new buoyant bubble is triggered after the previous buoyant bubble has moved sufficiently rearward that the downward branch of the vertical circulation associated with the bubble (to its right) no longer counteracts the rising branch of the vertical circulation produced at the leading edge of the cold pool.*

Figure courtesy of author.

Figure 4.71 *The southern end of a multicell convective storm that later evolved into a supercell in western Kansas, on May 24, 2021. View is to the northwest. The flow from the colder, more humid outflow under cloud base from the precipitation area to the right into the updraft, marked by the dark cloud base, is shown by the curved arrow. The amounts by which the updraft cloud base appears to have been lowered is also indicated by the double-arrowed, short, line segments.*
Photo courtesy of author.

During the springtime especially and late spring/early summer in midlatitudes and during midsummer at higher latitudes (where the polar jet is found), the deep-layer (from near the ground to the tropopause) vertical shear can become rather strong. These conditions can also occur in the fall, but with less frequency, and also during the winter, primarily at lower latitudes. Under these circumstances, the convective storms can become more organized, more intense, and produce severe weather that is more capable of doing damage and causing harm.

Convective-storm features such as growing cumuliform towers, updrafts, anvils, penetrating tops, gust fronts along the leading edge of cold pools with arcus/shelf clouds, downdrafts, flanking lines, lowered cloud bases, lightning and thunder, rain and hail, and strong surface winds and heavy precipitation are the hallmarks of most convective storms, though not all of them exhibit all the features just mentioned. The three common features that occur everywhere under all conditions are cumuliform, buoyant towers, updrafts, and downdrafts.

Figure 4.72 *Cloud-to-ground lightning in Norman, Oklahoma, on Aug. 21, 1979.*
Photo courtesy of author.

4.4   Supercells

In the late 1950s and early 1960s, radar meteorologists noticed that some convective storms, unlike ordinary-cell storms, seemed to last for a much longer period of time. Multicell storms can last for a long time, hours, but the individual cells themselves still persist for less than an hour (Ludlam 1963). A series of a succession of discrete cells in multicell convective storms sometimes occurs in such rapid succession that there is

Figure 4.73 *Cloud-to-cloud lighting over Norman, Oklahoma, on Sept. 2, 1979. There are also several cloud-to-ground lightning flashes in the distance on the left side of the image, partially hidden by precipitation.*
Photo courtesy of author.

a nearly *continuous* succession of updrafts or cells for a very long time, at least an hour and often even longer. Furthermore, they tend to move along not with the mean wind (averaged over the depth of the troposphere by mass) as ordinary cells do, but instead, owing to the formation of new cells adjacent to older cells. *propagate.* They thus give the *appearance* of motion, just as a gravity wave *appears* to move, but is really just propagating. Based on what I have seen in the scientific literature, there appears to have been an obsession with these types of storms in the 1960s. Multicell storms in an environment of vertical wind shear (both deep and shallow) were found to tend to move to the right of the mean winds and to be more intense or severe than those under the influence of only weak or non-existent shear. They were considered to be more organized in the sense that there was a place that cumuliform clouds were triggered, moved, and the location of new radar-echo cores with respect to earlier ones seemed to follow certain rules; precipitation that fell out from the convective storm tended not to fall out into updrafts and thereby weaken them. Chester Newton at NCAR was one of the researchers who early on was prominent in trying to explain the behavior of multicell convective storms.

A new animal in the convective-storm zoo was discovered about the same time and named by the pioneering, British meteorologist, Keith Browning, at the University of Reading, who analyzed an historically important severe hailstorm that struck just outside of London, at Wokingham,[29] in 1959, and who visited the National Severe Storm

---

[29] I remember once being on public transportation (probably a train), most likely en route to London from Reading, when I got very excited when passing through Wokingham, owing to its importance to the history

Figure 4.74 *Lightning in south-central Kansas on May 7, 2002. The flanking line (left), anvil, and penetrating/overshooting top are all visible.*
Photo courtesy of author.

Laboratory (NSSL) and documented a similar storm in western Oklahoma in 1963, both of which exhibited unusual characteristics (Browning and Donaldson 1963; Browning 1964). In the spirit of Superman, the Superbowl in the US, the Superball, and Super-glue, he named the special convective storm the "Supercell." In an e-mail he told me that he was originally derided for this, and like other brilliant meteorologists, such as Ted Fujita, he faced an uphill battle in getting the scientific community to accept his novel ideas, especially ideas that were based mainly on observations alone and without an as yet understood firm theoretical basis. The discovery of a new beast is exciting, but it can elicit condescension (as opposed to condensation) from the scientific establishment.

4.4.1   Dynamics of supercells: Why they don't move along with the wind, why they rotate, and what causes their unique architecture

The key to understanding long-lived multicells and supercells is recognizing the effect of deep vertical shear on a buoyant updraft as Chester Newton had first tried to do in the 1950s (Newton and Katz 1958; Newton and Newton 1959; Newton 1963; Newton and Fankhauser 1975). In long-lived multicells, shear promotes the persistent development of a series of discrete or nearly continuous new convective cells. Suppose that there is

of severe-storm meteorology. My attempt to pronounce it correctly failed and I was politely informed of the proper, local pronunciation by another passenger, native to the area.

Figure 4.75 *Anvil-to-cloud or ground lightning ahead of a multi-tiered supercell, southwest of Norman, Oklahoma during the early evening of April 3, 2003. The lightning may be striking the ground, but the flashes are hidden from view by clouds near the base of the storm.*
Photo courtesy of author.

an easterly wind at low levels and a westerly wind at upper levels of the troposphere, in the presence of a buoyant updraft. In the absence of horizontal turbulent mixing, momentum is conserved. Let's ignore horizontal turbulent mixing because it should be most significant only right at the edges of the updraft, but not well inside it. If the updraft is relatively wide, then the conservation of momentum is a pretty good assumption for most of the updraft. (What controls the width of the updraft is another story, perhaps not completely told yet. It might have something to do with the height of the cloud base: the higher the cloud base, the wider the updraft. Or perhaps, not.) The updraft will carry easterly momentum aloft (e.g., a piece of wood moving to the west, if carried upward, would continue to move to the west at the same speed as below), so inside most of the updraft the winds would be from the east. Suppose that the updraft moves with the mass-weighted environmental airflow, which is nearly stationary if the easterly wind below is of nearly the same speed as the westerly wind aloft (Fig. 4.76). If mixing of momentum inside the updraft with momentum in the environment, outside the cloud, is ignored, then air inside the updraft moves from east to west, as it also does at the base of the cloud. It follows then that an air parcel at the eastern edge of the updraft at midlevels must accelerate to the west as it enters the updraft. The only way the air can accelerate to the west is if it experiences a dynamic pressure-gradient force that acts from east to west. This pressure-gradient force is consistent with relatively low pressure on the eastern edge of the cloud, assuming that the pressure perturbations are negligible far to the east and far to the west of the updraft. At the surface, there are no accelerations and therefore there are no perturbation pressures at all. So, below midlevels there is an upward-directed dynamic-pressure-gradient force acting upward on the *downshear* side (in the direction of the vertical shear vector, which is directed from west to east, i.e., westerly vertical shear) of the updraft, to its east. It is easy to see that the opposite happens on the upshear side, to the west of the updraft. These opposite-direction vertical pressure-gradient forces therefore act to lift air and encourage new cell formation just to the east of the updraft and suppress new cell formation just to the west of the updraft.

The interaction between the updraft and the vertical shear thus promotes propagation of the convective storm in the downshear direction. So, the effect of unidirectional deep-layer shear is to keep the convective storm going by triggering new cells on the downshear direction, ahead of the old updraft. This effect is separate from whatever additional effect the low-level shear has on the development of new cells along the leading edge of the gust front. The effect of unidirectional shear is most prominent when the storms are lined up as a "squall line," that is, if they are two-dimensional. The effect of vertical shear on a buoyant updraft of promoting relatively low and high pressure on the downshear and upshear sides of an updraft, respectively, are "linear" because it can be shown mathematically that it depends on the vertical wind shear by itself, not on a product of the vertical shear with itself or with the wind.

The upward transport of momentum from low levels throughout the depth of the updraft can be used to explain why shear doesn't necessarily tear an updraft apart: the winds inside the updraft should remain nearly constant with height. When an updraft, however, is not strong, the vertical transport of momentum is slower, so that the wind inside the updraft aloft is more characteristic of the winds aloft, and consequently the updraft may indeed get ripped apart by strong shear (Fig. 3.39). This ripping apart of an updraft is reminiscent of a gruesome painting by an unidentified Flemish artist I have seen numerous times, with some horror, at the Museum of Fine Arts in Boston ("Martyrdom of Saint Hippolytus"), which showed an unfortunate individual being ripped apart by four horses "quartering" him in different directions.[30] Vertical shear in the atmosphere is like horses pulling an object (in this case, poor St. Hippolytus) in opposite directions.

---

[30] https://collections.mfa.org/objects/33760

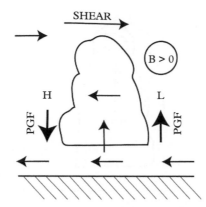

Figure 4.76 *Illustration of what happens when an updraft in a cumulus congestus interacts with vertical wind shear in its environment. The winds underneath cloud base in this idealized case are from right to left, or from the east. The winds at the top of the cloud are from the west, so the vector difference between the wind aloft and the wind near the surface is from the west . . . yielding* westerly *vertical wind shear. Inside the cloud, easterly momentum is transported upward by the updraft, so at midlevels the wind inside the cloud is from the east. Since there is no air motion at all in midlevels in the environment, air must accelerate to the west as it enters the cloud at midlevels on the right and decelerate as it exits the cloud on the left. The only way this can happen is if there is a dynamic pressure-gradient force acting from right to left (from the east) at the right edge of the cloud at midlevels and if there is a dynamic pressure-gradient force acting from the left to the right (from the west) at the left edge of the cloud at midlevels. If there are no pressure perturbations far away from the cloud, then there must be a relatively high dynamic pressure on the left side of the cloud at midlevels and a relatively low dynamic pressure on the right side at midlevels. Under cloud base the winds are uniform and from the east, so there are no accelerations and pressure perturbations or horizontal pressure-gradient forces. So, on the* downshear *side of the cloud there is an upward-directed dynamic pressure-gradient force (PGF) acting upward below midlevels, while on the* upshear *side there is a downward-directed dynamic pressure-gradient force acting downward. A new buoyant cell may be triggered if air is lifted to its LFC, on the downshear, eastern side. The convective storm will therefore propagate in the downshear direction, from west to east, while the storm as a whole does not move at all. (Note that we have not considered what happens if there is cold pool from precipitation. It would have a separate, additional effect)* Figure courtesy of author.

But this effect of vertical shear on a buoyant updraft in two dimensions is not the whole picture. When there is vertical shear and a buoyant updraft, the updraft *tilts* the horizontal vorticity (a measure of the tendency of the air to rotate about a horizontal axis) associated with the vertical shear of the wind in the environment, so that counter-rotating vertical columns of vortices form alongside the updraft. The tilting occurs when there is a gradient in updraft strength, which is found between the outer edges of the updraft (where the vertical velocity is zero) toward the center of the updraft (where the vertical velocity is the greatest). Updrafts in supercells can be as fast as 50 m s^{-1} or even more. In Figure 4.77, top panel, for example, we see how in westerly shear, a cyclonically rotating (in this case in a counterclockwise direction) region forms at midlevels at

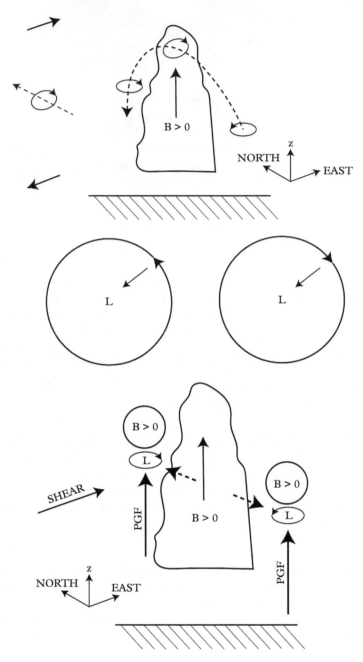

Figure 4.77 *Why supercells in an environment of uni-directional vertical shear split into symmetric parts that propagate away from each other. (top panel) Illustration of the idealization of a buoyant cloud tilting horizontal vorticity onto the horizontal plane, producing vertically oriented, counter-rotating vortices. The winds (arrows on the left) aloft are from the west (upper left) and from the east at low*

the equatorward edge of the updraft while an anticyclonically rotating (in this case in clockwise direction) region forms at midlevels at the poleward edge of the updraft. Each rotating column of air is characterized by relatively low dynamic pressure at its center at all levels. This is necessary for the air to rotate, by always accelerating air toward the left in the cyclonic vortex and to the right in the anticyclonic vortex (Fig. 4.77, middle panel).

Since at midlevels each vortex is characterized by relatively low dynamic pressure and the strength of the vortices is maximized at midlevels, where the combination of vertical shear and lateral variation in updraft strength is the greatest, there is an upward-directed dynamic pressure-gradient force on the right side of the shear vector, on the equatorward side of the updraft and also on the left side of the shear vector, on the poleward side of the updraft (Fig. 4.77, bottom panel). So, new cell growth tends to be forced, according this tilting mechanism, *both* to the right and left of the updraft. Thus, the convective storm cannot be two-dimensional when there is strong shear and buoyancy: there must be a three-dimensional aspect to it owing to the tilting of the vorticity associated with the vertical wind shear. It is thus noteworthy that vertical shear in one direction is responsible for adding a dimension to the convective storm in a direction *normal* to the shear vector.

Figure 4.77 (Continued) *altitudes (lower left); the vertical shear vector, the vector difference between the wind aloft and the wind below is from the west. If a paddlewheel were placed in the flow, it would rotate in a clockwise direction, as indicated on the left. The dashed arrow going through the circular, solid arrow indicates the sense of rotation (vorticity) that would occur indicated by the curl of one's right hand, with the dashed arrow pointing in the direction of one's thumb. A positively buoyant cloud (B > 0) forces an updraft (solid arrow pointing upward). If there is no vertical motion outside the cloud, then the dashed arrow marks the line in which the vorticity vector associated with the environmental vertical shear is bent by the updraft, so that the arrow points downward to the north and upward to the south, to the right of the vertical shear vector. The downward—directed dashed arrow is associated with the clockwise-turning winds of a vortex and the upward-directed dashed line is associated with the counter-clockwise-turning winds of a vortex. In the Northern Hemisphere, the former is an anticyclone and the latter is a cyclone. (middle panel) Illustration of how air in both counterclockwise (left) and clockwise (right) rotating vortices are accompanied by centers of low dynamic pressure. In this example, each circularly symmetric vortex is forced by a dynamic pressure-gradient force (arrows) acting toward the left (for the counterclockwise-rotating vortex) and right (for the clockwise-rotating vortex). (bottom panel) Illustration of how the convective storm in an environment of westerly vertical wind shear forces new cell growth both to the left and right of the vertical wind shear vector (to the north and to the south). Both the clockwise and counterclockwise vortices are associated with centers of low pressure, while at the ground there is no dynamic-pressure perturbation. So, there is an upward-directed dynamic perturbation pressure-gradient force (PGF) acting upward beneath each vortex. Rising air forced upward may reach its LFC, triggering regions of positive buoyancy (B > 0) to the north and south of the original, buoyant convective cloud. Thus, the original cell appears to split as it propagates both northward and southward (dashed arrows). The production of precipitation and a cold pool are not taken into account in this illustration.*

Figures courtesy of author.

The process of new cell growth on the flanks of the first updraft results in a split-ting of the updraft into two, symmetrical parts. Storm splitting was first documented in nature by conventional radar in the 1960s and was much puzzled about (e,g., Charba and Sasaki 1971). Early theories for the splitting involved using spinning baseballs or rotating solid cylinders as analogs, though incorrectly, because convective storms are *not* solid objects. This effect of vertical shear acting on a buoyant updraft is nonlinear because it turns out that it depends on the product of vertical vorticity with itself. This nonlinear effect turns out to be even more important than the linear effect in most cases. So, the nonlinear effect causes a convective storm to split into two symmetrical parts, as propagation occurs normal to the deep-layer shear, while the linear effect causes a convective storm to propagate in the downshear direction (i.e., in the direction of the vertical-shear vector) (Klemp and Wilhelmson 1978; Rotunno and Klemp 1982, 1985; Klemp 1987; Davies-Jones 2002). Storm splitting is most easily seen in a sequence of radar images from a conventional radar (Fig. 4.78), as the radar-echo core seems to split like a biological cell undergoing mitosis (Bluestein and Sohl 1979).

Sometimes there can be a series of multiple splits. The parent storm of a devastating tornado that leveled the town of Greensburg, Kansas on May 4, 2007 (Tanamachi et al. 2007), split three times before it spawned a number of tornadoes.[31] It is rare to be able see the separate cloud bases of a splitting storm, but I have seen them a number of times (Fig. 4.79). You can see in Figure 4.79 how the cloud features tend to be somewhat symmetrical about the line separating the two clouds. A short time later, the right mover had completely separated from its companion left mover (Fig. 4.80).

When storm splitting occurs in adjacent storms there can be collisions between indi-vidual cells that have evolved from the storm split; the collisions are like the opposite of splitting but I have yet to get any striking photographs of storm collisions, probably because precipitation often obscures the collisions. I have, however, seen many radar images that document storm collisions. When a line of closely-spaced storms is trig-gered, the left-moving splitting storm from one storm may collide with the right-moving splitting storm from an adjacent storm (Fig. 4.81).[32] Unlike billiard balls that go off in different directions when they collide and their subsequent motion is determined by the principles of momentum conservation, *convective storms are not solid objects* and are there-fore the dynamics of collisions are more complicated. Based on observations of storms, we have learned that collisions between storms may result in a weakening of the resulting storm merger or in a strengthening of the resulting storm merger, depending on many different environmental factors. The most important consideration is probably whether or not the new merged updraft still has access to air that is warm and moist enough to sustain itself and to promote the continuation of upward motion.

Now let's consider a more accurate, but also more complicated picture: one in which the vertical-shear vector changes direction with height; in the case we consider now, the shear vector changes by 180°. In our earlier example (Fig. 4.76), the vertical shear was

---

[31] Bluestein, 2009.
[32] Bluestein and Weisman, 2000.

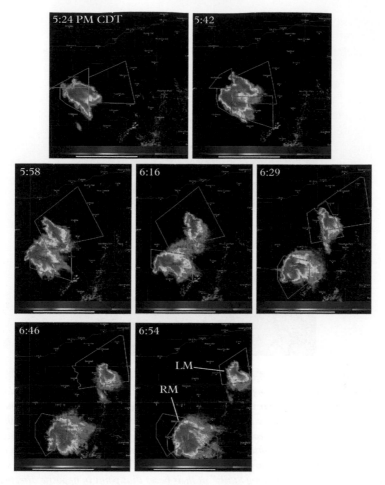

Figure 4.78 *Storm "mitosis:" An example of the radar echo from a convective storm that splits into a cell that moves to the right (RM—"right mover") and a cell that moves to the left (LM—"left mover") of the mean wind or vertical shear vector. Radar reflectivity imagery from the WSR-88D radar at Frederick, Oklahoma, in southwest Oklahoma, on May 7, 2020 at the times indicated. The storm elongated along the "deep" shear vector (not shown, but it was approximately from the west-northwest to east-southeast). The RM moved to the east-southeast and developed into a supercell and produced hail as large as 3.25 inches (8.25 cm) in diameter; the LM moved more rapidly off to the northeast and weakened. The National Weather Service issued severe thunderstorm warnings for both cells (yellow polygons).*

Figure courtesy of author.

unidirectional; it was always from west to east. First consider what happens at some arbitrary height in the environment of the convective storm. In the linear case, there is relative low pressure in the downshear direction and relative high pressure in the upshear direction. For the special case in which the winds vary with height as shown in Figure 4.82,

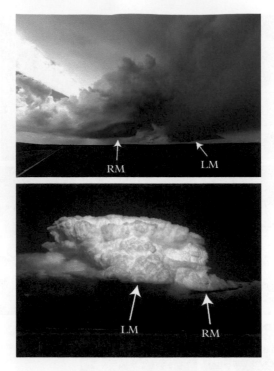

Figure 4.79 *Splitting storms. (top) A convective storm that has split into right—(RM) (left in photo) and left—(LM) (right in photo) moving parts, viewed to the west, from west of Limon, Colorado, on June 4, 2015. The two storms appear to be symmetric, as the right mover has a line of clouds (flanking line), which extends to the south, while the left mover has a tail at cloud base extending to the north. The right mover eventually produced a spectacular set of cyclonic and anticyclonic tornadoes near Simla, to the south. The view is to the west. (bottom) As in the top panel, but for a splitting storm in eastern Colorado on Aug. 16, 2009, viewed to the east., from Boulder, Colorado. Again, note the symmetric look to the two convective towers, having "tails" at their bases that point in opposite directions.*
Photos courtesy of author.

the shear vector rotates in a clockwise direction with height; it has completely reversed direction aloft from what it is near the surface. The shear is from the south below, from the west at midlevels, and from the north at high levels. An upward-directed pressure-gradient force forces air upward on the *right* side of the updraft, to the right of the mean vertical-shear vector and downward on the *left* side of the updraft. When this linear effect is combined with the nonlinear effect, we can speculate that when the vertical shear turns in a clockwise direction, convective storms may split, but if they do, the right-moving, cyclonically rotating cell is favored to persist while the left-moving, anticyclonically rotating cell is favored to weaken. The opposite happens in the rarer instances in which the vertical shear turns in a counterclockwise direction with height. These findings are backed up by numerical simulations (Weisman and Klemp 1984) and by observations. The relative rarity of unidirectional wind shear in nature and the

Figure 4.80 *Portrait of a supercell as a young storm: The right mover in the storm shown in Figure 4.98, top panel), but a short time later after it had separated from the split. The left mover is barely visible to its right. Very little, if any precipitation is visible from the right mover at this time, which looked like a very large, rotating, soup can.*

Photos courtesy of author.

ubiquity of clockwise turning of the vertical-shear vector with height explains why most severe convective storms don't split into perfectly symmetric storms, but in the presence of strong vertical shear propagate mainly to the *right* of the mean winds. As a result of surface friction (the interaction of airflow with the ground), which decreases with height just above the ground (the air is slowed down by friction the most right next to the ground), it turns out that the wind shear vector at low levels should turn in a clockwise manner in the Northern Hemisphere when the horizontal pressure gradient associated with synoptic-scale features does not vary with height and when there is at least an approximate balance among the horizontal pressure-gradient force, the Coriolis force, and friction, as is frequently observed. The relationship between the way the actual wind varies with height and the way the shear vector varies with height is complex and there is no need for the casual observer to consider the nuances of it. For example, it is possible for the direction of the wind to change with height, while the direction of the wind shear vector with height does not.

A brief summary of what we know about supercells from theory and observations follows: In an environment of strong deep-layer vertical wind shear, especially in the lower half of the troposphere (as demonstrated by numerical experiments and confirmed

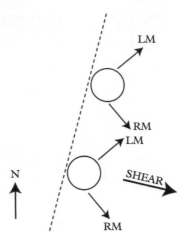

Figure 4.81 *Idealized example of how adjacent members of neighboring splitting storms may collide. Suppose two storms (circles represent the updrafts from two nearby storms) form along a surface boundary (shown by the dashed line) in an environment of vertical shear throughout the depth of the storms as indicated. If both storm cells split into right (RM) and left (LM) moving storms as shown, then the LM cell from the southern storm will collide with the RM cell from the northern storm. The RM of the southern storm and the LM cell from the northern storm may go off without experiencing any collisions. This figure just scratches the surface of how complicated neighboring cell interactions can be; and the behavior of cold pools from precipitation is not even included in this picture.*
Figure courtesy of author.

by observations), and of strong potential buoyancy, any convective storm that is triggered by a buoyant updraft develops rotation and has a component of propagation normal to the vertical-shear vector; when the updraft rotates and is long-lived (longer than about an hour), it is called a "supercell." In the Northern Hemisphere, most supercells have a cyclonically rotating main updraft (counterclockwise in the Northern Hemisphere) and propagate to the *right* of the mean wind in the troposphere. In a few, very rare, instances, supercells have an anticyclonically rotating main updraft and propagate to the left of the mean wind in the troposphere. In these cases, we find that the vertical shear turns anomalously with height in a counterclockwise direction. There are some supercells, especially during the night, that form when the vertical shear is extremely strong but the potential buoyancy is rather modest or even weak. Strong potential buoyancy and steep lapse rates seem to be less important than the vertical shear in producing rotating updrafts. Both multicells and supercells require vertical shear, but supercells require even stronger deep-layer vertical shear, especially in the lower half of the troposphere, from the surface to about 6 km AGL (e.g., Weisman and Klemp 1982).

It is nearly impossible to tell a supercell from a multicell just by looking at the cloud top features from such a great distance so great that we cannot see what is happening at cloud base. When viewed at relatively close range, when there are no intervening clouds or heavy precipitation, from a location approximately normal to the vertical-shear

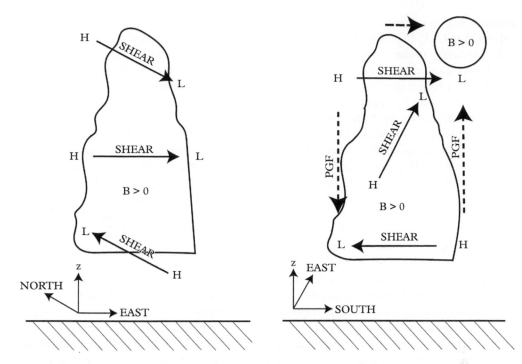

Figure 4.82 *Illustration of what happens when a positively buoyant (B > 0) updraft interacts with vertical wind shear in an environment in which the shear vector rotates completely around 180° from the base of the cumulus congestus cloud up to the top of the cumulus congestus cloud. (Compare with happens when the vertical shear vector does not change direction, in Fig. 4.76) (left panel) In this highly idealized case, the shear vector turns from southerly in the bottom portion of the cloud, to westerly at midlevels in the cloud, to northerly in the top portion of the cloud. The linear effect described in Figure 4.76 for a constant vertical shear vector acts to produce a relatively high dynamic pressure upshear of the updraft and a relatively low dynamic pressure downshear of the updraft. In the left panel, our view is to the north. (right panel) From a different vantage point, now looking toward the east, we see that the turning with height of the vertical shear vector acts to induce an upward-directed pressure-gradient force (PGF, dashed arrow) just south of the buoyant updraft and a downward-directed pressure-gradient force just north of the buoyant updraft. The result of this is to promote the formation of a new cell (B > 0) above the LFC on the south side of the updraft and to suppress the formation of any new cell on the north side. The component of motion due to propagation in this case is to the south (long-dashed arrow at the top), to the right of the direction of the mean vertical wind shear in the troposphere (from the west).*

Figures courtesy of author.

vector, on its right side, most supercells appear to have a unique architecture. As the American architect Louis Sullivan once proclaimed for buildings, "form follows function," for supercells arguably the same can be said. Just as bathrooms in houses tend to be inside or near bedrooms and dining rooms tend to be near kitchens, distinct cloud features within supercells are located in specific areas of the storm with respect to each other (Fig. 4.83) (e.g., Moller et al. 1994). In supercells, air enters them on the upshear side, especially under the flanking line, or from the downshear side ahead of the updraft, and leaves them either through the penetrating/overshooting top, anvil, or as downdrafts laden with precipitation on the downshear side, or on the upshear side. Wall clouds are found underneath the main updraft and tail clouds, when visible, are located between the wall cloud and a region of precipitation. The air entering wall clouds may consist of recycled air or some ambient air. Tornadoes tend to be found underneath wall clouds during their formation and mature stages. The idealized visual structure of supercells seen in the top panel of Figure 4.83 is based on a composite of many supercells; several different storms are seen in the bottom three panels to illustrate how in general the idealized depiction is and also to appreciate deviations in real life from the ideal composite. One is reminded again of how the supercell composite is like dinosaur skeletons in museums: The dinosaur skeletons actually consist of bones from many different dinosaurs.

A non-tornadic supercell strikingly illuminated at sunset is shown in Figure 4.84 to illustrate some of the variety and similarity of how they look.

Keith Browning synthesized what we had learned about convective storms and proposed that the basic building blocks of convective storms are either ordinary cells or supercells (Weisman and Klemp 1986). There sometimes is, however, some ambiguity in telling one from another, and ordinary cells sometimes evolve into supercells and later evolve back into a series of ordinary cells, multicells. Both multicells and supercells usually have anvils, some part of which may be "backsheared" (moving in the direction opposite to that of the ambient wind at high levels) owing to very strong updrafts shooting into the lower stratosphere, where the winds may be weaker than they are at the tropopause), mamma underneath the anvil, penetrating/overshooting tops above the anvil, flanking lines, and lowered cloud bases.

Multicells, however, have a number of distinct convective towers, each with separate, non-rotating updrafts and tend to have more extensive cold pools and stronger gust fronts. In supercells, on the other hand, there is usually just one main, long-lived updraft, which rotates. The intensity of the updraft and the degree of rotation may, however, fluctuate with time. While one may be impressed by a massive cumulonimbus with a penetrating top as see from a distance, one really can't tell whether the main updraft in the convective storm is rotating and long lived. In both multicells and supercells, from a distance one can often see a succession of penetrating/overshooting tops. It is also not possible to distinguish, looking below, from a satellite above, a supercell from a multicell. The main characteristic of a supercell is that *its main updraft rotates* and is long lived. The cyclonically (anticyclonically) rotating part of the air current is called a *mesocyclone* (*meso-anticyclone.*).

The radar view of a supercell (Fig. 4.85) is usually less ambiguous than an actual visual view of the cumulonimbus cloud of which it is composed. A distinct, concave-shaped edge to the core of the storm on its right flank (with respect to its movement) and a hook echo on the right-rear flank of the radar-echo mass are telltale hallmarks of a supercell. The hook echo is evidence of a circulation associated with a rotating

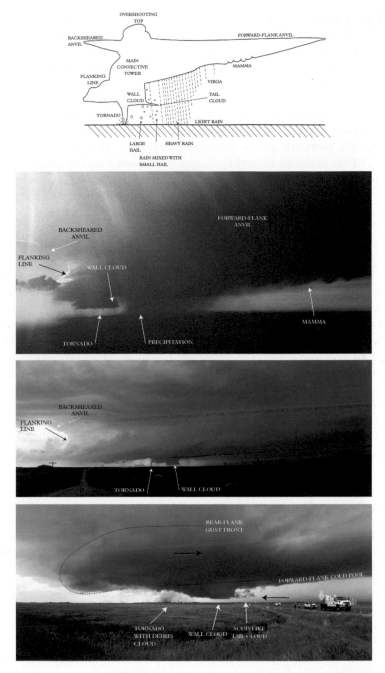

Figure 4.83 *Supercells with wall clouds and tornadoes. (top) Idealized depiction of features in a tornadic supercell (the Platonic ideal), as viewed to the right of the vertical shear vector and to the right*

updraft. The updraft in supercells can be so strong, with air moving upwards at speeds of 50 m s^{-1} (greater than 100 mph), that there is not enough time for a parcel of air to produce precipitation until it gets to very high altitudes in the storm. When a radar cuts a vertical cross-section through the updraft, a localized region of relatively weak echo, called the *weak-echo region (WER)* or *bounded-weak-echo region (BWER)*, if it is totally surrounded by higher reflectivity, is seen. This has also been referred to as the *vault* (Fig. 4.86). The BWER is often seen on a constant-altitude plane as a crescent-shaped region of relatively low radar reflectivity or as a hole in the reflectivity pattern (Fig. 4.87).

In a supercell, there are several prominent airstreams (Fig. 4.88). The "rear-flank" downdraft (RFD) is driven in part by negative buoyancy and in part by a downward-directed pressure-gradient force. The former is caused by cooling due to evaporation and sublimation and melting of precipitation as it falls into unsaturated air and air warmer than 0°C; the latter is thought to be caused by the interaction between the updraft and the vertical shear (Fig. 4.76) and/or by a weakening with height (becoming less cyclonic or more anticyclonic with height) of vorticity (Fig. 4.89). Precipitation loading may also contribute to the RFD. The RFD spreads out at the ground and terminates in a rear-flank gust front (RFGF), which is prominently viewed from a location to the right of the motion of the supercell along its flanking line. The RFGF marks the leading edge of the RFD as it impinges upon the ambient warm, moist air flowing toward the lower portion of the supercell. The RFD is not always steady, but sometimes comes in spurts; as a result, there are secondary surges which result in secondary RFGFs to the rear of the leading edge of the cold pool. There are instances, however, when the cool pool is weak, and even some instances when the air behind the RFGF is actually relatively warm, mainly due to subsidence (adiabatic warming as the pressure increases and the air in the downdraft is compressed). Warm RFDs occur if unsaturated air is driven by a downward-directed pressure-gradient force more than it is by negative buoyancy. Warm RFDs probably look relatively clear, with an absence of precipitation, though you would need to make actual temperature measurements to be certain; there could have

---

Figure 4.83 (Continued) *of the storm as it moves by. Features are not necessarily drawn to scale in order to show them more clearly. Inspired by work done by C. Doswell, which appeared in Moller (1978). (second row) Airborne view from the NOAA P-3 research aircraft of a tornadic supercell over southwestern Kansas, near Ulysses, on May 26, 1991, during COPS-91 (Cooperative Oklahoma[a] Profiler Studies—1991). View is to the northwest. Compare with the top panel. (third row) Tornadic supercell, viewed looking to the west and northwest, in southeastern Wyoming, in Goshen County, on June 5, 2009, during VORTEX2. The backsheared anvil (far left), flanking line (left), tornado (center left), wall cloud (to the right of the tornado), and opaque precipitation core (right) are all visible in this "classic" supercell. Compare with the top panel. (bottom) Panoramic view of a tornadic supercell over Selden, Kansas on May 24, 2021, while being probed by RaXPol (right). Black arrows denote the approximate direction of the airflow. Curved dotted line represents the approximate location of the rear-flank gust front. View is to the west. Compare with the top panel.*

[a] As noted earlier, despite the name of the field program, we "trespassed" into Texas and Kansas from time to time.

Figure and photos courtesy of author.

Figure 4.84 *Nontornadic supercell on the Oklahoma-Kansas border near sunset on April 15, 2017. The anvil is seen spreading out to the rear (left) of the main convective updraft tower, which is leaning in the opposite direction (to the right). View is to the north-northwest.*
Photo courtesy of author.

been copious amounts of precipitation earlier. Cold RFDs are most likely laden with precipitation.

Precipitation that falls from the anvil located downshear from the updraft evaporates or sublimates or melts, and cools. The air becomes negatively buoyant and hits the ground, where it spreads out and forms the forward-flank downdraft (FFD) (Fig. 4.88). There is not always a sharp boundary at the surface between the FFD and the ambient air flowing toward it and colliding with it. However, there is a temperature gradient created by the juxtaposition of cooler air in the FFD with the ambient warm, moist air approaching it. This diffuse boundary between the warm ambient air and the cool air under the FFD is of great importance. Along it, air acquires horizontal vorticity aligned approximately along the direction of the air flowing from the FFD, since air is accelerated downward in the negatively buoyant, cold pool, but not all in the environment (Fig. 4.88, right panel). This rotating, horizontal column of air (called the streamwise vorticity current, or SVC, especially if there is a smaller-scale horizontal vortex associated with a leading Kelvin–Helmholtz wave (Schueth et al. 2021)), which travels toward the main updraft and is stretched horizontally and tilted upward along its edge, so that a low-level, cyclonically rotating vortex is formed. This vortex is called a "low-level mesocyclone"

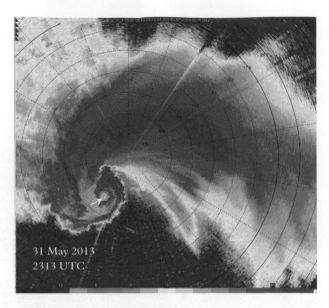

Figure 4.85 *Radar reflectivity factor in dBZ (color scale is at the bottom) in a tornadic supercell near El Reno, Oklahoma, on May 31, 2013, at 5° elevation angle. See the main tornado associated with this storm in Figure 5.18, bottom panel, in Chapter 5. The short white arrow points to the weak-echo hole (WEH) that marks the center of the tornado. North is up and east is to the right. Around the WEH is the hook echo. The main core of the precipitation is northeast of the tornado and is color-coded red. From RaXPol.*

(See also Bluestein et al., 2015.)
Figure courtesy of author.

and could be the parent vortex for a tornado. The low-level mesocyclone is distinct from the mesocyclone that forms at midlevels as a result of the tilting of horizontal vorticity associated with vertical shear in the environment. Although the two don't necessarily line up in the vertical because they owe their existence to two different physical processes, the downward-directed pressure-gradient force associated with the decrease in vorticity above it contributes to the RFD, as noted earlier.

The formation of the low-level mesocyclone requires a cold pool in the FFD region (if there is only weak or no horizontal vorticity in the environment, i.e., no or weak low-level wind shear). If the cold pool is too weak, then the rate of formation of horizontal vorticity (spin about the horizontal) is too slow to result in the formation of a low-level mesocyclone. If the cold pool is too strong, then the cold-air-driven RFD wraps around the mesocyclone as it forms and pushes the RFGF far ahead of the updraft and the updraft weakens as it is cut off from the supply of ambient, warm air. Or, instead, if there is a sharp discontinuity in temperature rather than a broad zone of horizontal temperature gradient, then the baroclinic zone behaves like a density current and flows outward, normal to the edge of the cool pool and effectively cutting off the supply of

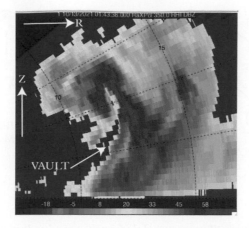

Figure 4.86 *Vertical cross-section of radar (RaXPol) reflectivity factor (dBZ) shown by color scale at the bottom, through the updraft in a supercell that had earlier produced a tornado in western Oklahoma during the evening of Oct. 12, 2021. The vault is seen as a narrow, curved band of low reflectivity (purple). At higher elevations it is completely surrounded by much higher reflectivity (red/brown) and is known as a bounded-weak-echo region (BWER). Z is the height of the radar beam above the ground and R is the distance (km) from the radar.*

Figure courtesy of author.

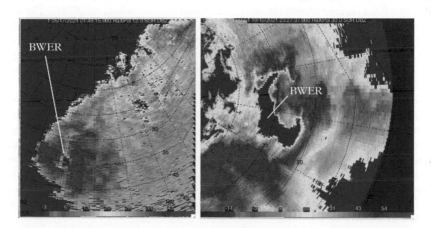

Figure 4.87 *Examples of BWERs from RaXPol radar reflectivity data (color scale in dBZ). (left) compact BWER 35 km away from the radar, at 7.4 km above the ground in West Texas during the evening of May 16, 2021, in an isolated low-precipitation (LP) supercell that had earlier produced a tornado; (right) crescent-shape BWER at closer range, 15 km, at 8.7 km above the ground, in southwestern Oklahoma during the evening of Oct. 10, 2021, in a high-precipitation (HP) supercell that had earlier produced a tornado. Photographs of the storm associated with the former are seen in Figure 4.108.*

Figure courtesy of author.

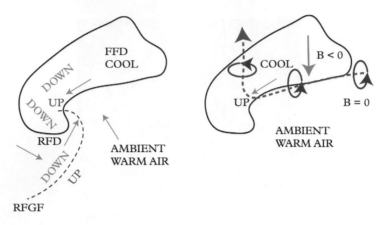

Figure 4.88 *Aspects of supercell structure and the formation of a low-level mesocyclone. (left panel) Idealized depiction looking down on the storm near the ground, of the locations of the forward-flank downdraft (FFD), which is relatively cool owing to the melting, sublimation, and evaporation of precipitation; the rear-flank downdraft (RFD), which is due in part to cooling and precipitation loading and part to a downward-directed dynamic perturbation pressure-gradient force; the rear-flank gust front (RFGF), which marks the boundary between air that has descended in the RFD and the ambient, warmer (usually, but not always; the RFD air may be cooler, the same temperature as, or even warmer than the air in the environment) air in the environment; at this interface (dashed black line), air is lifted upward (UP). Air is also lifted upward by the main updraft is also marked by "UP." Air is descending (DOWN) in a band curving around the main updraft and also just behind the RFGF. The solid black line marks, approximately, the outer boundary of the region of precipitation (what would be detected by radar). The green arrows denote salient parts of the wind field near the ground. If there were a tornado, it would be located near the tip of the RFGF, near the region where the main updraft abuts against the RFD. (right panel) Idealized explanation for the formation of a low-level mesocyclone in a supercell. The solid line, as in the left panel, represents the outer edge of most of the precipitation in the storm. "UP" marks the location of the main updraft. The air near the ground inside the precipitation core is relatively cool and negatively buoyant (B < 0; blue vector represents buoyancy), while the warm air in the environment of the storm is neutrally buoyant (B = 0). A paddlewheel placed along the region where there is a temperature gradient pointing outward from the precipitation core will begin to rotate as shown, with upward motion outside the core and sinking motion within it. The red dashed line represents the orientation of the vorticity vector if you curl your right hand about the circulation with your thumb pointing along the red dashed line. Air coming from the right (green arrow represents the low-level wind) acquires spin as shown and is then tilted upward at the main updraft, where a mesocyclone at low altitude is formed.[a]*

[a] How the horizontally spinning air is tilted upward is a bit more complicated than shown in this figure. The air may actually descend a little as it wraps around the updraft before it is tilted upward.
Figures courtesy of author.

warm, moist air into the updraft. When viewing a supercell and the RFGF moves way out ahead of the storm the chances of seeing a tornado are much diminished. There must therefore be some cold pool, but one neither too strong nor too weak, and the baroclinic zone must not move way out ahead of the storm.

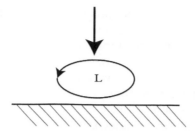

Figure 4.89 *Illustration of how a cyclone at low levels in a convective storm induces a downward-directed perturbation pressure-gradient force. A cyclone near the ground is associated with a center of low dynamic pressure. If the cyclone weakens with height and even disappears at some level aloft, then the pressure deficit at the center of the cyclone decreases with height, resulting in a pressure-gradient force that is directed downward. The cyclone effectively sucks air downward into its center.*
Figure courtesy of author.

Figure 4.90 *Cyclonically rotating wall cloud being probed by RaXPol (right) in Northwest Kansas, east of Selden, which had just been hit by a tornado (see Figs. 1.1, top left panel; 5.9, middle panels), on May 24, 2021. There is a short tail cloud feeding into the wall cloud from the north (right). This wall cloud is rather ragged in appearance, suggestive of turbulent mixing.*
Photo courtesy of author.

Low-level vertical wind shear in the direction of the FFD boundary also plays an important role. If the shear is too strong, then the air flowing toward the updraft along the zone of temperature gradient is too rapid, so while horizontal vorticity may be generated, air does not experience this rate of generation of vorticity long enough to be significant. If the shear is too weak, the horizontally spinning air takes too long to reach the updraft. There must therefore be some low-level vertical shear, but neither too little nor too much. Low-level vertical wind shear normal to the FFD boundary may also play a role, according to RKW theory, in affecting how rapidly the FFD boundary moves normal to itself.

There may also be some contribution to the formation of the low-level mesocyclone from vertical wind shear in the environment, not associated with processes going on in the storm., i.e., that associated with vorticity not produced by the storm, but rather by

Figure 4.91 *More wall clouds and their formation. (top, left) Wall cloud forming under an updraft base in northwest Texas on May 25, 1999. The cloud formed as scud below cloud base and connected to the cloud base above. (top, right) A supercell in the Texas Panhandle, viewed to the southwest, on May 21, 2010. The cloud base associated with the updraft is flared out, flat, and apparently free of precipitation, while curtains of precipitation are seen falling to its northeast (right). There is a second ring of elevated, ragged-looking convection (left) above the laminar base. (bottom, left) Scud forming underneath a cloud base in a supercell near Elk City, in western Oklahoma, on May 31, 2022. (bottom, right) As in the bottom, left panel, but a short time later (several minutes) as the scud became attached to the cloud base above, becoming a wall cloud.*

Photos courtesy of author.

Figure 4.92 *Tail clouds feeding into wall clouds. (top) Tail cloud feeding into a supercell in south-central Oklahoma on May 4, 2020. Viewed to the north. (bottom) Wall cloud with a tail cloud (right), viewed to the west, from east of Boulder, Colorado, on June 26, 2020. This supercell, seen over the Foothills, formed at a higher elevation, just east of the Continental Divide, near Nederland, and produced large hail as it propagated southeastward off the mountains. There must have been some interaction of the storm's airflow with Bear Peak, which was right in the way of the wall cloud.*
Photos courtesy of author.

Figure 4.93 *Wall clouds with tail clouds. (top) Wall cloud and tail cloud, coming into the wall cloud from the forward flank (right of the wall cloud) in a supercell in West Texas, early in the evening on May 9, 2017. View is to the north-northwest. There is another, rare, tail cloud feeding into the wall*

vorticity that is already there in the environment anyway and is drawn into the storm. Air that moves in the same direction as the axis of rotation associated with the spinning motion (which may be produced by the storm or imported from the environment) is known as "streamwise vorticity" (Davies-Jones 1984). Airflow that has large amounts of streamwise vorticity traces out a helix and is said to have a lot of "helicity." In the reference frame of the moving storm the helicity is known as "storm-relative helicity" (SRH). Since air from both the environment and inside the storm, where precipitation is falling, enters the rotating updraft, *both* environmental shear and the SVC-induced shear probably play roles in producing the low-level mesocyclone.

As the low-level mesocyclone forms, it is accompanied by a decrease in the dynamic component of pressure. This decrease in pressure can result in a slight lowering of the condensation level and cloud base. This lowering of cloud base owing to a drop in pressure will be considered in more detail in the next chapter, on tornadoes. More significant to the lowering of the cloud base under the updraft is the ingestion of cooler, but more humid air from the FFD region. This ingestion of more humid air leads to the further lowering of cloud base into what Ted Fujita named the "wall cloud" (Fig. 4.90 and 4.91). Ted Fujita came up with this name to describe the vertical face on one side of the lowered cloud base in a tornadic supercell that struck Fargo, North Dakota back in 1959 (Fujita 1960). A meteorological (fluid) rock climber would be challenged with this side of a wall cloud, sometimes extending many kilometers upward. Remarkably, Fujita's nomenclature was based on an analysis of photographs taken of just this one storm, yet it has stood the test of five decades of time. The lowering of the cloud base is frequently preceded by the formation of scud clouds under the cloud base, which then get attached to the cloud base above, effectively lowering the cloud base (Fig. 4.91, bottom panels). The updraft region of supercells can take on many appearances. Sometimes the air near the ground is relatively dry, the cloud base relatively high and flat. Other times, the air is relatively humid and the wall cloud can be low to the ground. A band of cloud extending from rain-cooled air in the FFD is sometimes seen connecting to the wall cloud as a "tail cloud" (Figs. 4.92 and 4.93).

Tail clouds may be mistaken for funnel clouds or tornadoes when they are highly tilted and extend down to or almost touch the ground (Fig. 4.96).

Underneath a low-level mesocyclone there is an upward-directed dynamic-pressure-gradient force, which may be strong enough to overcome any negative buoyancy there is if the air flowing into the updraft is relatively cool. When this happens, the air can be

---

Figure 4.93 (Continued) *cloud from the rear flank (left of the wall cloud). (middle) Wall cloud (left) and tail cloud (right) to the northwest, being closely monitored by my former graduate student Trey Greenwood, west of Selden, Kansas, on May 24, 2021, prior to the formation of a tornado. (bottom) Panoramic photograph of a rotating wall cloud (right), tail cloud feeding into the wall cloud from the north (far right), and rear-flank gust front (RFGF) cloud (center and left), looking to the west (left) and to the north (right), on May 24, 2021, before the Selden, Kansas tornado formed a bit later, to the west of Selden. A schematic of the flow viewed from above is given in Figure 4.94 The tail cloud and the RFGF cloud look as if they coil up at the wall cloud. The tornado subsequently formed on the far, rear, western side (left side) of the wall cloud near the clear slot.*
Photos courtesy of author.

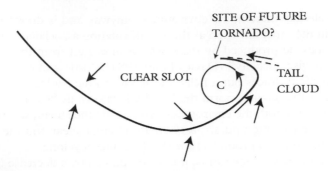

Figure 4.94 *View looking down of the features shown in the photograph in Figure 4.93, bottom panel. The solid line marks the arcus cloud lowered from cloud base above; the curved, solid line marks the boundary between the ambient air (flow denoted by arrows) and the air in the "clear slot," coming from the RFD, that is, the RFGF. The tail cloud (dashed line) is under cloud base to the right of the low-level mesocyclone denoted by "C." The site of possible future tornadoes at the cusp of the RFGF is noted. See also Figure 4.95 for funnel clouds in the same relative location.*
Figure courtesy of author.

lifted stably up to the LFC, below which the edge of the cloud may look like a smooth orographic wave cloud (Figs. 1.1, bottom, right panel; 4.53, 4.97). This laminar appearance is particularly striking when viewed from the ahead of or to the right side of the motion of the updraft. Supercells like these have been described by some storm chasers as "mother ships,"[33] owing to their resemblance to giant flying saucers, the icons of many science-fiction movies of the 1950s. In many instances, the cloud bases are neatly flared out. One can often see striations along the edges of the cloud, like those in some lenticular clouds (see Chapter 2). These striations sometimes give one the sense of rotation though it is not clear that they are caused by rotation; instead, they could be caused simply by vertical inhomogeneities in moisture as originally hypothesized by Richard Scorer in 1972 and verified quantitatively by M. Hills and Dale Durran[34] at the University of Washington for lenticular clouds. Clouds are not like thick liquids that are stirred, or like pottery that is being spun and shaped. The smooth striated bases are frequently seen in the early evening as the surface temperature falls and the boundary layer becomes stable, though they can be seen earlier in the day also if there is a strongly stable layer. Sometimes the smooth bases are markedly flared out, as in Figures 4.97 and 4.98.

In the following set of photographs, tornadoes and funnel clouds are seen having various spatial relations to other cloud features in a supercell (Figs. 4.99–4.102).

The rear flank of a supercell often has a clear slot, marking the rear-flank downdraft (Figs. 4.103 and 4.104, bottom panel).

One can, on rare occasions, get a remarkable view of an entire supercell cumulonimbus with a tornado, viewed from the rear, when one is far enough away (Figs. 4.100,

---

[33] Strictly speaking, though, mother ships are simply those that yield smaller ships or act as bases.
[34] Hills and Durran, 2014.

Figure 4.95 *Tilted funnel clouds in supercells along with a wall cloud. (top) Highly tilted funnel cloud (left) pendant from the wall cloud of the supercell seen slightly earlier in Figure 4.93, bottom panel, on May 24, 2021. View is to the north. It is likely that the tilt is caused by outflow from the storm flowing at low levels from the right to the left. While the funnel cloud seems to be symmetrically located with respect to the tail cloud seen at the right, it is functionally different. (bottom) Funnel cloud in a supercell in the Texas Panhandle on May 20, 1999, viewed from the south-southeast. Note how the funnel cloud leans to the right with height as does the one seen in the top panel. In this case, however, no tail cloud is evident. The wall cloud is mostly smooth looking and flat. The tilt of this funnel cloud may also be caused by outflow from the storm flowing from right to left.*

Photos courtesy of author.

Figure 4.96 *Tail cloud touching the ground on the northern side (right) of a lowered cloud base associated with the updraft of a developing tornadic supercell near the border of eastern Colorado and western Kansas, west of Tribune, on May 25, 2010, during VORTEX2. This cloud feature is not to be mistaken for a funnel cloud or tornado; it does not display any signs of rotation.*
Photo courtesy of author.

4.104, 4.105). When this happens, the powerful tornado looks like a highly concentrated, tiny area of energy, since it is so small in comparison with the size of the storm, both in its depth and width. Seeing an *entire* storm with a tornado within it is an awe-inspiring sight, almost as great as being close enough to see the debris cloud and sense the rotation of the tornado and its parent cloud. I am often content, especially when using a mobile radar, to sit tight and let a storm approach the radar crew. That way, I may get to see the entire storm first, and later, when a tornado forms, have it close by. And, furthermore, my students and I can safely document the formation of a tornado beginning well before the tornado has formed.

4.4.2 Precipitation and its effect on the visual appearance of supercells

How supercells appear visually depends a lot on how much precipitation is being produced. When a supercell approaches you from the upshear direction, you first see (Fig. 4.83, top panel, moving from left to right) the anvil coming toward you. An anvil that

Figure 4.97 *Supercells with smooth bases that have striations in the columns of cloud above. (top, left) Isolated, non-tornadic supercell, viewed to the west, in northwest Kansas, north of Hays, on June 10, 2008, at sunset. The cloud base is flared and smooth with some striations. This storm produced large hail. I chased this storm serendipitously, as we were migrating to Boulder from Norman for the summer, and came upon it as we stopped for dinner and a layover in Hays. The sun is purely by chance visible on the western horizon just on the southern edge of region of precipitation; no adjustment by the photographer was necessary to see the sun in this position. (top, right) Supercell in southwest Oklahoma, near Mangum, on March 18, 2012. The bottom portion of the cloud is laminar and striated and probably stable, while the upper portion is bubbly and buoyant. (middle, left) Supercell viewed to the north from southwest of Lubbock, Texas, on May 17, 2021. The cumuliform tower at the upper right is*

looks crisp, with a sharp edge, probably means the main updraft is vigorous. A mushy, soft-edged anvil or one that does not seem to be diverging may be a sign that the updraft is not very strong. Light rain begins to fall from the anvil and progressively becomes more and more intense; it gets darker, as the precipitation core becomes less translucent. Then it may get mixed with small hail. As the hail gets larger and larger and rain is less and less mixed with it, the visibility increases and sunlight (if it is daytime, of course) increases, so it gets brighter. As the hail continues to get larger and larger, it becomes less and less concentrated and it may appear as if there is no precipitation falling at all. Don't be fooled! At this point you are probably right near the updraft and may be at a location that is about to get hit by a tornado.

After chasing many supercells, other chasers and I have noticed that sometimes very little precipitation can be seen at all near the main convective-cloud tower, leading me to have named these storms "low-precipitation" supercells or "LP" storms. These storms look like the skeletons of supercells, since there are no opaque precipitation cores blocking one's view (Figs. 4.106–4.109).

That there is no precipitation at all is probably an illusion; one often finds large hail in these storms, so that the precipitation core is likely to be translucent if the hailstones are widely separated. If there is large hail only, then it is unlikely there will be a significant cold pool, which would argue that strong low-level mesocyclones and tornadoes are not as likely, unless there is horizontal vorticity in the environment associated with low-level vertical wind shear to replace or augment the missing, baroclinically generated horizontal vorticity. However, many LP supercells do, in fact, produce tornadoes. We need in situ measurements of temperature and precipitation in them to verify our theories, as visual observations alone are insufficient. From a great distance, one sees large anvils streaming out, giving the illusion that the storms are very large. However, upon

---

Figure 4.97 (Continued) *folding back on itself as a result of the buoyancy gradient at the edge of the cloud (highly buoyant inside the cloud, neutrally buoyant outside the cloud), as in the spherical-vortex-like flow about a buoyant bubble. The clear slot is not visible at this time from this vantage point. As in the previous panels, the bottom portion of the cloud is laminar, while the upper portion is bubbly-looking. (middle, right) Supercell in southwest Kansas on June 9, 2009, during VORTEX-2, viewed to the west-northwest. The convective tower is laminar and striated; the striations are stacked, looking like some orographic wave clouds (e.g., Fig. 1.1, bottom, right panel). A small tail cloud is seen on the north side (right) of the cloud base. I was being interviewed live on the Weather Channel while I was photographing this storm and had to turn my back on it, a very anxiety-producing situation of having to ignore a potentially tornadic storm, though at it this point a tornado did not appear to be imminent. (bottom, left) As in the middle, right panel, but a short time later, but as a narrow updraft convective tower that is dissipating. The midsection is laminar with helical striations and the bottom is flared out, with evidence of a clear slot underneath. There are a few discrete inflow bands on the southern side (left) of the tower, above the cloud base. (bottom, right) A supercell in the northern Texas Panhandle, viewed to the west, on May 24, 2000. The lower cloud base is flared out, laminar, with striations. There is a mid-level layer of cumuliform convective towers wrapping around from the southwest and south (left).*

Photos courtesy of author.

Figure 4.98 *Supercells with smooth, flared-out bases. (top) Bell-shaped or spinning-top-shaped, smooth, cloud base portion of a convective tower in a supercell west of Oklahoma City, on May 26, 2004, viewed to the west. There is an elevated inflow tail aloft on the southern side (left) of the convective tower and vertical, parallel bands of clouds, of unknown origin, on the north side (right) of*

getting closer and closer to them, one finds that there is only one narrow convective tower feeding the massive anvil and that the storm is actually rather compact. On radar, one sees mainly a small radar-echo core. When some LP supercells dissipate before ever evolving into classic supercells and producing significant amounts of precipitation near their updraft cloud base, the diameter of the cloud base shrinks until it disappears (Fig. 4.109), through a process I call a "downscale transition" (Bluestein 2008), which is the opposite of what happens when a rain-cooled downdraft spreads out, triggering new cells along the outflow edges, causing the storm to get wider and move "upscale."

It was also noted that the opposite extreme happens, that is, there is so much precipitation that under the updraft cloud base there is an opaque, dark region, which might even hide a tornado within it. These convective storms are called "high-precipitation" (HP) supercells (Figs. 4.110 and 4.111). Their precipitation cores, as they appear on radar, are relatively large. Because there is so much precipitation, one would expect that a cold pool associated with an HP supercell might be more intense; that depends, however, on how humid the low-level air is. If the humidity is relatively high below the cloud base, then evaporation and cooling should be tempered and the cold pool not necessarily extremely cold.

In between the LP and HP storms on a spectrum of storm types, we find the "classic" supercells (e.g., Fig. 4.83), in which there is heavy (and visibly opaque) precipitation underneath the forward flank, and relatively precipitation-free, clear air underneath the updraft and wall cloud. There are a number of factors that can control the precipitation efficiency of supercells. These include the dryness of the environmental air mass, the concentration of condensation nuclei, the type of aerosols suspended in the air, and the vertical shear. Sometimes classic and LP supercells can coexist in the same general environment. In this case, we sometimes find that an LP supercell is the equatorward-most storm, while there are classic supercells poleward of the LP storm. The equatorward-most storm is sometimes known as the "tail-end Charlie" storm, a British military term that referred to the last aircraft in a group of aircraft during World War II. It might be

---

Figure 4.98 (Continued) *the tower. The precipitation from this storm was falling so far from the cloud base (to the rear of the photographer), that the location of the storm was being given by the National Weather Service as many miles downshear from the location of the updraft base shown in this photograph. The most-heavy precipitation was therefore far removed from this cloud base. (middle) The supercell, on April 22, 2020, near Springer, Oklahoma (also seen in Fig. 4.103, top panel; Fig. 4.104, and on the cover) on the eastern side of the interstate highway seen in the foreground (I-35), viewed to the northeast, with a flared out, bell-shaped, laminar cloud base, with roiling cumuliform convective clouds overhead, prior to the appearance of a tornado. The motion at the top is like that of the spherical vortex described about a thermal bubble (Fig. 3.6) and seen at the top of the cloud shown in Fig. 3.7 as a roiling, curled up cloud mass casting a shadow. (bottom) Tornado in a supercell in south-central Oklahoma, near Sulphur, on May 9, 2016, viewed to the west. Note the similarity of the flared-out base north (right) of the tornado to the flared-out bases seen in the top two other panels, but viewed from a different relative vantage point.*

Photos courtesy of author.

Figure 4.99 *Broad, laminar, lowerings/funnel clouds in supercells, viewed from ahead of the storm. (top) Supercell in southwest Oklahoma, near Mangum, on March 18, 2012, viewed to the west. A wall cloud (and possible funnel cloud) is seen (right center) associated with an older mesocyclone, while a lowered cloud base is about to become a wall to its north (right). (bottom) Supercell with a wall cloud lowering, looking to the west, from southeastern Wyoming, on June 5, 2009, in a new supercell, separate from the Goshen County tornadic supercell seen in Figure 4.83, third row. Note the similarity of the storm structure seen here to that on March 18, 2012, in the top panel.*
Photos courtesy of author.

Figure 4.100 *Supercells with tornadoes, viewed from ahead of the storm. (top) Supercell with a tornado, in the northern Texas Panhandle, near Spearman, on May 31, 1990, viewed to the west. The region to the right of the tornado is marked by a tall vertical wall of cloud and a translucent area, probably populated by widely spaced, large hailstones or large raindrops. (bottom) A tornadic supercell in eastern Colorado, near, viewed to the southwest, on June 10, 2010, during VORTEX2. There are several stacked levels to the cloud on its southern side (left).*

Photos courtesy of author.

Figure 4.101 *Two in a series of 13 tornadoes produced in a supercell in southwest Kansas, south of Dodge City, on May 24, 2016, viewed to the west. Two of my former graduate students (Kyle Thiem and Zach Wienhoff) are seen at the lower right also photographing the storm and in a sense of symmetry act as a pair of observers for the two tornadoes. RaXPol, not seen in this photograph (to the rear of the photographer), is scanning the storm at this time. My students are literally "outstanding in their field." In this image, the tornadoes are on the left side of the cloud base, rather than on the right, as in Figure 4.100 (Wienhoff et al., 2020).*
Photo courtesy of author.

that the anvil from the tail-end Charlie LP storm drops ice crystals onto the next storm down the line, seeding it so that precipitation becomes more intense. In trying to target a potentially tornadic supercell, I have tried to game which storm to target: the tail-end Charlie storm or the storms poleward of it, without any consistent results. Sometimes the tail-end Charlie is the only one that produces tornadoes, while other times it's the storm next in line poleward that produces tornadoes. It's often a good strategy to target the storm poleward of the tail-end Charlie storm first, however, but move on down the line to the tail-end Charlie storm if the storm you are looking at does not produce a tornado or if it does, but then it dissipates. On many occasions one can leave an old storm to pick up a newer storm down the line, which may also eventually spawn a tornado. If we were to target the tail-end Charlie storm first, and then find out that we made a

Figure 4.102 *More funnel clouds and tornadoes seen near wall clouds with tail clouds. (top, left) Tornado (left center), wall cloud (center), and tail cloud (right) in a supercell, viewed to the west from east of Lake Thunderbird, Norman, Oklahoma, on May 19, 2013. Frame from a video. (top, right) As in the top, left panel, but during a multiple-vortex phase of the same tornado. (middle) Multiple-vortex tornado (left) in a supercell over Selden, Kansas, on May 24, 2021. A tail cloud is visible to the right. View is to the west. From a frame of an iPhone video. Note the similarity to the top, right panel. (bottom) Wall cloud, tail cloud (right) and funnel cloud (left) in north-central Oklahoma on May 1, 2008. View is approximately to the north-northwest.*
Photos courtesy of author.

mistake and have to target the next storm on the poleward side, we might have some difficulty in catching up to it since it would likely have a component of motion *away* from us, while if we were target the tail-end Charlie storm it would take less time to

Figure 4.103 *Clear slots in supercells. (top) The updraft region of a supercell forming in south-central Oklahoma, west of Springer, on April 22, 2020, viewed to the south. There appears to be a circular hollow region at the tail end of the developing storm. After the storm crossed Interstate 35 (left), it produced a tornado (Fig. 4.104 and cover). (bottom) A supercell in south-central Kansas, near Geuda Springs, on May 14, 2018, viewed to the north. The clear slot is visible to the north (left). A tornado subsequently formed just north of the clear slot. RaXPol is seen probing the storm and one of my former graduate students, Dylan Reif, is keeping a close eye on the storm. This view is what one probably would have seen from the other side of the storm in the previous panel, in which the clear slot was viewed to the south.*

Photos courtesy of author.

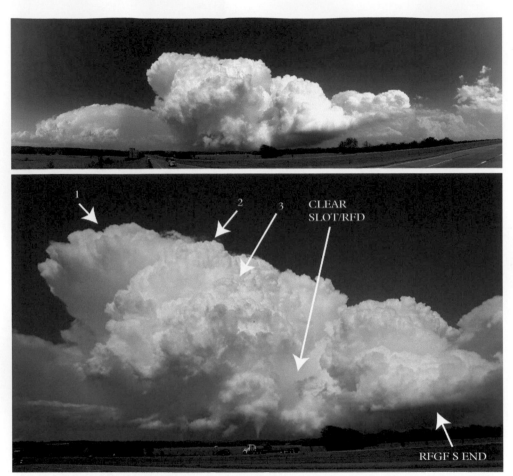

Figure 4.104 *The entire cloud of tornadic supercells, as viewed from the rear of the storm. (top) As in Figure 4.103, top panel, on April 22, 2020, but a short time later (perhaps 5 minutes or so) and with a view after the storm had crossed the highway; a wide, panoramic view from north-northwest (left) to southeast (right) of a supercell with a tornado (center left), and another supercell farther away to the north (far left) and yet another one farther away to the southeast (far right). I-35 is seen extending from the north (left) to the south (right). (bottom) As in the top panel, but a narrower view, a few minutes later, looking to the northeast through east-southeast, with the tornado in view near the center. Note how small the tornado appears in relation to the size of the parent cumulonimbus. The dry/clear slot and rear-flank downdraft (RFD) is seen as an indentation of clear air just to the southeast (right) of the (white) tornado, which extends up to the top of the cloud. The cloud base to the southeast (right) is probably located at the southern edge of the rear-flank gust front (RFGF). There are three separate updraft pulses, labeled "1," "2," and "3," from the first, second, and third pulses, respectively. Updraft pulse 1 is about to spew out an anvil. A closer view of this tornadic supercell is seen on the cover.*
Photos courtesy of author.

Figure 4.105 *Supercell with a dissipating, rope-like, white tornado leaning to the north with height (left) and its parent updraft base, viewed from a distance, well to the northeast, in eastern Colorado, south of Keenesburg, on June 19, 2018. A new updraft base is seen to the south (right), as cyclic mesocyclogenesis is occurring. It is not often that one gets to see an unobstructed view of a tornado and all the surrounding cloud features as here.*
Photo courtesy of author.

reach because it would have a component of motion *toward* us. In some instances, LP supercells evolve into classic (e.g., those in Fig. 4.106 evolved into classic supercells) or even HP supercells.

The structure of most of the supercells I have seen in the Plains regions of the US is relatively easy to see and the storms are tall. There are supercells, some tornadic, that are found underneath upper-level cyclones, where the tropopause is relatively low (or anywhere else where the equilibrium level [see Chapter 3] is relatively low). Because these supercells are not as tall as most, they have been referred to as "mini-supercells"[35] or "low-topped supercells." These supercells may not even have extensive anvils, owing to relatively weak winds near the tropopause.

Supercells also occur in landfalling tropical cyclones, particularly in the right-front quadrant of the tropical cyclone (with respect to the motion of the tropical cyclones)

---

[35] Kennedy et al. 1993; Suzuki et al. 2000; Davies, 2006.

Figure 4.106 *Low-precipitation (LP) supercells. (top) LP supercell in southwest Kansas, viewed to the west, on 9 June 2009, during VORTEX-2. There is some precipitation falling from the anvil, but much of the region underneath the storm is translucent. This supercell evolved into a "classic" supercell later (Fig. 4.97, middle right, bottom left panels). There is an inflow band on the southern side of the cloud base (left) and also one at midlevels. (middle) LP supercell developing off the dryline in far West Texas, southwest of Lubbock, on 17 May 2021. Later on, it evolved into a classic supercell (Fig. 4.97, middle left). (bottom) Supercell in southwestern Oklahoma on May 23, 2011. This supercell produced hail as wide as 6 inches (15 cm) in diameter. Precipitation curtains are seen underneath the anvil, but there is some translucence, giving the appearance of an LP supercell. View is to the west and northwest.*
Photos courtesy of author.

Figure 4.107 *Wide views of LP supercells viewed a distance from the front of the storms. (top) LP supercell in western Kansas, west of Hays, on June 10, 2008, viewed to the west. The region underneath the updraft cloud base (center, near the horizon) and underneath the anvil (right) is mainly translucent, indicative of little precipitation falling, or possibly widely spaced hailstones or raindrops.*

in outer rainbands (Fig. 4.112), in very humid environments, where intervening precipitation severely limits visibility. I've never seen one myself and have only rarely seen photographs, mostly posted on social media by others. The environment of these supercells has strong vertical shear, especially at low altitude, owing to the strong winds at low and midlevels in hurricanes and typhoons. The buoyancy in these supercells tends to be much less than that in the Great Plains supercells.[36]

4.5 Mesoscale convective systems

Mesoscale convective systems (MCSs), mentioned many times earlier, are convective storms that are organized on the mesoscale. They may form ready-made when convective storms are triggered along a linear boundary separating two distinct air masses such as a surface front, outflow boundary from earlier convection no longer in existence, or a dryline, etc. (Purdom 1976). They are "ready-made" in the sense that they begin as a long line of cells, on the mesoscale (tens to hundreds of kilometers) immediately, not on the convective scale (~ only 10–30 km).[37] Or, MCSs can form from just one ordinary cell, on the convective scale, which then evolves into a series of multicells. In the latter case, the outflow boundaries/gust fronts from cold pools produced by each cell conglomerate and "grow upscale" so that the boundaries along which subsequent cells

---

Figure 4.107 (Continued) *There is an inflow band on the southern side of the cloud base (left) and also one at midlevels. The area of the anvil appears massive compared to the area of the updraft base. This supercell developed into a "classic" supercell later on (Fig. 4.97, top left). Approaching a storm like this from a distance can be disappointing, since it looks so massive from a distance, but appears tiny from close range. Sometimes these storms dissipate quickly, while other times they develop more precipitation and are longer lived. (middle) Developing tornadic, low-precipitation supercell in northeast Colorado viewed to the west, over the Pawnee National Grassland, on July 21, 2000. At this early stage in the storm's life, virtually no precipitation is visible, save for some shafts on the northern (right) side of the cloud base. An "inflow tail" is visible on the southern (left) side of the cloud base (but is not a tail cloud because it is not feeding air into a wall cloud), while scud fragments are seen feeding into the storm area to its left. A line of low, cumuliform/scud-like clouds tails out on the northern (right) side of the cloud base and may be associated with a surface boundary along which the storm is forming. The tornadic stage of this supercell is seen in the bottom panel. (bottom panel) As in the middle panel, but later on (on July 21, 2000), when a tornado was occurring. Note the copious amount of light passing through the storm, both to the north (right) of it, under the base near the tornado (center), and to the south (left) of it. There is some virga falling from the bottom of the anvil (far left), which produced an apparent microburst at the surface later. The haze is probably caused by dust particles and perhaps some widely spaced precipitation. The view is to the southwest.*
Photos courtesy of author.

[36] McCaul, 1991.
[37] Bluestein and Jain, 1985.

Figure 4.108 *LP supercell viewed to the south, in West Texas, south of Olton, early in the evening of May 16, 2021 (top). A wall cloud is visible underneath the updraft base (right) and mamma are seen underneath the anvil emanating from the updraft cloud. Although the region beneath the anvil is translucent, there actually was precipitation detected by the surveillance WSR-88D Doppler radar at Lubbock. This storm had produced a tornado earlier in the evening. The eastern (left side) of the cloud base is flared outward, as in many supercells (e.g., Fig. 4.97). A cloud band is seen extending to the west (right) of cloud base, probably marking the leading edge of the RFGF (middle) as in the top panel, but the anvil and two convective storms in the distance to the southeast and are illuminated by the setting sun and the mamma are no longer illuminated; (bottom) at sunset, with brilliant orange/red coloring. This sequence of cloud images is in the spirit of Claude Monet's paintings of the cathedral at Rouen under different lighting conditions.*

Photos courtesy of author.

Figure 4.109 *Sequence of photographs (left to right, then down) of a dissipating LP supercell northwest of Limon, Colorado, on June 4, 2015, viewed to the north, through a "downscale transition." The first photograph (upper, left) shows a tail extending to the left of the cloud base, probably associated with a rear-flank gust front, which dissipates, and then a ragged-looking tail cloud appears briefly to the right (middle, right). Some precipitation is evident falling from a high cloud base downshear from the storm, to the right. The cloud base ultimately (bottom panel) looks like a ragged funnel cloud leaning in the downshear direction as it disappears.*
Photos courtesy of author.

form yield a much larger area of convection.[38] During their early evolution they may be composed of isolated supercells, or multicells of both.

[38] Weisman and Klemp, 1984.

Figure 4.110 *High-precipitation (HP) supercells viewed from the front. (top) An HP supercell, viewed to the west, in southeastern Colorado, west of Lamar, on June 11, 2009, during VORTEX-2. The base of the storm is flared out at the bottom. There is an opaque precipitation region underneath the cloud base and a small tail cloud on the north (right) side. While it is bright to the north, it is not optically translucent. (middle) Wide-angle view of a high-precipitation (HP) supercell, viewed to the west, near Okarche, Oklahoma, on May 29, 2012. The wall cloud appears on the right side, with the precipitation curtains making it look as if the wall cloud is housed in a cage; this cage is sometimes referred to by storm chasers as the "bear's cage." (bottom) High-precipitation (HP) supercell in western Oklahoma, near Calumet, on May 29, 2004. This storm produced a strong, damaging mesocyclone and an anticyclonic tornado.*

Photos courtesy of author.

Figure 4.111 *More HP supercells as seen visually, but also by radar. (top) Panoramic view of an HP supercell in the northern Texas Panhandle on May 16, 2016, viewed to the west (left) to the northeast*

4.5.1 Squall lines: Quasi-linear (mesoscale) convective systems (QLCSs)

The line-storm clouds fly tattered and swift,
  The road is forlorn all day,
Where a myriad snowy quartz stones lift,
  And the hoof-prints vanish away.
The roadside flowers, too wet for the bee,
  Expend their bloom in vain.
Come over the hills and far with me,
  And be my love in the rain.

The birds have less to say for themselves
  In the wood-world's torn despair
Than now these numberless years the elves,
  Although they are no less there:
All song of the woods is crushed like some
  Wild, easily shattered rose.
Come, be my love in the wet woods; come,
  Where the boughs rain when it blows.

There is the gale to urge behind
  And bruit our singing down,
And the shallow waters aflutter with wind
  From which to gather your gown.

Figure 4.111 (Continued) *(right). The updraft base is striated, there is some greenish tint along one of the striations, and mamma are pendant from the anvil on the right. It appears dangerously dark under cloud base when contrasted with the relative brightness surrounding it. (second row) HP supercell with a funnel cloud approaching Oklahoma City's Will Rogers Airport, viewed from Moore, Oklahoma, on April 23, 2022. Viewed to the northwest. The RFGF is enshrouded in heavy precipitation and a tail cloud is visible. (third row) Panoramic photographic of the precipitation region of an HP supercell in northwest Texas, on May 8, 2015, viewed to the west; the precipitation has a greenish tint, which is discussed in more detail at the end of this chapter. (bottom row, left) Radar reflectivity factor dBZ) at 2° elevation angle, east of Tulia, Texas, in the Texas Panhandle, on May 7, 2019, of an HP supercell, as seen by RaXPol. Note how a much broader area of precipitation is detected by radar around the mesocyclone than there is in the classic supercell seen in Figure 4.85. The tiny weak-echo hole marking a very small vortex near the center of the low-level mesocyclone is completely hidden from view by a thick barrier of heavy precipitation. The weak-echo notch is found, when looking from the right-rear quadrant of the HP supercell, just to the right of the cloud base, as a region that is brighter than the dark area where the precipitation surrounding the mesocyclone is located. (bottom row, right) As in the left panel, but for Doppler velocity (m s^{-1}). The low-level mesocyclone is enclosed by the circle. The deep-purple-coded Doppler velocities are directed toward the radar at 40 m s^{-1} and greater.*
Photos and figure courtesy of author.

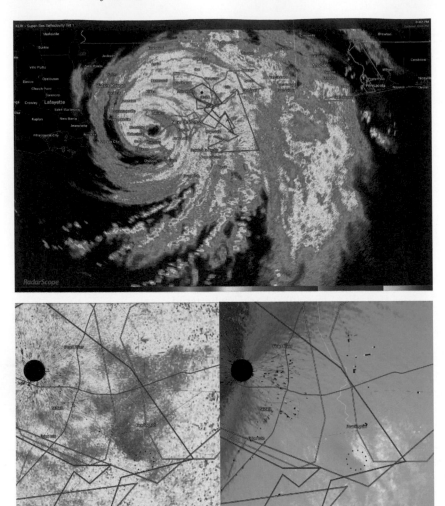

Figure 4.112 *Radar depiction of a supercell in a landfalling hurricane. WSR-88D radar images from KLIX (Slidell, Louisiana) on Aug. 29, 2021 shortly after Hurricane Ida made landfall. The narrow, black circle marks the location of the radar. (top) Radar reflectivity showing the eye west-southwest of New Orleans, with tornado warnings (within areas defined by red lines) issued by the National Weather Service, at 6:42 p.m. CDT. (left, bottom) Zoomed in radar reflectivity at 6:45 p.m. CDT, showing the supercell, one of several (red and orange color codes), surrounded by moderate rain (yellow color code). The radar is marked by a broad, black circle. Since the supercell was surrounded by moderate rain, it was probably very difficult to see detailed cloud and storm structure. The rear end of the supercell, which was rapidly moving to the west-northwest, had a curved band of heavy precipitation connected to it, similar to the band seen along the RFGF in a typical Great Plains supercell. (right, bottom) As in the left, bottom panel, but for Doppler velocity. The approximate location of the tornadic vortex signature (TVS) is indicated by a circular dotted line (blue-white to dark green color-coded Doppler velocities) and marked also in the left, bottom panel. The supercell was located approximately just 20 km from the radar.*

What matter if we go clear to the west,
    And come not through dry-shod?
For wilding brooch shall wet your breast
    The rain-fresh goldenrod.

Oh, never this whelming east wind swells
    But it seems like the sea's return
To the ancient lands where it left the shells
    Before the age of the fern;
And it seems like the time when after doubt
    Our love came back again.
Oh, come forth into the storm and rout
    And be my love in the rain.

Robert Frost, *A Line-storm Song*
(In the Public Domain)

The term squall line is still used today and is rooted in history. When I grew up in the Boston area, I remember squall lines coming in from the northwest late in the afternoon on some days during the summer, having formed in the Berkshires of western Massachusetts or in the mountains and higher terrain of southern New Hampshire. It got dark, the winds shifted to the northwest, it got cooler, there was distant thunder, and, if we were lucky, it would eventually rain, but frustratingly often the storms dissipated before reaching my house. Squall lines are today considered a special case of MCSs. Because they are comprised of a solid, broken (mostly solid, but with some breaks), or slightly wavy line of convective storms, they are now called "quasi-linear convective systems (QLCSs)."

They have a very specific architecture (Figs. 4.113 and 4.115), just as supercells do. As they continue to produce precipitation, an extensive cold pool builds up at the surface. New cells are triggered along the leading edge of the cold pool, along the gust front (Fig. 4.65), when the rate of import of horizontal vorticity associated with the low-level shear in the environment ahead of the storms is nearly opposite to that produced at the leading edge of the cold pool, a lá RKW theory (Fig. 4.69d). In nature, we tend to see winds "veering" with height (turning in a clockwise direction, say from southeasterly to southerly to southwesterly to westerly. In this case, if the leading edge of the cold pool is oriented in the north-south direction, then the wind at the surface has an easterly component, while the wind aloft has a westerly component: There is thus westerly shear, which induces rotation (vorticity) opposite to that produced at the leading edge of the cold pool where buoyancy creates a downward force adjacent to the region east, ahead of the cold pool where there is no buoyancy.

There is usually a "leading convective line," which produces a long line of intense radar echoes. As more and more precipitation falls below into unsaturated air, the cold pool builds up in strength (it gets colder and deeper); if the low-level shear in the environment does not change, then the rate at which horizontal vorticity is produced along the leading edge of the cold pool exceeds the rate at which horizontal vorticity of the

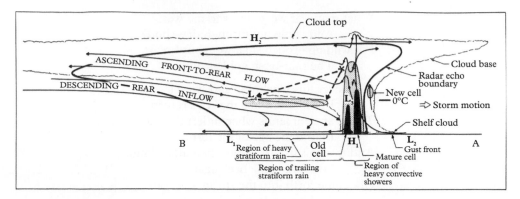

Figure 4.113 *Idealized vertical cross-section through a QLCS/squall line during its mature stage (from Houze et al., 1989, their Fig. 1. © American Meteorological Society. Used with permission). See also Figure 4.61, top panel, which shows a vertical cross-section through a real QLCS as seen by RaXPol. The thick black line marks the edge of the storm as it would appear on a radar. The light, scalloped line represents the edge of the cloud associated with the storm. The stippled regions represent moderate radar reflectivity and moderate rain along the "leading convective line," while the solid back regions represent heavy precipitation (rain and possibly some hail). The narrow, horizontal stippled region to the rear (left) of the leading edge of the storm represents the "bright band," a strip of enhanced radar reflectivity found at the "melting level," where there are water-coated ice particles. A mesoscale region of stratiform precipitation is found there. L_2 marks a mesoscale area of low hydrostatic pressure at the surface, which is sometimes found ahead of the line of convective storms, where air subsides and warms. A mesoscale, hydrostatic low-pressure area, a "wake low" (L_4) may be found at the rear (left) of the mesoscale area of precipitation, as dry air sinks and warms from aloft (coming in the "descending rear inflow"), as in a heat burst. A mesoscale, hydrostatic high-pressure area (H_1) is found underneath the leading edge of convection. The asterisk at the upper-rear portion of the leading convective line represents the location where ice particles flow from the top of the deep convective cloud at the leading edge, rearward, and fall slowly; they grow as water vapor condenses onto them and become large aggregates as they warm and fall. After they melt, they become moderate or heavy, trailing, stratiform rain.*

opposite sign is imported and a vertical circulation is produced that leans with height toward the rear of the QLCS (Fig. 4.114).

When the vertical circulation leans toward the rear, ice crystals, supercooled water droplets, and other mixed-phase particles are carried aloft toward the rear of the deep convection at the leading edge of the QLCS to produce a trailing anvil and a mesoscale, "stratiform precipitation area." The airflow is described as front-to-rear flow. The trailing anvil is often a place to look at sunset, especially when the anvil is overhead and extending to the west, for spectacular mamma (e.g., Fig. 4.30). It is prudent, when storm chasing, to "let" a mature MCS progress eastward or southeastward to avoid getting stuck in heavy precipitation and to await the mamma display at sunset (as mentioned earlier). I view the extensive regions of brilliant, colorful mamma at sunset as a prominent characteristic of the post-MCS/QLCS sky (e.g., Fig. 4.30).

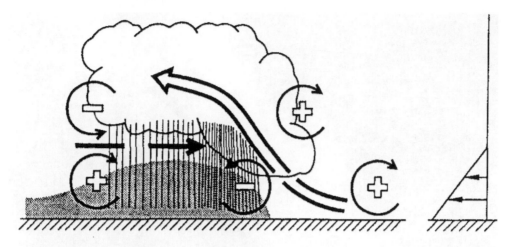

Figure 4.114 *This idealized vertical cross-section through a QLCS represents the next stage to that shown in the illustrations shown earlier in Figure 4.69. The vertical profile of the storm-relative winds is shown at the bottom right. Rotation (vorticity) is induced in the clockwise direction (indicated by the "+" inside the curved streamline) near the ground just ahead (to the right) of the QLCS by the vertical shear. The leading edge of the cold pool induces rotation (vorticity) of the opposite sense ("-;" counterclockwise direction). However, by now, unlike in Figure 4.69d, the cold pool has become so strong that the rotation (counterclockwise) in the opposite sense overwhelms the rotation induced by the vertical shear in the environment (clockwise) and the air flowing up and over the leading edge of the cold pool leans to the rear of the system with respect to height (large arrow leading double solid line). By this time, rotation induced at the rear of the cloud, owing to buoyancy in the cloud above, but no buoyancy outside the cloud to the rear, is in the counterclockwise direction, while rotation induced at the rear edge of the precipitation and cold pool (aloft) is in the clockwise direction. As a result of these two superimposed areas of rotation, the flow is enhanced from the rear into the storm, as indicated by the two thick, solid arrows. This feature is called the "rear-inflow jet."*
(From Weisman, 1993, their Fig. 14c. © American Meteorological Society. Used with permission.)

In between the stratiform precipitation and the leading line of deep convection there is a region of weaker precipitation, which is called the "transition zone" (Fig. 4.115). The larger particles fall out early near the leading convective line, while the smaller particles remain aloft longer and seed the clouds to the rear, producing the enhanced, stratiform precipitation area to the rear of the leading convective line, resulting in a lull in precipitation in between.

4.5.2 Bow echoes: Extreme QLCSs

If QLCSs persist for a number of hours, then counter-rotating vortices can develop along the opposite sides of the system, where horizontal vorticity generated at the edge of the gust front/cold pool is tilted onto the vertical at the opposite ends of the where the updraft flowing in over the cold pool line vanishes (Fig. 4.116). These mesoscale vortices

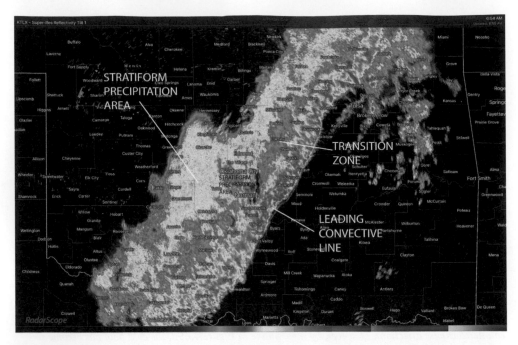

Figure 4.115 *An example of the radar reflectivity depiction of a QLCS from the WSR-88D radar near Oklahoma City, on Oct. 13, 2021 at 6:54 a.m. CDT.*
Figure courtesy of author.

were named "book-end vortices" by Morris Weisman at NCAR. With time, the cyclonic member on the poleward side may increase in intensity to become a "mesoscale convective vortex (MCV)." The cyclonic (counterclockwise in the Northern Hemisphere) member is favored when there is upward motion aloft and convergence near the surface. This convergence acts to increase the absolute vorticity. The absolute vorticity is the vorticity due to the storm's airflow relative to the ground plus Earth's vorticity, which is associated with the Coriolis force. On the other hand, the magnitude of the absolute vorticity in the anticyclonic vortex is less than that in the cyclonic vortex, because the relative vorticity and Earth's vorticity are of the opposite sign; in this case, the absolute vorticity is *less* than the Earth's vorticity. The punch line is that QLCSs, with time, may take on an "asymmetric-" looking structure and appear on radar to have a comma shape (Fig. 4.117), a mesoscale version of the synoptic-scale comma cloud we will look at in Chapter 5. MCVs, if they persist, may force new QLCSs the next day, and, when they are over a warm ocean, develop into a tropical wave or tropical cyclone if the vertical shear is not too strong.

We note, for later reference in Chapter 5, that when the gust front as depicted in Figure 4.116 is the RFGF of a supercell, that anticyclonic vorticity, clockwise rotation in the Northern Hemisphere, is produced at the far end (southern in the case of Fig. 4.116) (Fujita 1981) and is likely the source of vorticity for some anticyclonic tornadoes.

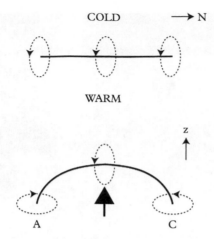

Figure 4.116 *Schematic showing how book-end vortices ("A" is the anticyclonic, clockwise-turning vortex and "C" is the cyclonic, counterclockwise turning vortex, in the Northern Hemisphere) are produced at the opposite ends of a north–south oriented, mature[a] gust front. In the top panel, rotation is produced owing to the gradient of buoyancy, from negative buoyancy behind the gust front, to no buoyancy in the warm air ahead of the gust front to the east. In the bottom panel, counter-rotating vortices are produced by tilting of the vorticity (rotation such that sinking motion is found just behind, to the west, of the gust front and rising motion just ahead of the gust front to the east) at the far ends of the gust front by an updraft (thick arrow). This updraft is driven by air that is forced up and over the gust front, lifted to its LFC, resulting in a buoyant updraft.*

[a] On the other hand, early on, before much precipitation has fallen and before a cold pool has been produced, these vortices can form when a buoyant updraft in the flanking line acts on low-level easterly shear brought in from the environment (particularly when there is a "low-level jet" of air from the east above the surface), ahead of the flanking line.
Figure courtesy of author.

One other aspect of long-lasting QLCSs bears mention. At the rear edge of the QLCS, horizontal vorticity of opposite signs is produced near the base of the cloud (Fig. 4.114): The gradient of rising, buoyant air in the cloud with air having no vertical motion at all outside the cloud is juxtaposed above a gradient of sinking, negatively buoyant air in the cold pool below with no vertical motion at all outside the cold pool, to yield a net flow of air from outside the convective system into it. This airflow into the convective storm complex is called the "rear-inflow jet" (Weisman 1992). It extends from the rear of the storm toward the front of the storm. When the rate at which horizontal vorticity is produced in one way at the rear above the cold pool is less than produced the other way below the cold pool, the rear-inflow jet descends to the ground, resulting in very strong, damaging winds at the leading edge of the gust front. This situation requires the cold pool to be relatively strong. The radar-echo pattern at the leading edge then takes on a bow shape and is therefore called a "bow echo" (Fig. 4.117). Ted Fujita coined the phrase "bow echo." When these windstorms may produce a long swath of damage because they are long-lived and travel great distances, they are called "derechos" (Johns

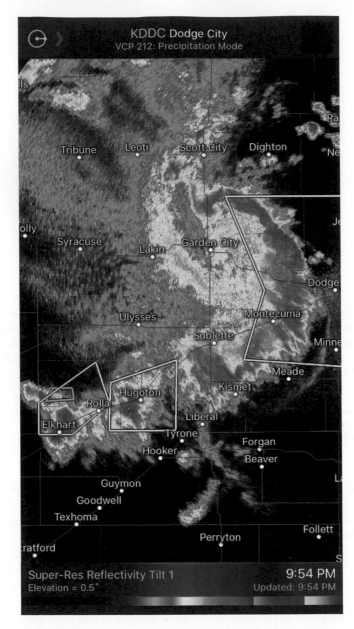

Figure 4.117 *WSR-88D radar depiction of an asymmetric MCS on May 6, 2019, with a comma shape on the northern end, a fine line (color-coded light blue) marking an eastward-bulging gust front ahead of the leading line, and isolated cells on the southwestern end, in far southwestern Kansas. The tip of the comma-shaped radar echo was rotating cyclonically (in a counterclockwise direction). The time shown is in CDT. The two cells at the southwestern end are severely warned (yellow quadrilaterals). The bulge in the radar echo at the leading edge of the MCS is called a "bow echo."*

and Hirt 1987). QLCSs with bow echoes can produce awe-inspiring shelf clouds at their leading edge such as the one shown in Figures 4.55 and 4.57, bottom panel.

4.6 Overall summary of convective storms

When convective clouds with buoyant updrafts extend high up in the troposphere, precipitation may form. Precipitation falling into unsaturated air and melting of precipitation, if the top of the cloud is well below 0° C, both result in the production of a low-level cold pool and an ordinary-cell convective storm. Low-level vertical shear in the environment associated with horizontal vorticity, when imported at approximately the same rate at which horizontal vorticity is generated at the leading edge of the cold pool, can foster periodic growth of convective cells associated with a multicell convective storm. If the vertical shear in the lowest half of the troposphere is also large, rotating updrafts are forced in the direction normal to the shear by upward-directed dynamic perturbation-pressure-gradient forces associated with the interaction of the updrafts with the vertical shear. These storms become supercells, which are much longer lived than ordinary cells and multicells, and may generate low-level mesocyclones that originate in large part from the region of horizontal temperature gradient along the edge of the FFD. Even when the deep-layer shear is not strong enough for supercells, a conglomeration of cold pools may lead to the formation of an MCS with a trailing region of stratiform precipitation and possibly the development of an asymmetric structure with book-end vortices, an MCV, and a bow-echo structure.

Figure 4.118 *Green cloud (left) in precipitation behind the rear-flank gust front of an HP supercell in western Oklahoma on May 31, 2022. View is to the north. The color reproduced here is more of an aqua tint, rather than pure green.*
Photo courtesy of author.

To me, supercells are the holy grail of convective storms, owing to their beauty, their ability to produce tornadoes and large hail, and their relative rarity compared to ordinary-cell and multicell cumulonimbus clouds. They remind me of my boyhood search for seashells on the shoreline of South Miami Beach, where I found that the supercell of shells was the rarer and more beautiful (in my mind) univalves such as left-handed whelk shells and cone shells, while the ordinary cell–multicell of shells was the ubiquitous and the less beautiful bivalves such as turkey wing shells and clam shells.

One postscript to our look at convective storms in general, regardless of whether they are supercells, multicells, or MCSs: A greenish hue, not easily reproduced in photographs (Fig. 4.111, top and third rows; 4.118), is sometimes seen. Some people have postulated that green clouds are associated with large hail, but my experience has been that green clouds are rather associated with lots of rain and perhaps some small hail, and also possibly a thinness in the clouds as, for example, seen in some mamma though in the latter figure the color is more aqua than green. The thinness allows some sunlight to pass through. Craig Bohren and Alistair Fraser at Pennsylvania State University in 1993 hypothesized that green clouds do not actually "reflect" (my pun; my addition of quotes) the color of the clouds, but rather are a result of selective scattering or the absorption by hydrometeors of light late in the day.[39] Bill Beasley, a former colleague of mine at OU, once did a study with one of his students, Frank Gallagher, in which they used a spectrophotometer and found that green clouds are indeed green.[40]

References

Achtemeier, G. L., 1991: The use of insects as tracers for "clear-air" boundary-layer studies by Doppler radar. *J. Atmos. Ocean. Technol.*, 8, 746–765.

Atkins, N. T., and R. M. Wakimoto, 1991: Wet microburst activity over the southeastern United States: Implications for forecasting. *Wea. Forecasting*, 6, 470–482.

Atkins, N. T., R. M. Wakimoto, and T. M. Weckwerth, 1995: Observations of the sea-breeze front during CaPE. Part II: Dual-Doppler and aircraft analysis. *Mon. Wea. Rev.*, 123, 944–969.

Atlas, D. (ed.), 1990: Radar in meteorology. *Amer. Meteor. Soc.*, 21, 806 pp.

Barzun, J., 2000, *From Dawn to Decadence*, p. 209, Harper Perennial, 912 pp.

Bedka, K. M., E. Murillo, C. R. Homeyer, B. Scarino, and H. Mersiovski, 2018: The above anvil-cirrus plume: An important severe weather indicator in visible and infrared satellite imagery. *J, Appl. Meteor.*, 33, 1159–1181.

Benjamin, T. B., 1968: Gravity currents and related phenomena. *J. Fluid Mech.*, 31, 209–248.

Biggerstaff, M. I., L. J. Wicker, J. Guynes, C. Ziegler, J. M. Straka, E. N. Rasmussen, A. Doggett IV, L. D. Carey, J. L. Schroeder, and C. Weiss, 2005: The Shared Mobile Atmospheric Research and Teaching Radar: A collaboration to enhance research and teaching. *Bull. Amer. Meteor. Soc.*, 86, 1263–1274.

Bluestein, H. B., 2008: On the decay of supercells through a "downscale transition:" Visual documentation. *Mon. Wea. Rev.*, 136, 4013–4028.

---

[39] Bohren and Fraser, 1993.
[40] Gallagher III et al., 1996.

Bluestein, H. B., 2009: The formation and early evolution of the Greensburg, Kansas supercell on 4 May 2007. *Wea. Forecasting*, **24**, 899–920.

Bluestein, H. B., and M. H. Jain, 1985: Formation of mesoscale lines of precipitation: Severe squall lines in Oklahoma during the spring. *J. Atmos. Sci.*, **42**, 1711–1732.

Bluestein, H. and C. Sohl, 1979: Some observations of a splitting severe thunderstorm. *Mon. Wea. Rev.*, **107**, 861–873.

Bluestein, H.B. and W. P. Unruh, 1989: Observations of the wind field in tornadoes, funnel clouds, and wall clouds with a portable Doppler radar. *Bull. Amer. Meteor. Soc.*, **70**, cover and 1514–1525.

Bluestein, H. B., and A. L. Pazmany, 2000: Observations of tornadoes and other convective phenomena with a mobile, 3-mm wavelength, Doppler radar: The spring 1999 field experiment. *Bull. Amer. Meteor. Soc.*, **81**, 2939–2951.

Bluestein, H. B., and M. L. Weisman, 2000: The interaction of numerically simulated supercells initiated along lines. *Mon. Wea. Rev.*, **128**, 3128–3149.

Bluestein, H. B., M. M. French, I. PopStefanija, R. T. Bluth, and J. B. Knorr, 2010: A mobile, phased-array Doppler radar for the study of severe convective storms: The MWR-05XP. *Bull. Amer. Meteor. Soc.*, **91**, 579–600.

Bluestein, H. and Coauthors, 2014: Radar in the atmospheric sciences and related research: Current systems, emerging technology, and future needs. *Bull. Amer. Meteor. Soc*, **95**, 1850–1861.

Bluestein, H. B., J. C. Snyder, and J. B. Houser, 2015: A multi-scale overview of the El Reno, Oklahoma tornadic supercell of 31 May 2013. *Wea. Forecasting*, **30**, 525–552.

Bluestein, H. B., D. T. Lindsey, D. Bikos, D. W. Reif, and Z. B. Wienhoff, 2019: The relationship between overshooting tops in a tornadic supercell and its radar-observed evolution. *Mon. Wea. Rev.*, **147**, 4151–4176.

Bluestein, H. B., F. C. Carr, and S. J. Goodman, 2022: Atmospheric observations of weather and climate. *Atmos. – Oceans*, DOI: 10.1080/07055900.2022.2082369. 39 pp.

Bohren, C. F., and A. B. Fraser, 1993: Green thunderstorms. *Bull. Amer. Meteor. Soc.*, **74**, 2185–2193.

Brown, R. A., L. R. Lemon, and D. W. Burgess, 1978: Tornado detection by pulsed Doppler radar. *Mon. Wea. Rev.*, **106**, 29–38.

Browning, K. A., R. J. Donaldson, Jr., 1963: Airflow and structure of a tornadic storm. *J. Atmos.Sci.*, **20**, 533–545.

Browning, K. A., 1964: Airflow and precipitation trajectories within severe local storms which travel to the right of the winds. *J. Atmos. Sci.*, **21**, 634–668.

Byers, H. R., and R. R. Braham, Jr., 1949: *The Thunderstorm*. U. S. Gov't Printing Office, 287 pp.

Charba, J., 1974: Application of gravity current model to analysis of squall-line gust front. *Mon. Wea. Rev.*, **102**, 140–156.

Charba, J., and Y. Sasaki, 1971: Structure and movement of the severe thunderstorms of 3 April 1964 as revealed from radar and surface mesonetwork data analysis. *J. Meteor. Soc. Japan*, 191–213.

Clarke, R. H., R. K. Smith, and D. G. Reid, 1981: The morning glory of the Gulf of Carpentaria: An atmospheric undular bore. *Mon. Wea. Rev.*, **109**, 1726–1750.

Christie, D. R., 1992: The morning glory of the Gulf of Carpentaria: a paradigm for non-linear waves in the lower atmosphere. *Aust. Met. Mag.*, **41**, 21–60.

Crum, T. D. and R. L. Alberty, R. L., 1993: The WSR-88D and the WSR-88D operational support test facility. *Bull. Amer Meteor. Soc.*, 74, 1669–1687.

Davies, J. M., 2006: Tornadoes with cold core 500-mb lows. *Wea. Forecasting*, 21, 1051–1062.

Davies-Jones, R., 1984: Streamwise vorticity: The origin of updraft rotation in supercell storms. *J. Atmos. Sci.*, 41, 2991–3006.

Davies-Jones, R. 2002: Linear and nonlinear propagation of supercell storms. *J. Atmos. Sci.*, 59, 3178–3205.

Donaldson, R. J., Jr., 1970: Vortex signature recognition by a Doppler radar. *J. Appl. Meteor.*, 9, 661–670.

Doviak, R. J., and R. Ge, 1984: An atmospheric solitary gust observed with a Doppler radar, a tall tower, and a surface network. *J. Atmos. Sci.*, 41, 2559–2573.

Doviak, R. J., and D. S. Zrnić, D. S., 2006: *Doppler Radar and Weather Observations (2nd ed.)* Dover, San Diego, California, 562 pp.

Dowell, D. C., H. B. Bluestein, and D. P. Jorgensen, 1997: Airborne Doppler radar analysis of supercells during COPS-91. *Mon. Wea. Rev.*, 125, 365–383.

Droegemeier, K. K., and R. B. Wilhelmson, 1987: Numerical simulation of thunderstorm outflow dynamics. Part I: Outflow sensitivity experiments and turbulence dynamics. *J. Atmos.Sci.*, 44, 1180–1210.

Fabry, F., 2015: *Radar Meteorology: Principles and Practice.* Cambridge University Press, Cambridge, U.K., 256 pp.

Fovell, R. G., and P. – H. Tan, 1998: The temporal behavior of numerically simulated multicell-type storms. Part II: The convective cell life cycle and cell regeneration. *Mon. Wea. Rev.*, 126,551–577.

French, M. M., H. B. Bluestein, I. PopStefanija, C. A. Baldi, and R. T. Bluth, 2013: Reexamining the vertical development of descending tornadic vortex signatures in supercells. *Mon. Wea. Rev.*, 141, 4576–4601.

Fromm, M., and coauthors, 2010: The untold story of pyrocumulonimbus. *Bull. Amer. Meteor. Soc.*, 91, 1193–1209.

Fujita, T. T., 1960: *A Detailed Analysis of the Fargo Tornadoes of June 20, 1957.* Research Paper to the U. S. Weather Bureau, No. 42, University of Chicago, 67 pp.

Fujita, T. T., 1963: Analytical mesometeorology: A review. *Meteor. Monogr.*, 5, 77–128.

Fujita, T. T., 1982: Principles of stereoscopic height computations and their applications to stratospheric cirrus over severe thunderstorms. *J. Meteor. Soc. Japan*, 60, 355–368.

Fujita, T. T., 1981: Tornadoes and downbursts in the context of generalized planetary scales. *J. Atmos. Sci.*, 38, 1511–1534.

Gallagher, III, F. W., W. H. Beasley, and C. F. Bohren, 1996: Green thunderstorms observed. *Bull. Amer. Meteor. Soc.*, 77, 2889–2897.

Haghi, K. R., D. B. Parsons, and A. Shapiro, 2017: Bores observed during IHOP_2002: The relationship of bores to the nocturnal environment. *Mon. Wea. Rev.*, 145, 3929–3946.

Heymsfield, G. M., L. Tian, A. J. Heymsfield, L. Li, and S. Guimond, 2010: Characteristics of deep tropical and subtropical convection from nadir-viewing high-altitude airborne Doppler radar. *J. Atmos. Sci.*, 67, 285–308.

Heymsfield, G. M., L. Tian, L. L., McLinden, M., and J. I. Cervantes, 2013: Airborne radar observations of severe hailstorms: Implications for future spaceborne radar. *J. Appl. Meteor. Climatol.*, 52, 1851–1867.

Hills, M. O. G., and D. R. Durran, 2014: Quantifying moisture perturbations leading to stacked lenticular clouds. *Quart. J. Roy. Meteor. Soc.*, **140**, 2013–2016.

Hitschfeld, W., 1960: The motion and erosion of convective storms in severe vertical wind shear. *J. Meteor.*, **17**, 270–282.

Hjelmfelt, M. R., 1988: Structure and life cycle of microburst outflows observed in Colorado. *J. Appl. Meteor.*, **27**, 900–927.

Houze, R. A., Jr., and C. – P. Cheng, 1977: Radar characteristics of tropical convection observed during GATE: Mean properties and trends over the summer season. *Mon. Wea. Rev.*, **105**, 964–980.

Houze, R. A., Jr., S. A. Rutledge, M. I. Biggerstaff, and B. F. Smull, 1989: Interpretation of Doppler weather radar displays of midlatitude mesoscale convective systems. *Bull. Amer. Meteor. Soc.*, **70**, 608–619.

Hutson, A., C. Weiss, and G. Bryan, 2019: Using the translation speed and vertical structure of gust fronts to infer buoyancy deficits within thunderstorm outflow. *Mon. Wea. Rev.*, **147**, 3575–3594.

Idso, S. B., R. S. Ingram, and J. M. Pritchard, 1972: An American haboob. *Bull. Amer. Meteor. Soc.*, **53**, 930–935.

Isom, B., Palmer, R., Kelley, R., Meier, J., Bodine, D., Yeary, M., Cheong, B.-L., Zhang, Y., Yu, T.-Y. & Biggerstaff, M. I. (2013). The Atmospheric Imaging Radar: Simultaneous volumetric observations using a phased-array weather radar. *J. Atmos. Oceanic Technol.*, **30**, 655–675.

Johns, R. H., and W. D. Hirt, 1987: Derechos: Widespread convectively induced windstorms. *Wea. Forecasting*, **2**, 32–49.

Kanak, K. M., and J. M. Straka, 2002: An unusual reticular cloud formation. *Mon. Wea. Rev.*, **130**, 416–421.

Kanak, K. M., J. M. Straka, and D. M. Schultz, 2008: Numerical simulation of mammatus. *J.Atmos. Sci.*, **65**, 1606–1621.

Kennedy, P. C., N. E. Wescott, and R. W. Scott, 1993: Single-Doppler radar observations of a mini-supercell tornadic thunderstorm. *Mon. Wea. Rev.*, **121**, 1860–1870.

Kessler, E. (ed.), 1986: *Thunderstorm Morphology and Dynamics, 2nd ed.* University of Oklahoma Press, Norman, 411 pp.

Klemp, J. B., 1987: Dynamics of tornadic thunderstorms. *Ann. Rev. Fluid Mech.*, **19**, 369–402.

Klemp, J. B., and R. B. Wilhelmson, 1978: Simulations of right- and left-moving storms produced through storm splitting. *J. Atmos. Sci.*, **35**, 1097–1110.

Kristovich, D. A. R., N. F. Laird, and M. R. Hjelmfelt, 2003: Convective evolution across Lake Michigan during a widespread lake-effect snow event. *Mon. Wea. Rev.*, **131**, 643–655.

Kumjian, M. R., and A. V. Ryzhkov, 2008: Polarimetric signatures in supercell thunderstorms. *J. Appl. Meteor. Climatolog.*, **47**, 1940–1961.

Kummerow, C. D., W. Barnes, T. Kozu, T. Shiue, and J. Simpson, 1998: The Tropical Rainfall Measuring Mission (TRMM) sensor package. *J. Atmos. Oceanic Technolog.*, **15**, 809–817.

Lemon, L. R., 1976: The flanking line, a severe thunderstorm intensification source. *J. Atmos. Sci.*, **33**, 686–694.

Ludlam, F. H., 1963: Severe local storms: a review. *Meteor. Monogr.*, **5**, Amer. Meteor. Soc., Boston, Mass., 1–30.

Ludlam, F. H., 1990: *Clouds and Storms: The Behavior and Effect of Water in the Atmosphere.* Penn. State University Press, State College, Penn., 488 pp.

Marks, Jr., F. D., and R. A. Houze, Jr., 1984: Airborne Doppler radar observations in Hurricane Debby. *Bull. Amer. Meteor. Soc.*, **65**, 569–582.

McCaul, E. W., Jr., 1991: Buoyancy and shear characteristics of hurricane-tornado environments. *Mon. Wea. Rev.*, **119**, 1954–1978.

Markowski, P., and Y. Richardson, 2010: *Mesoscale Meteorology in Midlatitudes*. Wiley, 407 pp.

Moller, A. R., 1978: The improved NWS storm spotter's training program at Ft. Worth, TX. *Bull.Amer. Meteor. Soc.*, **59**, 1574–1582.

Moller, A. R., C. A. Doswell III, M. P. Foster, and G. R. Woodall, 1994: The operational recognition of supercell thunderstorm environments and storm structures. *Wea. Forecasting*, **9**, 327–347.

Nesbitt, S. W., and E. J. Zipser, 2003: The diurnal cycle of rainfall and convective intensity according to three years of TRMM measurements. *J. Climate*, **16**, 1456–1475.

Newton, C. W., 1963: Dynamics of severe convective storms. *Meteor. Mongr.*, **5**, Amer. Meteor. Soc., Boston, Mass., 33–58.

Newton, C. W., and S. Katz, 1958: Movement of large convective rainstorms in relation to winds aloft. *Bull. Amer. Meteor. Soc.*, **39**, 129–136.

Newton, C. W., and J. C. Fankhauser, 1975: Movement and propagation of multicellular convective storms. *Pure Appl. Geophys.*, **113**, 748–764.

Newton, C. W., and H. R. Newton, 1959: Dynamical interactions between large convective clouds and environment with vertical shear. *J. Meteor.*, **16**, 483–496.

Niziol, T. A., W. R. Snyder, and J. S. Waldstreicher, 1995: Winter weather forecasting throughout the Eastern United States. Part IV: Lake effect snow. *Wea. Forecasting*, **10**, 61–77.

O'Neill, M. E., L. Orf, G. M. Heymsfield, and K. Halbert, 2021: Hydraulic jump dynamics above supercell thunderstorms. *Science*, **373**, 1248–1251.

Parsons, D. B., and R. A. Kropfli, 1990: Dynamics and fine structure of a microburst. *J. Atmos. Sci.*, **47**, 1674–1692.

Pazmany, A. L., J. B. Mead, H. B. Bluestein, J. C. Snyder, and J. B. Houser, 2013: A mobile, rapid-scanning, X-band, polarimetric (RaXPol) Doppler radar system. *J. Atmos. Ocean. Technol.*, **30**, 1398–1413.

Peterson, D. A., E. J. Hyer, J. R. Campbell, J. E. Solbrig, and M. D. Fromm, 2017: A conceptual model for development of intense pyrocumulonimbus in western North America. *Mon. Wea. Rev.*, **145**, 2235–2255.

Purdom, J. F. W., 1976: Some uses of high-resolution GOES imagery in the mesoscale forecasting of convection and its behavior. *Mon. Wea. Rev.*, **104**, 1474–1483.

Rauber, R. J., and S. W. Nesbitt, 2018: *Radar Meteorology: A First Course*. Wiley, 496 pp.

Reif, D. W., H. B. Bluestein, T. M. Weckwerth, Z. B. Wienhoff, and M. B. Chasteen, 2020: Estimating the maximum vertical velocity at the leading edge of a density current. *J. Atmos. Sci.*, **77**, 3683–3700.

Roberts, R. D., and J. W. Wilson, 1989: A proposed microburst nowcasting procedure using single-Doppler radar. *J. Appl. Meteor.*, **28**, 285–303.

Rottman, J. W., and J. E. Simpson, 1989: The formation of internal bores in the atmosphere: A laboratory model. *Quart. J. Roy. Meteor. Soc.*, **115**, 941–963.

Rotunno, R., and J. B. Klemp, 1982: The influence of the shear-induced pressure gradient on thunderstorm motion. *Mon. Wea. Rev.*, **110**, 136–151.

Rotunno, R., and J. B. Klemp, 1985: On the rotation and propagation of simulated supercell thunderstorms. *J. Atmos. Sci.*, **42**, 271–292.

Rotunno, R., J. B. Klemp, and M. L. Weisman, 1988: A theory for strong, long-lived squall lines. *J. Atmos. Sci.*, **45**, 463–485.

Rust, D. W., and D. R. MacGorman, 1998: *The Electrical Nature of Storms.* Oxford University Press, 422 pp.

Schueth, A., C. Weiss, and J. M. L. Dahl, 2021: Comparing observations and simulations of the streamwise vorticity current and the forward-flank convergence boundary in a supercell storm. *Mon. Wea. Rev.,* **149,** 1651–1671.

Schultz, D. M. and Coauthors, 2006: The mysteries of mammatus clouds: Observations and formation mechanisms. *J. Atmos. Sci.,* **63,** 2409–2435.

Serafin, R. J., and Coauthors, 2003: *Radar and Atmospheric Science: A Collection of Essays in Honor of David Atlas,* R. M. Wakimoto and R. C. Srivastava, eds., American Meteorological Society, pp.

Simpson, J. E., 1997: *Gravity Currents in the Environment and the Laboratory, 2nd ed.* Cambridge University Press, 244 pp.

Snyder, J. C., H. B. Bluestein, D. T. Dawson II, and Y. Jung, 2017: Simulations of polarimetric, X-band radar signatures in supercells. Part II: Z_{DR} columns and rings and K_{DP} columns. *J. Appl. Meteor. Climatol.,* **56,** 2001–2026.

Srivastava, R., 1987: A model of intense downdrafts driven by the melting and evaporation of precipitation. *J. Atmos. Sci.,* **44,** 1752–1773.

Suzuki, O., H. Niino, H. Ohno, and H. Nirasawa, 2000: Tornado-producing mini supercells associated with Typhoon 9019. *Mon. Wea. Rev.,* **128,** 1868–1882.

Tanamachi, R. L., H. B. Bluestein, J. B. Houser, S. J. Frasier, and K. M. Hardwick, 2012: Mobile, X-band, polarimetric Doppler radar observations of the 4 May 2007 Greensburg, Kansas, tornadic supercell. *Mon. Wea. Rev.,* **140,** 2103–2125.

Tory, K. J., and J. D. Kepert, 2021: Pyrocumulonimbus firepower threshold: Assessing the atmospheric potential for pyroCb. *Wea. Forecasting,* **36,** 439–456.

Trapp, R. J., 2013: *Mesoscale-Convective Processes in the Atmosphere.* Cambridge University Press, 377 pp.

Wakimoto, R. M., 1985: Forecasting dry microburst activity over the High Plains. *Mon. Wea. Rev.,* **113,** 1131–1143.

Wakimoto, R. M., and N. T. Atkins, 1994: Observations of the sea-breeze front during CaPE. Part I: Single-Doppler, satellite, and cloud photogrammetry analysis. *Mon. Wea. Rev.,* **122,** 1092–1114.

Wakimoto, R. M., and V. N. Bringi, 1988: Dual-polarization observations of microbursts associated with intense convection: The 20 July storm during the MIST project. *Mon. Wea. Rev.,* **116,** 1521–1539.

Wakimoto, R. M., and R. Srivastava, 2003: *Radar and Atmospheric Science: A Collection of Essays in Honor of David Atlas.* Amer. Meteor. Soc., Boston, 270 pp.

Wakimoto, R. M., W. – C. Lee, H. B. Bluestein, C. – H. Liu, and P. H. Hildebrand, 1996 :ELDORA observations during VORTEX 95. *Bull. Amer. Meteor. Soc.,* 77, 1465–1481.

Wakimoto, R. M., N. T. Atkins, K. M. Butler, H. B. Bluestein, K. Thiem, J. Snyder, and J. Houser, 2015: Photogrammetric analysis of the 2013 El Reno tornado combined with mobile X-band polarimetric data. *Mon. Wea. Rev.,* **143,** 2657–2683.

Weisman, M. L., 1992: The role of convectively generated rear-inflow jets in the evolution of long-lived meso-convective systems. *J. Atmos. Sci.,* **49,** 1826–1847.

Weisman, M. L., 1993: The genesis of severe, long-lived bow echoes. *J. Atmos. Sci.,* **50,** 645–670.

Weisman, M. L., and Klemp, 1986: Characteristics of isolated convective storms. *Mesoscale Meteorology and Forecasting,* P. Ray, ed., Amer. Meteor. Soc., Boston, 331–358.

Weisman, M. L., and J. B. Klemp, 1982: The dependence of numerically simulated convective storms on wind shear and buoyancy. *Mon. Wea. Rev.*, **110**, 504–520.

Weisman, M. L., and J. B. Klemp, 1984: The structure and classification of numerically simulated convective storms in directionally varying wind shears. *Mon. Wea. Rev.*, **112**, 2479–2498.

Wienhoff, Z. B., H. B. Bluestein, D. W. Reif, R. M. Wakimoto, L. J. Wicker, and J. Kurdzo, 2020: Analysis of debris signature characteristics and evolution in the 24 May 2016 Dodge City, Kansas, tornadoes. *Mon. Wea. Rev.*, **148**, 5063–5086.

Wilson, J. W., and C. K. Mueller, 1993: Nowcasts of thunderstorm initiation and evolution. *Wea. Forecasting*, **8**, 113–131.

Wilson, J. W., and R. M. Wakimoto, 2001: The discovery of the downburst: T. T. Fujita's contribution. *Bull. Amer. Meteor. Soc.*, **82**, 49–62.

Wurman, J., J. M. Straka, and E. N. Rasmussen, 1996: Fine-scale Doppler radar observations of tornadoes. *Science*, **272**, 1774–1777.

Zhang, G., 2016: *Weather Radar Polarimetry*. CRC Press, Boca Raton, Florida, 304 pp.

Zrnić, D. S., A. V. Ryzhkov, 1999: Polarimetry for weather surveillance radars. *Bull. Amer. Meteor. Soc.*, **80**, 389–406.

5
Clouds Influenced by Rotation

Aloft all hands, strike the top-masts and belay;
Yon angry setting sun and fierce-edged clouds
Declare the Typhon's coming.
Before it sweeps your decks, throw overboard
The dead and dying—ne'er heed their chains Hope,
Hope, fallacious Hope!
Where is thy market now?

—J. M. W. Turner (1812)
(from his unpublished poem "Fallacies of Hope")[1]

5.1 Clouds associated with vortices

A vortex is an airflow pattern that is characterized by a circular array of wind vectors. If you were to follow a parcel of air, you would travel around in a circle. It is an atmospheric feature that is found on many different space and time scales. Some vortices consist of vertical columns of rotating air (like tornadoes and waterspouts), while others are tilted columns or even horizontal columns of rotating air (as we noted in Chapter 2 in rotors and in Chapter 4 in circulations (rolls) produced along outflow boundaries). We first consider the biggest vortices.

If one looks at Earth from space, one sees evidence of some of these vertical columns of rotating air. For example, on the planetary scale, the winds tend to blow from west to east in midlatitudes (~35–55° N and S) at high altitudes in the troposphere (~5–10 km AGL). This circumpolar vortex is known as the *polar jet* (the term *jet stream* is also used to describe a localized, concentrated region of airflow within the polar jet), a meandering ring of airflow around the North Pole and another one around the South Pole. Each ring of airflow can be thought of as a planetary-scale vortex (Fig. 5.1) when viewed looking down from above, at either pole. The *subropical jet* is also found, but at lower latitudes (~20–35° N and S). Both jets persist for periods of many days. When averaged over a season for many years, the polar jet and subtropical jet show up as localized westerly wind maxima at some latitude and height above the ground.

The polar jet (or subtropical jet) is not necessarily continuous around the entire globe, and may contain undulations on the synoptic scale (i.e., having time scales of

---

[1] From a panel on view at the Museum of Fine Arts, Boston, Massachusetts, May 28, 2022.

The Architecture of Clouds. Howard B. Bluestein, Oxford University Press. © Howard B. Bluestein (2024).
DOI: 10.1093/oso/9780198870548.003.0005

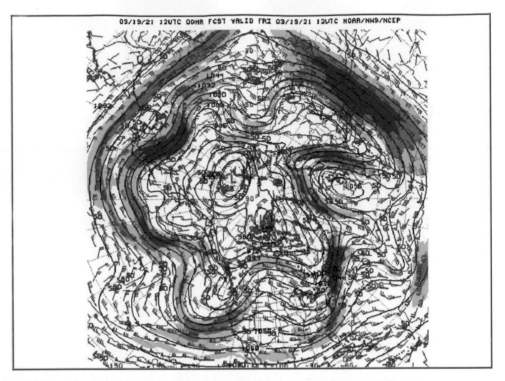

Figure 5.1 *Map of winds and heights at 250 hPa (~10–11 km altitude) at 1200 UTC on March 19, 2021 in the Northern Hemisphere, but outside the tropics. Extremely strong winds are color-coded purple, very strong winds are coded dark blue, and just strong winds are coded light blue. The lines connecting the barbs sticking out from them point in the direction from which the wind is blowing; half barb, whole barb, flag indicate 5, 10, and 50 kts (2.5, 5, and 25 m s^{-1}), respectively.* (From NOAA.)

days and space scales of thousands of kilometers), some of which form bulges poleward or equatorward (Figs. 5.1 and 5.2). The bulges can be described as the superposition of the jet with synoptic-scale vortices, producing wavelike patterns in the wind flow that are known as troughs and ridges[2] (the most equatorward locations and the most poleward locations, respectively, in the pattern swept out by each streamline representing the wind field). These troughs and ridges during portions of their lifetimes exhibit some tilt with height. The troughs and ridges, which appear to move (propagate actually, since they are not solid objects) generally from west to east, are responsible for our everyday weather in midlatitudes and its changes.

Downstream from troughs (and upstream from ridges) there tends to be rising motion and cooling to the saturation point, resulting in clouds; upstream from troughs (and downstream from ridges) there tends to sinking motion and clear skies. However,

---

[2] They are most significantly described as troughs and ridges in the pressure field; troughs are zones of relatively low pressure and ridges are zones of relatively high pressure. The corresponding waviness in the wind field is due to the relationship between the geostrophic wind and the pressure field.

some of the clouds produced downstream from troughs may spill over the downstream ridges, while the clear regions upstream from troughs (sometimes called the "dry slot") may intrude into the cloudy region downstream from troughs. This pattern looks like the yin-yang symbol (Fig. 5.3), which is indeed appropriate, since the symbol represents opposite forces that are interconnected: sinking and rising air motions; clear, subsaturated air and cloudy, saturated air. The pattern of clouds seen on a satellite image often take the shape of a comma, and is therefore referred to as a *comma cloud* (Carlson 1980) (Fig. 5.3), which may "give one pause" as to what it means. Some of the bulges may be so extreme that there is a complete disruption in the east-to-west flow and the flow is said to be "blocked." When there are clouds associated with the jet, they tend to be high-altitude cirrus (where it is very cold) and may be carried thousands of kilometers from their source (i.e., where there is rising motion and enough moisture for air to be lifted to its LCL). We noted in Chapter 2 that these cirrus clouds are sometimes referred to as "forerunners." Evidence of atmospheric mischief (stormy weather) may thus be carried far from the "crime scene" (i.e., where they were conceived) by the jet stream.

The polar jet tends to be associated with strong pole-to-equator temperature gradients, which are concentrated within the confines of a belt of latitudes. It is thus relatively cold on the polar side of the jet and relatively warm on equatorward side of the jet, especially at low to middle altitudes. The subtropical jet is found at lower latitudes and is associated, unlike the polar jet, with strong pole-to-equator temperature gradients mainly at higher altitudes. Winds tend to increase in westerly speed (or decrease in easterly wind speed) with height when the temperature decreases toward the poles and increases in easterly wind speed (or decreases in westerly wind speed) with height when the temperature increases toward the poles. Above the mid-latitude tropopause, in the lower stratosphere, the temperature gradient reverses, so that is actually warmer toward the poles and cooler toward the equator. Thus, there is not only a narrow belt of westerly winds, but the winds also tend to be restricted by altitude, reaching their maximum intensity near the tropopause. Much of the debate about the effects of anthropogenic warming on climate change involves how the pattern of the jet will change overall if the pole-to-equator temperature gradient changes in response to warming at high latitudes. If it warms more at high latitudes than at low latitudes, then the pole-to-equator temperature gradient will decrease, and the strength of the jet will, on average, decrease; the distribution of waves within the jet will likely change; and this will have an effect on how clouds will be distributed in space.

Embedded within the *polar jet*, we find that surface fronts (e.g., cold fronts, warm fronts, and stationary fronts), very narrow zones of locally enhanced temperature gradient at the ground, tend to form in association with the troughs and ridges at higher altitudes; they form as the air is initially deformed by the synoptic-scale wind field such that horizontal temperature gradients are enhanced when relatively warm air is scrunched closer to relatively cold air in the presence of a rotating Earth. While the fronts are forming, we tend to see rising zones of air and parallel sinking zones of air as a dynamic and thermodynamic response to the tightening of the temperature gradients. These circulations are examples of horizontally oriented vortices whose axes elongated and may extend for hundreds or even thousands of kilometers, while their vertical extent

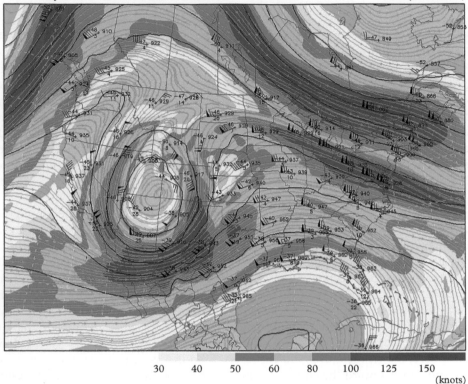

300 mb rawinsonde data 00z Sun 14 Mar 2021

300 mb Heights (dm) / Isotachs (knots)

0–hour analysis valid 0000 UTC Sun 14 Mar 2021   RAP (00z 14 Mar)

30   40   50   60   80   100   125   150

(knots)

Figure 5.2 *Weather map at the 300 hPa level late in the afternoon/early in the evening on March 13, 2021. At each station, wind direction is indicated by flags, and half barbs, whole barbs, and triangles indicate wind speeds of 5, 10, and 25 m s^{-1}; height of the 300 hPa surface is indicated in tens of meters (around 9 km). The wind speeds are color-coded in kts using the scale at the lower right. Each channel of strong winds, the jets, are coincident with zones of strong horizontal temperature gradients below. There is a trough over portions of Arizona, New Mexico, Utah, and Colorado. There is a ridge over the upper Midwest, extending southward to Louisiana.*
(From NCAR/RAL.)

is only 5 to 10 km: They are skinny, long vortices that may bend slightly. On satellite images, we frequently see curved lines of clouds triggered by the rising branch of the horizontal vortex associated with them.

The main focus of this chapter, however, is on clouds influenced by rotation that are much smaller in size, so small, in fact, that although we can see them visually with our own eyes as separate entities themselves, on satellite images they are not generally resolvable because they are either too small or are hidden below by opaque layers of clouds above them. (The exception to be looked at later is the quasi-circular eyewall

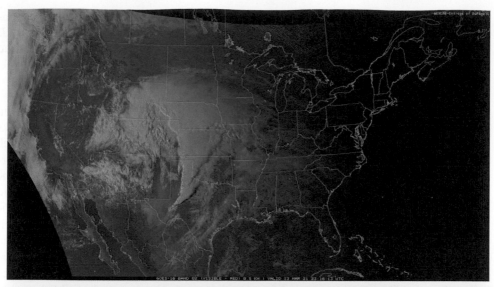

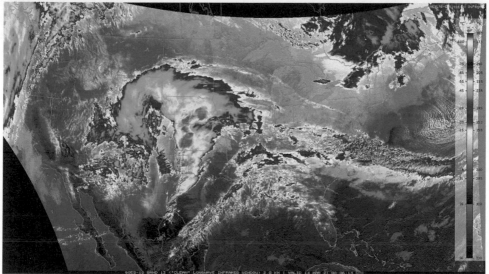

Figure 5.3 *Comma cloud over the central portion of the US during the late afternoon of March 13, 2021. A severe weather outbreak was occurring over the Texas Panhandle, while snow was falling over the Colorado Rockies. (top) Visible image from the GOES-East NOAA geosynchronous satellite. A 1, 000-km wide vortex was centered at middle and upper levels of the troposphere roughly over portions of Colorado, New Mexico, Arizona, and New Mexico, to the west of or near the western portion of the comma cloud. The cirrus clouds seen over eastern Nebraska, southern South Dakota, and western Iowa may be interpreted as "forerunners." The clear areas of west Texas and eastern New Mexico lie in the dry slot. (bottom) As in the top panel, but in the infrared channel. The highest (coldest) cloud tops are located in thunderstorms in red/orange. The streak of clouds running from southeast Texas, northeastward into the mid-Atlantic states and then east-southeastward (with some breaks) into the Atlantic are associated with rapidly flowing upper-level jets of air, in excess of 100 mph in some places at some altitudes.*

(Courtesy of College of DuPage Meteorology.)

of a tropical cyclone, which can also be seen in satellite images when the eye is completely clear and nicely seen from aircraft that penetrate the eye.) These rotating clouds appear the way they do not only because the air movement in vortices transports cloud material (water droplets, ice crystals, etc.) around in semi-circular paths (but typically not perfectly circular paths because some of the motion is due in part to the circular motion in vortices and some is due to the apparent motion of the vortices themselves) as in the comma cloud, but also for another reason involving the relationship between pressure and rotation. It must be recognized that vortices are *not* solid objects like blocks of wood being pushed around by various forces. In the case of blocks of wood, which travel around by themselves intact, the amount of material in them doesn't change. Vortices, on the other hand, are composed of *patterns* of air parcels hopping aboard the vortex, swirling partway around or all the way around, maybe even more than once, and then exiting the merry-go-round of air. Vortices (among other structures defined by the flow of air) are not distinct pieces of material, but are composed of circulating parcels of air; air enters somewhere, remains in the vortex for a period of time, and is then expelled. A vortex can be defined only if air moves around a center of rotation so that at any given instant in time, one can visualize circular or quasi-circular airflow. A vortex is therefore not like a separate piece of matter; it is an abstraction made visible by many parcels of air all in motion. To say that a vortex *moves* is only partially correct; it just *appears* to move, i.e., it propagates, disappearing on one side and appearing on the other side. We noted this phenomenon before in orographic wave clouds and in convective storms.

The time scale of a vortex is at least as long as the time it takes an air parcel to orbit around the vortex. Air traveling completely around the globe by the jet stream takes a very long time; air travels around a tornado much more quickly. So, the time scale of the planetary-scale polar jet around the Earth is very long, many days, while the time scale of a tornado is much shorter, a matter of only seconds or tens of seconds. This latter deduction may be vividly verified by watching a tornado evolve. Funnel clouds appear to move up, down, appear, and disappear, sometimes in just a matter of seconds.

5.2   Funnel clouds, tubular clouds, and debris clouds

We will now look at what is arguably the most purely rotation-influenced cloud on Earth, the "funnel cloud" (vividly displayed in Fig. 5.4 and also known by the meteorological cognoscenti by its Latin name, *tuba*, for tube, or trumpet). Funnel clouds form when the pressure is lowered enough that the air is cooled to the dew point, but unlike in many other types of clouds such as cumulus clouds or many stratiform clouds, the pressure is lowered *not primarily* because air is being lifted up (or warm, moist air is being mixed upward), expanding into a region of lower pressure and cooling, but rather because *the pressure at the center of a vortex is lower than that of its surroundings* at the same altitude. (However, in the portions of a tornado where there is strong rising motion, certainly some of the cooling must be due to the expansion of air as it rises.) When a vortex

Figure 5.4 *Two observers admiring a large tornado moving away from them, on April 14, 2012, in north-central Oklahoma. From a video frame shot by the author while he and his graduate students were in hot pursuit of the tornado.*

forms, the pressure at the center must drop. To understand why this happens, consider the case of a cyclonic (turning in the same sense as the Earth is rotating) vortex in the Northern Hemisphere (where it turns in a counterclockwise direction). As it forms, air must always be diverted or forced in a direction to the left of the air motion at any given instant if the trajectory of the air is circular. If the air travels in a circle, there must therefore be a pressure-gradient force always acting to the left of the air motion (Fig. 4.77, middle panel). If this is the case, then the pressure must be lower at the center and increase radially outward from the center to the pressure in the environment, where it becomes equal to the pressure there would be if there were no vortex present. We looked at this effect when we considered how supercells behave.

When the rotation is strong enough and there is sufficient water vapor in the air, then condensation occurs if the pressure is lowered to the saturation pressure. Since the air is saturated already at cloud base, saturation occurs first in a circular pattern just under cloud base if rotation develops in a column at cloud base. Since the lowest pressure is at the center of the vortex, the cloud one sees comes to a point below cloud base and looks like a funnel[3] or cone (Fig. 5.5[4]). Some tornadoes rotate anticyclonically, in which case the air curves to the right (in the Northern Hemisphere).

---

[3] Speed Geotis, the radar engineer at MIT when I was there as a student, used to, with a twinkle in his eye, call funnel clouds "funeral clouds," in deference to the potential danger they present.

[4] It is noted that the funnel cloud over the Indian Peaks of Colorado is just one of many that have been observed over mountainous terrain (e.g, Bluestein 2000) and that tornadoes do occur over mountainous and high-elevation terrain.

Figure 5.5 *Funnel clouds. (top, left) Cone-shaped funnel cloud viewed to the northwest, on May 5, 2001, from just east of Cordell, Oklahoma. Although there is no visible debris cloud in this photograph, a tornado was reported around or just after this time. This funnel formed in a supercell along the dryline. The yellowish haze is associated with the sun hitting lots of airborne dust near sunset. My students and I at the time of this photograph were scanning the tornadic storm with the University of Massachusetts mobile, W-band, Doppler radar. (top, right) Hose/tubelike-like funnel cloud just viewed to the north, on I-70 in western Kansas, near Hays, on August 19, 2006, during the afternoon. I was driving back to Norman, Oklahoma from Boulder, Colorado at the time; this funnel cloud was totally unexpected. It was not possible to see the ground to find out if there was a debris cloud, but a tornado was not reported. The funnel cloud was pendant from a line of cumulus congestus and not associated with a mesocyclone. It looked very much like the funnel clouds one sees in developing waterspouts in the Florida Keys (See Fig. 5.31, left panel, and Fig. 5.32, top panel for a comparison). The reader is referred to Bluestein (2008) for more meteorological details on weather conditions relating to the funnel cloud. View is approximately to the north. (middle, left) Funnel cloud pendant from a convective cloud over the Indian Peaks, west of Ward, Colorado, near Lake Isabelle, on June 17, 2013. This funnel*

It does not, however, always look like a cone or funnel, but rather sometimes looks like a long garden hose, tube, or elephant's trunk with a funnel shape only at its lower tip (Fig. 5.5, top right, bottom right)). "Funnel" clouds should probably be referred to as "tubular" clouds (Bluestein 2005), since they frequently do not come to a point and look like funnels at all. The reason for this may be that while the horizontally averaged pressure in the atmosphere decreases with height, the amount of the drop in pressure necessary to produce condensation decreases with height because the humidity increases with height. That is, it takes only a small pressure drop to lower the cloud base just a bit, but it takes a lot greater pressure drop to lower cloud base all the way to the ground. But, if the amount of water vapor in the air decreases with height, then it takes a greater pressure drop near cloud base to lower it and it takes less of a pressure drop to produce saturation near the ground. In this case, the term "funnel" cloud may not be the best descriptor because the funnel shape is seen only at the bottom end of the tubular cloud. Funnel clouds (also known as "condensation funnels") herald the formation of a tornado when they occur beneath a "boundary-layer" based cloud, which is at relatively low altitude, but they may also appear pendant from clouds at much higher altitudes (to be shown later) and which have little hope of ever producing a vortex that will make contact with the ground.

If the vortex associated with the funnel cloud is intensifying, it will appear as if the funnel cloud descends from cloud base (Fig. 5.6), but it is not *really* descending; it is propagating downward. If the air is very dry so that even a modest pressure drop will not cause the air to become saturated, then the first sign that a tornado is forming is the appearance of a column of dust or debris near the ground and rotating cloud elements in the cloud base above (Fig. 5.6, top, left). As the vortex intensifies, one may see both a funnel cloud pendant from cloud base above, a clear break below the funnel cloud, and a rotating dust/debris column below (Fig. 5.6, top right panel). Just because there is no visible funnel cloud at all, or if there is a funnel cloud, but it does not extend all the way down to the ground, does not mean that there is no tornado. As long as a strongly rotating column of air makes contact with the ground, technically there is a tornado, but it might

---

Figure 5.5 (Continued) *cloud lasted only several minutes and there were no indications that it had ever "touched down" and produced a tornado. View is to the northwest. It's not often one gets to see a funnel cloud when there is snow on the ground. (middle, right) Very ragged-looking funnel cloud in a supercell near Russell, Kansas, on May 25, 2012. A strong cyclonic vortex signature was noted in RaXPol data, coincident with the location of the funnel cloud. View is to the northwest. (bottom, left) Long, tilted, smooth-looking, white funnel cloud in a dissipating tornado on June 19, 2018 in northeastern Colorado, south of Keenesburg. The vortex circulation was being undercut by a gust front moving from left to right. The view is to the northeast. (bottom, right) White funnel cloud with a shadow on its right side, in a dissipating tornado near the Colorado–Kansas border, in a supercell north of Tribune, Kansas, on May 25, 2010, during VORTEX 2. The view is to the north-northeast.* Photos courtesy of author.

Figure 5.6 *Life cycle of a tornado, viewed to the west, near Attica, Kansas, on May 12, 2004. (top, left) A debris cloud appears underneath a small funnel cloud, beneath a wall cloud. A "clear slot" is seen wrapping around the wall cloud, to its left. (top, right) As in the top, left panel, a short time layer, when the funnel cloud has "descended" about a third of the way to the ground and the debris cloud has become more pronounced. This view is slightly to the right (north) of the view in the previous photograph. Note the location of the clump of trees just to the right (north) of the debris cloud. (middle, left) As in the top, right panel, but a short time later, when the funnel cloud has broadened and disappears inside the debris cloud rising from the ground half way up to cloud base. Cars probably bearing storm chasers are seen to the right, north of the tornado, just barely outrunning it. In this photograph, the clump of trees seen in the previous image is to the left (south) of the debris cloud. (middle, right) Note the headlights from vehicles with storm chasers trying to keep ahead of the tornado and the vivid debris cloud. This photograph was taken shortly after the one in the middle, left panel, as the tornado was narrowing and about to dissipate. (bottom) Just after the photograph in the middle, right panel, when the tornado condensation funnel was continuing to narrow, while the debris cloud near the ground was becoming wider and shallower (Bluestein et al. 2007).*

Photos courtesy of author.

be invisible. Some wild-eyed, eager storm chasers inflate the number of tornadoes they claim to have seen by counting every momentary display of whirling dust or debris (even flying leaves or twigs) near the ground. One must be cautious though, in accepting some of these claims, since a wind speed of only 20 m s^{-1} is sufficient to lift some dust, a wind speed that is obviously not dangerous. On the other hand, if the ground surface is free from loose soil or other small particles, a strongly rotating column of air will not produce any visible sign of a dust or debris column. In this case, the number of tornadoes may actually be underreported. Trees falling in the forest are not noticed by observers outside the forest.

5.3   Visual appearance of tornadoes: The debris is blowin' in the wind

In this section we'll look at a sample of the many different shapes and sizes tornadoes and their over-water versions, waterspouts appear. We'll first look at the condensation funnel and the debris cloud (Fig. 5.7).

The column of rotating dust or debris in the form of vegetation (e.g., loose dirt, leaves, tree branches, etc.) or pieces of structures built by humans (e.g., tiles from a roof, pieces of wood from a house, etc.) is called a "debris cloud." These are often beautiful to behold, especially when colorful red or brown dirt is lofted and contrasted with surrounding green vegetation (Fig. 5.8).

Debris clouds, when silhouetted against a bright background produce a stark contrast, especially when there is clear air just above the debris cloud because there is no funnel cloud directly above it (Fig. 5.9). A tiny condensation funnel making it all the way to the ground, with a debris cloud in a "secondary vortex," is seen in Figure 5.10.

5.4   Pressure in a tornado and its relationship to its funnel cloud

One can compute how low the pressure must be for a given wind speed. It turns out that in a tornado, well above the ground where the effects of the air rubbing against the Earth's surface (slowing down the air as a result of frictional drag) are not felt, the force due to the acceleration of air normal to the airflow (radially inward) is given by the inward-directed pressure-gradient force. In the reference frame of the ground, it appears as if the air is always accelerated (in this case just the change in wind direction, not a change in wind speed) normal to its motion. The force due to the continuous turning of the air parcel is the *centripetal force*. Alternatively, in the reference frame of the air itself, the inward-directed pressure-gradient force is balanced by an outward-directed *centrifugal force*. If you sit in a car and it turns to the left, your body will keep moving straight ahead and you will feel as if you are being forced to the right, outward from the car. If you are Toto, Dorothy's dog in the movie *The Wizard of* Oz, and are carried off by a tornado, you must feel as if you are, for reasons of safety, being thrown out from the tornado.

Figure 5.7 *Tornadoes with broad condensation funnels and debris clouds near the ground. (top) Silhouette of a tornado with a broad debris cloud, viewed to the west, south of Dodge City, Kansas, on May 24, 2016. This was one of over a dozen tornadoes spawned by the same supercell. (bottom) Large tornado near Verden, Oklahoma, on May 3, 1999 (Burgess et al. 2002). This tornado tracked toward and struck Moore and Oklahoma City, to the northeast, a short time later, with devastating consequences.*

Photos courtesy of author.

Figure 5.8 *Tornadoes with long funnel clouds and colorful debris clouds. (left) Tornado with large debris cloud (brown) being probed by a portable Doppler radar from the Los Alamos National Laboratory and operated by my graduate students, near Hodges, Texas, on May 13, 1989. View is to the west. (right) Tornado in far southwestern Nebraska on June 10, 2004, which is seen as a well-defined funnel cloud that is tapered up to cloud base, while narrowing to a thin vertical rope, with a debris cloud hovering at and above the ground.*
Photos courtesy of author.

This effect will be shown later to be important when using Doppler radars to map out the wind field in tornadoes. The centrifugal (and centripetal) force is proportional to the square of the swirling wind speed about the vortex and inversely proportional to the distance from the center of the vortex. Using calculus and assuming that the swirling wind speed varies from zero at the center and linearly increases with distance from the center, we can estimate the pressure drop from the edge of the vortex to its center. In this case, the vortex is said to be in "solid-body" rotation since it is as if there were a solid (not a fluid) object rotating about its axis (like passengers on a Ferris wheel). This technique actually turns out to be pretty good, as borne out by photogrammetric analyses of debris (tracking the motions of identifiable pieces of debris from frame to frame in movies and videos) and mobile Doppler-radar measurements.

Figure 5.9 *Tornadoes with condensation funnels not reaching the ground or with a break between the them and the debris clouds. (top, left) Tornado in far northeastern Colorado, viewed to the south, on June 10, 2004. If it weren't for the small debris cloud seen at the ground, one would not know that there was a tornado actually being produced underneath the funnel cloud. (top, right) Funnel cloud west of Lubbock, Texas, on May 17, 2021. The red dirt lofted below is not associated with a tornado, but rather from strong surface winds associated with the rear-flank gust front. View is to the north or northwest. (middle, left) Tornado over Selden, Kansas, on May 24, 2021. The funnel cloud narrows to one ropelike funnel at the center, while the debris cloud near the ground is much wider than the condensation funnel. Some of the debris appears to have been lofted almost up to cloud base. View is to the west. (middle, right) As in the middle, left panel, but the funnel cloud may actually be composed of four smaller, secondary funnel clouds. (bottom) Anticyclonic tornado on the southern edge of the rear-flank gust front in a supercell in central Oklahoma, near El Reno, on April 24, 2006.*

Suppose now, for example, that in a tornado vortex the swirling wind speed varies from zero at the center to 75 m s^{-1} (a strong tornado) at a radius of 150 m. Then the rotation rate of the "solid-body" vortex 75 m from the center is [75 m s^{-1}/150 m] = 0.5 radians s^{-1} [1 radian is 57.3° or (180/π)°]. The pressure drop from the environment to the center turns out to be around 30–35 hPa. For a violent tornado, with maximum swirling wind speeds of 130 m s^{-1} and a *core radius* (the distance from the center of rotation at which the linear increase of wind speed ceases, levels off, and decreases) of 500 m[5] which in the case of solid-body rotation is also the radius of maximum wind (RMW), the pressure drop is about ~100 hPa. To get a feeling for how large this pressure drop is, note that it represents more than the typical range of pressure at the surface of the Earth (~950 hPa to 1, 050 hPa). If this pressure change occurs on synoptic time scales of a day or longer, you don't notice it; if occurs in 10 minutes, then your ears may "pop."

If cloud base is at around 900 hPa and the surface pressure is around 1, 000 hPa, then the condensation funnel may extend all the way to the ground. This estimate, however, neglects the increase in hydrostatic pressure as one gets lower in altitude and also variations with height of water-vapor content. For the case in which the pressure drop is around 30–35 hPa, the funnel cloud may extend only around 1/3 of the way to the ground. Scientists have estimated the wind speeds in tornadoes using the "funnel cloud" method, in which they use photogrammetry to measure how far below cloud base a funnel cloud extends, and then a nearby sounding in which the variation with height and pressure of temperature and humidity in the environment of the tornado are known and taken into account.[6]

The theoretical estimates of the pressure drop in tornadoes have been roughly verified by actual in situ measurements by instruments. Such measurements are very difficult and dangerous to obtain for obvious reasons. One doesn't blithely walk into a tornado holding a barometer! Over the years there have been a number of serendipitous measurements made by both amateur and professional meteorologists near or in tornadoes (e.g., Karstens et al. 2010). Most of these were from near misses, but a measurement

---

Figure 5.9 (Continued) *This tornado formed while a cyclonic tornado was dissipating to the north. See Bluestein et al. (2016) for more details. There is a well-defined condensation funnel pendant from the cloud base above and a hollow-looking debris cloud, underneath of which there is a broader debris cloud. This tornado struck a small airport south of El Reno and inflicted substantial damage to a hangar, which was documented in a spectacular fashion by an Oklahoma City television station (Channel 9) helicopter pilot.[a]*

[a] I have a copy of this video, but I do not have permission to post it; I was not able to locate it on any public website.
Photos courtesy of author.

[5] This is a description qualitatively of a "mile-wide" tornado, since the diameter is greater than 1 km and the tornado extends beyond the core radius.
[6] Dergarabedian and Fendell, 1970.

Figure 5.10 *Helical striations in a secondary vortex in a tornado near Verden, Oklahoma, on May 3, 1999. The view is to the south. When we returned the next day to look for damage, to our surprise, we could not find any.*
Photo courtesy of author.

was made of a pressure drop of as high as 82 hPa in a tornado in St. Louis in 1896. In the early 1980s my graduate students and I, using an instrument developed and built by Al Bedard and Carl Ramzy in Boulder, Colorado at NOAA's Wave Propagation Laboratory, tried to make, for the first time ever to the best of our knowledge, *deliberate* measurements of pressure, temperature, humidity, and wind in tornadoes. The device we used we called TOTO, after Dorothy's dog in the *Wizard of Oz*. TOTO (TOtable Tornado Observatory) was rather cumbersome, weighing around 400 lbs and having to be deployed from the back of a pickup truck on a ramp (Bluestein 1983). Efforts were only slightly successful and only one measurement near a tornado was made by Lou Wicker at National Severe Storms Laboratory (NSSL) in 1985. On one occasion, in

1982, we had a tornado in sight coming directly toward us from the southwest; after we deployed TOTO, it dissipated before reaching the instrument (we of course had planned to get out of the way, but didn't have to). A second tornado then formed to our west, but it moved to the northwest, eluding us. After our experiences, a number of other smaller, lightweight devices were developed by others. During the first Verification of the Origins of Rotation in Tornadoes Experiment (VORTEX) (Rasmussen et al. 1994) in 1995, Bill Winn from the Langmuir Laboratory/New Mexico Institute of Mining and Technology, measured a pressure drop of 50 hPa at the edge of a large tornado in the Texas Panhandle on June 8. At the time, I was in an NCAR (National Center for Atmospheric Research) Electra aircraft equipped with a Doppler radar (ELDORA—ELectra DOppler RAdar), looking at the rather formidable-looking tornado under the guidance of my colleague Roger Wakimoto, then at UCLA (Fig. 5.17, center panel). Much more successful measurements were made by Tim Samaras, an amateur meteorologist and engineer at Applied Research Associates, Inc., based near Denver, Colorado, beginning in the late 1990s, first using tiny, much more lightweight instruments developed by Frank Tatom of Engineering Analysis, Inc. He later used instruments developed by himself, that could be dropped in a linear array at the side of the road, ahead of an approaching tornado, thus increasing the chances for a close encounter of the tornadic type. The inspiration for this approach came from one of my colleagues, Fred Brock, at the University of Oklahoma, using what he called "Turtles." Tim's measurements indeed showed pressure drops as high as 100 hPa directly in the path of a tornado on June 24, 2003 in South Dakota. His exploits were well documented in *National Geographic* magazine. Tragically, Tim tragically lost his life while attempting to make measurements in a violent tornado in El Reno, Oklahoma on May 31, 2013 (see Figs. 4.85; 5.16, bottom panel; 5.21; Figs. 5.38–5.40) (Bluestein et al. 2019).

Measurements near the ground, however, cannot be compared precisely with the theoretical measurements mentioned previously because they don't take into account surface friction. Pete Sinclair from Colorado State University used a mobile, instrumented tower as high as 10 m to make measurements in dust devils (to be mentioned later) in the desert in southern Arizona from 1960 to 1963 and found typical pressure deficits of just several hPa. Stirling Colgate, from the Los Alamos National Laboratory and New Mexico Institute of Mining and Technology, attempted unsuccessfully to launch instrumented rockets into tornadoes below cloud base from a Ccssna 210 aircraft in 1981. Verne Leverson and Pete Sinclair, along with Joe Golden at NOAA in Boulder (mentioned later), penetrated some waterspouts (also mentioned later) in the Florida Keys in 1974 and found pressure deficits as high as 10 hPa. More recently, Reed Timmer and his group successfully launched a probe into a large tornado in Kansas in 2019 using a small rocket.[7]

---

[7] Timmer et al., 2023

5.5   More visual characteristics of tornadoes and waterspouts

Tornadoes are defined as "violently" rotating columns of air that make contact with the ground and which can cause damage. They are connected to the base of clouds in convective storms and are driven by buoyant updrafts in them, but, as noted previously, do not necessarily have to have a funnel cloud. This definition is somewhat deficient because if the ground is barren and there is nothing available to be destroyed, what evidence does one have that there is a tornado? Without actual wind measurements, we don't know just by looking at a tornado. There have been tornadoes in which wind measurements were made by Doppler radar that were much higher than the winds expected to have caused the extent of damage actually found (Snyder and Bluestein 2014). One of our ground-based mobile Doppler radars (Bluestein 2022) measured wind speeds as high as 135 m s^{-1} (~300 mph) in a tornado near El Reno, Oklahoma in 2013, while not much was damaged, and the rating of the tornado was therefore consistent with that for much lower wind speeds. (Wurman et al. 2007 also reported similar maximum windspeeds in a tornado in Bridge Creek, Oklahoma, on May 3, 1999.) It is possible that the winds measured by the radar were not right at the surface and therefore the actual windspeeds at the ground could have been less (Bluestein et al. 2019). In general, though, we would like to include some objectivity to the definition of what is a tornado: The wind speeds need to be at least 30 m s^{-1} (assuming we can measure them), though some structures are not damaged by wind speeds this low while some trees may be toppled by wind speeds of only 20–25 m s^{-1}, if, for example, a tree is old and weak and if the ground is wet and saturated, or some structures of dubious structural integrity are struck by the wind. When a violent tornado strikes a populated area, the resulting damage can be devastating (Fig. 5.11).

My first encounter, indirect or otherwise, with tornadoes was during my childhood near Boston, Massachusetts, when a large tornado struck Worcester, just 40 miles to the west. My mother called me in when I was playing outside and told me that tornadoes take little boys and suck them up into the air. I did not believe her, but was forced to come inside anyway, though I'm sure I wanted to stay outside and watch. Perhaps this piqued my interest, which remained latent until years later. I also recall when a tornado severely damaged a fast-food restaurant in Miami, just to my west near a causeway, when I lived in Miami Beach as a youngster. We viewed the damage the next day and that spectacle also left a deep impression on me, especially seeing a car overturned on the road.

I saw my first tornado during a storm chase in southwestern Oklahoma on May 20, 1977 as a participant in the NSSL/OU Severe Storm Intercept Project under the leadership of Bob Davies-Jones at NSSL and Jeff Kimpel at OU. This tornado was a large wedge embedded in precipitation and was hard to see owing to poor contrast (Fig. 5.12).

Figure 5.11 *Extreme tornado damage. (top) Severe damage still evident on May 8 from a tornado that struck Moore, Oklahoma (south of Oklahoma City and north or Norman), on May 3, 1999. In this case it looks as if a pickup truck had been wrapped around the remnants of a tree. From this and the other panels, it appears as if Moore, Oklahoma might be a dangerous place to live. This photograph is reminiscent of a cartoon I saw in* The New Yorker *magazine, in which a television reporter stands in what appears to be a suburb or rural area just devasted by a tornado (a car is overturned, someone is draped over a denuded tree branch, furniture is strewn around) and says into the camera "But the weather looks great for the rest of the week."[xi] This is certainly a Panglossian attitude, given the circumstances. (middle) Panoramic view of a neighborhood of Moore, Oklahoma, several days after*

My first academic job was as a visiting, non-tenure track faculty member at the University of Oklahoma (OU) in Norman during the beginning of the late summer of 1976. I was enticed there by the founder of the NSSL, Edwin Kessler, who had been next door to me on a sabbatical at MIT, when I was a graduate student there. I came to Norman to spend a few years learning about severe convective storms and tornadoes, and intended to leave, but have remained there ever since. I had arrived by pure chance at the right place at the right time: While people had been storm chasing on a limited basis since the 1950s, it was not until the early 1970s, just several years prior to my arrival in Oklahoma, that scientists and students had begun storm chasing in an organized manner, under the organization of NSSL in cooperation with OU, with my late colleague and friend Jeff Kimpel at OU in charge administratively, along with Bob Davies-Jones at NSSL. Most, but not all of the tornado photographs in this book I took during organized storm chases, mostly supported by the National Science Foundation, during which I have had the exceptional opportunity to watch tornadoes with never-ending awe and study many of them along with simultaneous radar data (Bluestein 1999a, b).

While tornado vortices are typically vertical columns of rotating air, they sometimes become tilted, especially during the latter part of the life of the supercell tornado (Fig. 5.13), as the rear-flank gust front upends the vortex.

It has been reported that the famous tornado in *The Wizard of Oz* movie released in 1939 was simulated as a large, tapered, muslin sock blown by "wind machines" (probably large fans) and tilted when a gantry held the top of the sock was moved in a different direction than a rod, connected at the bottom, was moved. The tornado looked like a tilted rope snaking around and has stuck in the imagination of many as having the classic look of a "midwestern"[8] tornado. In some instances, a segment of the tornado condensation funnel may actually be horizontally oriented (Fig. 5.14). This extreme tilt is probably caused when the rear-flank gust front in supercells or the ordinary gust front in ordinary-cell convective storms moves the tornado away from the parent storm by a significant amount. When some tornadoes dissipate, they do so in a "top-down"

Figure 5.11 (Continued) *(May 25) the violent tornado that hit it on May 20, 2013. It looked as if a powerful bomb had destroyed everything in sight. (bottom) A view from a helicopter a few days later (May 22) of damage from the May 20, 2013 tornado that ravaged Moore, Oklahoma. It appears as if the damage path was almost five blocks wide. The sharp gradient in damage, indicative of a sharp gradient in wind speed, is apparent, as some houses with little, if any, damage are seen next to totally obliterated houses. (The pilot was Mason Dunn and the helicopter was supported by PBS NOVA, Channel 4 in the UK, and the Discovery Channel. I am grateful for the opportunity to have been able to photograph the damage on this flight.)*

[a] From *The New Yorker* magazine, June 5, 2000, p. 57. Frank Cotham.
Photos courtesy of author.

[8] Oklahoma, at center of tornado country, is in the Southern Plains, *not* in the Midwest as some mistakenly believe.

Figure 5.12 *Large wedge tornado (the author's first tornado) in southwestern Oklahoma, near Tipton, on May 20, 1977. View is approximately to the northwest.[a]*

[a] This is my first publication of this old photograph, which was not possible years earlier (e.g., in *Tornado Alley*), because I did not have a digital version or software that could enhance the contrast and correct for incorrect exposure.
Photo courtesy of author.

manner, leaving behind a visible funnel cloud near the ground, with a break aloft. It therefore may appear as if the funnel cloud is not even connected to the parent storm above it (Fig. 5.14, bottom panels).

To this day, probably as a result of the *Wizard of Oz* movie, I continue to be intrigued not only by the tornado phenomenon itself, but also by the semi-barren, stark landscape of the High Plains and Plains of the US, the frequent stage on which tornadoes make an appearance. The wildness of the Texas Panhandle, western Oklahoma, the Oklahoma Panhandle, eastern Colorado, western Kansas, southeastern Wyoming, and the Nebraska Panhandle have been my favorite backdrops for tornadoes. One advantage of studying tornadoes in these areas is that they are sparsely populated so that the likelihood of a tornado striking a residential area is relatively small; they are therefore good places to conduct field experiments. Georgia O'Keeffe, the prominent American painter, lived and taught in Canyon, Texas, in the Texas Panhandle, for a few years, and according to Anne Dingus, writing in *Texas Monthly* in 1987, wrote, "It is absurd the way I love this country." I am also intrigued by the regionalist paintings of the Plains and Midwest made during the Depression era in the US. The regionalists' view of the landscape along

Figure 5.13 *Tilted condensation funnels. (top, left) Silhouette of a tilted, tornado condensation funnel near irrigation equipment, south of Dodge City, Kansas, on May 24, 2016. View is to the west. (top, right) As in the top, left panel, but in southeastern Wyoming northwest of Pine Bluffs, on June 12, 2017. (second row, left) A highly tilted tornado in northeastern Colorado, south of Keenesburg, on June 19, 2018. View is to the northeast. (second row, right) A highly tilted tornado/funnel cloud in south-central Kansas, near Geuda Springs, on May 14, 2018. View is to the east. (bottom) A tilted condensation funnel in a dissipating tornado near Purcell, Oklahoma on Sept. 2, 1992.*

Photos courtesy of author.

Figure 5.14 *Tilted funnel clouds having horizontal segments. (top) A nearly horizontally oriented tornado snaking around in its dissipating stage in central Kansas, southwest of Russell, on May 25, 2012. The condensation funnel is nearly vertically oriented, however, right near the ground, at the left. View is to the east. (bottom, left) A nearly horizontal funnel cloud in a tornado in south-central Oklahoma, near Paul's Valley, on March 26, 2017. The funnel cloud/tornado vortex does become vertically oriented near the ground and the debris cloud, hidden by the tree, is underneath the funnel cloud. View is to the southeast. (bottom, right) Debris cloud and bottom portion of a remnant funnel cloud in a dissipating tornado near Bassett, Nebraska on June 5, 1999. View is to the south. This tornado looks somewhat like a ghost of a tornado, suspended in the air.*
Photos courtesy of author.

with tornadoes were combined in a painting, *The Prelude*,[9] made by John Steuart Curry, of John Brown, the abolitionist. This painting appears on the wall of the Kansas State Capitol in Topeka, which I tried to see during VORTEX-2 in 2009 when we spent a night in Topeka, but alas was unable to view because that part of the Capitol in which

[9] https://en.wikipedia.org/wiki/Tragic_Prelude

the painting was displayed was being renovated. Tornadoes over highly populated areas don't fascinate me as much. They bring horrific destruction and they are typically apparent only in tightly localized sections of photographs, owing to power lines, and buildings, and other nearby urban structures. If you want to see debris flying around and filling the air, you may chase tornadoes in an urban area, and good luck . . . but I don't recommend it, mainly for safety reasons.

Tornado funnels appear not only in many different contexts, rural, urban, etc., but they themselves take on many different kinds of appearances. Some funnel clouds are ragged-looking (Fig. 5.5, middle right panel), while others are smooth and laminar (Figs. 5.7, 5.8, and 5.13). The funnel may be very narrow and rope-like in appearance (Fig. 5.14, top panel), pendant from a relatively high cloud base (Fig. 5.8), or very wide, pendant from a relatively low cloud base (Fig. 5.12). Tornadoes in their dissipating stages frequently look like long, tilted ropes of cloud (Fig. 5.14). Very wide tornadoes look like broad cylinders and have been described as "wedges" (Figs. 5.15, 5.16, and 5.17, middle panel). Nothing excites the imagination of a storm chaser more than the appearance

Figure 5.15 *Large, wedge tornado near Binger, Oklahoma, on May 22, 1981, viewed to the northwest.*
Photo courtesy of author.

Figure 5.16 *Wedge tornadoes. (top) Large, wedge tornado, viewed to the south, southwest of El Reno, Oklahoma, on May 24, 2011 (Houser et al., 2015). A film crew from the PBS television program NOVA was filming us during field operations on this day, but the video of this tornado never made it on the program. (middle) A wedge tornado, viewed to the west, in south-central Oklahoma, near Sulphur, on May 9, 2016. (bottom) Large, violent, wedge tornado in central Oklahoma, near El Reno, Oklahoma, on May 31, 2013, viewed to the southwest.*

Photos courtesy of author.

Figure 5.17 *Wide tornadoes. (left) Broad tornado in east-central Kansas on May 8, 2003. View is to the northwest. It was extremely hazy on this day, owing to smoke from fires in Mexico, so I had to increase the contrast using Adobe Photoshop. (middle) Airborne view of a wedge tornado in the Texas Panhandle, on June 8, 1995, during VORTEX. I was aboard NCAR's Electra aircraft in which the ELDORA (ELectra DOppler Radar) radar was probing this tornado (Dowell and Bluestein, 2002a). I am fortunate that Roger Wakimoto, who was in charge of this mission, allowed me to participate as an observer on this particular day. (right) Airborne view of a broad tornado over northwest Texas on May 29, 1994, from a NOAA P-3 aircraft.*
Photos courtesy of author.

of a wedge tornado. However, the visual appearance of a tornado funnel does not necessarily reveal the intensity of the tornado. For example, if the air below cloud base is relatively humid, it takes much less of a pressure drop to become saturated and produce a wide cloud than if the air is relatively dry. So, some "wedge" tornadoes in a humid environment, such as in the southeastern US, may actually be weaker than some rope-like tornadoes in a dry environment, such as the High Plains of the US. But, granted, the wedge tornado looks fearsome. Some tornadoes may evolve from ropes into wedges and back to ropes again, before dissipating.

Tornadoes can occur all year round in the US, but they occur mostly during the warm season, except in the southeastern states and west coast, when they can occur during the late winter and midwinter, respectively. My students and I once viewed a rare autumn tornado on Nov. 7, 2011, west of Norman (Fig. 5.18). Around this time there were also earthquakes, one of which was accompanied by a thunderstorm and called a "thunder quake." The gods must have been very angry at this time.

The funnel cloud itself sometimes appears to be hollow, especially in some waterspouts (Figs. 5.29, 5.31, right panel, in a waterspout) so that when it is, it is a circular curtain rather than a "solid" cylinder of cloud droplets. Since the mass of liquid cloud droplets, or raindrops, or flying debris is much greater than the mass of air molecules, the strongest forces in the radial direction are the centrifugal force and drag; the inward-directed pressure-gradient force is less significant, so that the most massive particles are centrifuged radially outward (Snow 1984), so that a *radar* detects an eye-like feature called a "weak-echo hole" (WEH) (Fig. 5.19) (Dowell et al. 2005). This feature is different from the WER or BWER produced in the updraft region of a supercell, in which there is a lack of radar reflectivity because there is not enough time for precipitation to form at low and midlevels in the storm, owing to a strong updraft.

Figure 5.18 *Rare, autumn tornado on Nov. 7, 2011, in central Oklahoma, near Carnegie and Ft. Cobb. View is to the northwest.*
Photo courtesy of author.

I fortunately cannot claim that actually I have been an "eye" witness to attest to the actual hollowness of tornado funnel clouds. When considering the radar echo throughout the vertical, not just at one elevation angle or altitude, the vortex appears on radar as a weak-echo column (Fig. 5.19, right panels and 5.20).

The WEH, though it has been detected, remarkably, extending all the way to the top of the parent storm (Tanamachi et al. 2012), fills in near the ground, as a result of surface friction (Fig. 5.19, right panels; 5.20)). There have been reports of someone who opened the door of his/her underground storm shelter during a tornado and was able to peer upward through the funnel cloud. Without any measurements of wind, the presence of a WEH is indicative of rotation, but not necessarily of a vortex strong enough to be considered a tornado. We note parenthetically that the WEH that occurs in small-scale vortices such as tornadoes, waterspouts, and dust devils, is *not* produced in the same way an eye is produced in a strong tropical cyclone, or in an intense extratropical cyclone.

While most tornadoes are cyclonic, some are anticyclonic (Fig. 5.9, bottom panel; 5.21, and 5.22). In supercells, as we noted earlier in Chapter 4, anticyclonic vortices sometimes are found along the tail end of the rear-flank gust front, possibly as horizontal vorticity is tilted onto the vertical at the end of the line of convective clouds (Fig. 4.116; these anticyclonic vortices are probably the source of anticyclonic tornadoes in cyclonically rotating, right-moving supercells[10] (Figs. 5.21 and 5.22). I have been fooled

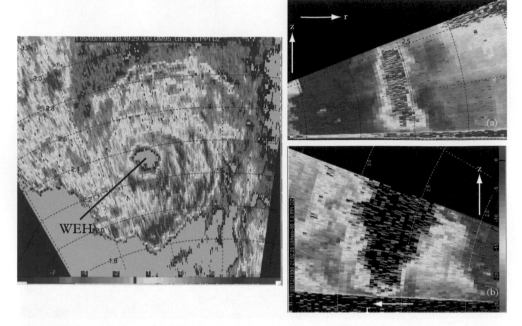

Figure 5.19 *Peering inside a tornado with a radar. (left panel) Close-up view (from 1.8 to 2.6 km range) of radar reflectivity of the tornado seen in Figure 5.7, bottom panel (May 3, 1999). The radar used was the University of Massachusetts W-band (3-mm wavelength) mobile Doppler radar, having a very fine-beam width of 0.18°, which corresponds to a spatial resolution of only 7 m at the center of the tornado. This image might represent the finest scale measurements ever made in a tornado using a radar. The weak-echo hole (WEH) is about 160 m across. This radar depiction of a tornado looks similar to the radar depiction of a tropical cyclone (see Fig. 4.112, top panel), but on a much smaller scale. The strongest reflectivity is color-coded yellow, brown, and red. They are not equivalent to the reflectivity measured by X-band, C-band or S-band radars because the scattering is not in the Rayleigh range (the scatterers are not small compared to the wavelength of the radar). The spacing between adjacent range rings is only 200 m (range rings are marked every 0.2 km). Figure courtesy of author. (right panels) Vertical cross sections (height on the ordinate; distance from the radar on the abscissa) of equivalent[a] radar reflectivity (dBZ$_e$) through the approximate center of two tornadoes showing a WEH extending to the top or nearly to the top of the shallow radar domains. From the U. Mass. W-band radar (top, right panel) on May 5, 2002, near Happy, Texas (Texas Panhandle) at 1849:18 CDT (similar to that in Bluestein and Pazmany 2000; Bluestein et al., 2004, Fig. 9 and Bluestein 2013, Fig. 6.50a) and (bottom, right panel) on 12 May 2004, near Attica, Kansas at 2002:15 CDT (similar to that in Bluestein et al., 2007, Fig. 13 and Bluestein, 2013, Fig. 6.50a). Note that in both panels the WEH disappears as the radar echo closes up right near the ground as a result of surface friction, while aloft the WEH is wider, possibly due to a maximum in centrifuging owing to higher azimuthal wind velocities, or more likely due to a horizontal circulation along the edge of the tornado, such that echo-free air from the center of the vortex is advected radially outward along the radially outward flowing branch of the circulation (while echo-rich air is brought radially inward by the radially inward flowing branch of the circulation).*

[a] Because the radar wavelength is not large compared to the size of the scatterers, the scattering is of the Mie variety, not Rayleigh scattering, so that the reflectivity cannot be interpreted in the same way the reflectivity that would be if it had been measured by an X-band, C-band, or S-band radar. When the radar reflectivity is treated as if it were measured for Rayleigh scattering by water droplets, it is known as the *equivalent radar reflectivity factor*.

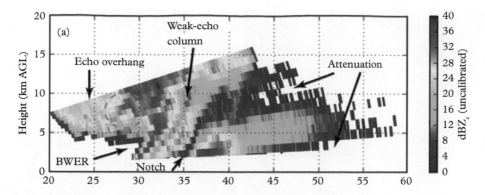

Figure 5.20 *Vertical cross section through a tornado in a supercell near Greensburg, Kansas on May 4, 2007, from a synthesis of data collected at constant elevation angles by the University of Massachusetts X-band polarimetric mobile Doppler radar. The weak-echo column is seen leaning to the right (north) with height.*

From Tanamachi et al., 2012, their Fig. 7a. © American Meteorological Society. Used with permission.

many times while watching a cyclonic tornado or cyclonically rotating wall cloud or funnel cloud, oblivious to the formation of an anticyclonic tornado off in another direction, toward the other end of the rear-flank gust front.

I know of documented instances of anticyclonic tornadoes in anticyclonically rotating, left-moving supercells, but they are incredibly rare and I have never seen one myself. Some anticyclonic tornadoes are produced by non supercell storms. Some anticyclonic tornadoes, regardless of how they formed, like cyclonic tornadoes are associated with weak-echo holes, spiral bands of reflectivity, anticyclonic vortex shear signatures, and low co-polar cross-correlation coefficients when debris is kicked up at the ground, as seen in Figure 5.21. Anticyclonic swirls are seen in a frame of a video looking straight up while the anticyclonic vortex at the tail end of a gust front in a cyclonically rotating, right-moving supercell passed by us, nearly overhead (Fig. 5.22).

Lighting also can significantly affect how a tornado appears. When a tornado is directly illuminated by the sun, it can appear to be white, producing the classic "white tornado" appearance (Fig. 5.23). This moniker was made famous in the US in 1970 by a television commercial for Ajax all-purpose cleaner which was purported to clean "like a white tornado." I guess a white tornado must be a clean one! However, doesn't flying debris make it dirty? Tornadoes look white when viewed from the rear side of the storm late in the day, where the sun may be shining, or from the south of the storm (in the Northern Hemisphere) (Fig. 5.13, middle panels; 5.44). On the other hand,

---

[10] My student Jake Margraf found very recently that a strong anticyclonic vortex/weak anticyclonic tornado in the Selden, Kansas tornadic supercell of 24 May 2021, formed along the tail end of a rear-flank gust front when a secondary surge behind the RFGF impinged upon the leading, RFGF (Margraf, 2023).

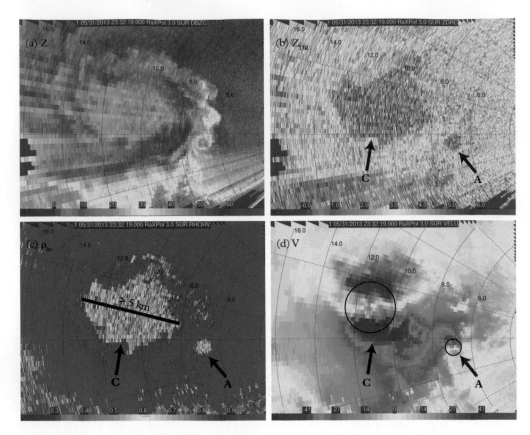

Figure 5.21 *RaXPol radar imagery from a large cyclonic tornado (C) and a simultaneously occurring anticyclonic tornado (A), near El Reno, Oklahoma, on May 31, 2013, at 2332:19 UTC. (a) radar reflectivity factor (corrected for attenuation) in dBZe; (b) differential reflectivity (ZDR) in dB: (c) co-polar cross-correlation-coefficient (ρ_{hv}); and (d) de-aliased Doppler velocity in m s^{-1}; the red-purple couplet (larger circle) marks the cyclonic vortex signature, and the purple-yellow couplet (smaller circle) marks the anticyclonic vortex signature.*

From Bluestein et al., 2015, their Fig. 19. © American Meteorological Society. Used with permission.

when viewed as a silhouette, the tornado funnel is black and ominous looking (Fig. 5.9, middle left panel).

Rainbows and tornadoes are sometimes seen simultaneously; in rare instances, they may coincidentally mimic the same shape and tilt as the tornado funnel (Fig. 5.24), of course without any physical basis for the similarity.

At night, tornadoes may be visible only as silhouettes when there is lightning behind them (Fig. 5.25). It's like seeing a photograph with lightning acting as a flash, though usually illuminating the tornado from the rear, not from the front.

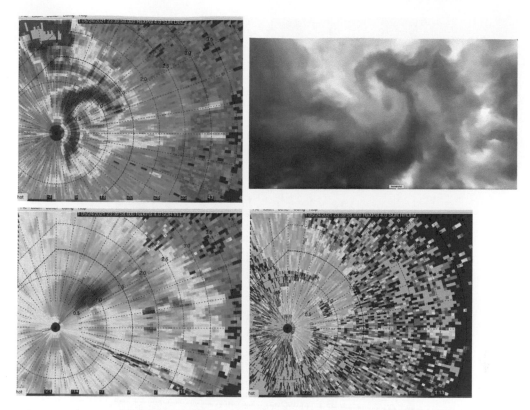

Figure 5.22 *Data collected by RaXPol near a strong anticyclonic vortex/weak anticyclonic tornado E of Selden, KS, on May 24, 2021. Radar reflectivity factor (dBZ, upper left); the echo pattern is spiral, with a WEH in the middle; frame grab from a video, looking straight up, courtesy of S. Emmerson, a member of my radar crew, of the anticyclonic vortex (shown as a spiral of darker cloud like a mirror image of the reflectivity pattern, which is viewed from above, not from below as the photograph is) as it moved by overhead (upper right); Doppler velocity (m s^{-1}, lower left) the velocity pattern (red next to green) indicates anticyclonic shear; and co-polar cross-correlation coefficient (ρ_{hv}, lower right); the hole (purple) surrounded by yellow/brown colors is a debris signature. Range rings plotted in km.*

When there is a lot of precipitation, especially rain, falling near a tornado, it may become completely encased and hidden by the rain. Storm chasers often report tornadoes as embedded in rain (or precipitation, in general). Sometimes it takes a lot of imagination to see something that is mostly hidden in rain, but I must admit that one can often identify tornadoes and funnel clouds even when they are encased inside precipitation cores (Fig. 5.26).

A tornado that occurs over water is called a *waterspout*. Waterspouts are seen over oceans, lakes, and rivers, and are obvious hazards to boats and ships. The "debris" cloud, in this case, is seen as a column of spray extending upward from the water surface (Fig. 5.27).

Figure 5.23 *White (actually whitish-tinted, relatively bright, illuminated—not in a shadow) tornadoes. (top) Tornado in far southwestern Nebraska on June 10, 2004. The tornado, which has a whitish color, appears to be embedded within a cylinder of cloud above that marks the parent mesocyclone. (middle) As in the top panel, but later on, and appearing white. (bottom) The same tornado seen in the other panels, but a bit later on as it became contorted and "roped out" as it dissipated. Note how, like some of the tornadoes seen in Figure 5.13, middle, left, and 5.14, the part near the cloud base is horizontal.*

Photos courtesy of author.

Figure 5.24 *Rainbow and tornado appearing to mimic each other in south-central Kansas, on May 12, 2004.*
Photo courtesy of author.

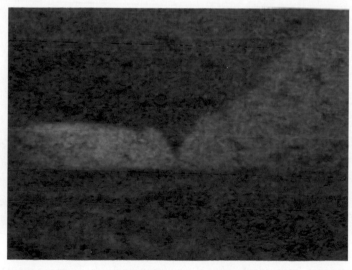

Figure 5.25 *Tornado silhouetted by in-cloud lightning from behind, at night in a supercell in north-central Nebraska on June 4, 1999. View is to the north-northeast. From a grainy video frame taken with a hand-held, older-technology, video camera.*
Photo courtesy of author.

Figure 5.26 *"Rain-wrapped" tornadoes. (left) Funnel cloud from a tornado embedded in rain, viewed to the west, in central Oklahoma, south of Piedmont, on May 29, 2012. (right) Tornado embedded in precipitation, viewed to the southwest, on June 13, 2010, in the Oklahoma panhandle, west of Slapout, during VORTEX-2.*

Photos courtesy of author.

Figure 5.27 *Waterspout east of Key Biscayne, Florida, on May 27, 1975, along a line of large cumulus congestus that developed into a line of cumulonimbus clouds. There is a spray ring (analogous to the debris cloud in a tornado) at the surface of the ocean and a wider funnel aloft within which there is a narrower funnel cloud extending from the ocean surface to midway between cloud base and the ocean surface. I took this photograph when I was a graduate student, attending my first professional conference, one on tropical meteorology run by the American Meteorological Society.*

Photo courtesy of author.

A disturbance in the ocean is sometimes visible in the aftermath of a waterspout (Fig. 5.28), possibly owing to the upwelling of water from below. In Figure 5.29, the wake is seen as a short, white, line segment, perhaps composed of water with dissolved bubbles of air. An incipient waterspout may be seen as a dark spot in the water (Fig. 5.29, upper right).

Although the oceanic regions at low altitude are typically more humid than the corresponding areas over the land, waterspouts mostly appear to be narrow and rope-like, perhaps because they are typically relatively weak (Fig. 5.30).

Waterspouts sometimes make landfall and become tornadoes and vice versa. Just a few of many instances of waterspouts making landfall and inflicting damage have been documented at the marina on Dinner Key, Florida, on June 7, 1968 and in Myrtle Beach, South Carolina, on July 6, 2001. The latter caused $8 million in damage. Landfalling waterspouts have obviously scared the hell out of beachgoers. Many waterspouts and funnel clouds form under a line of cumulus congestus, where there may be a number of them, all visible at once (Fig. 5.31, left panel). Some of the best photographs I have ever seen of multiple waterspouts in a line have come from the Mediterranean Sea in the fall.

Joe Golden at NOAA was the first to document meticulously the life cycle of waterspouts over the lower Florida Keys in the late 1960s and early 1970s (Golden 1974).

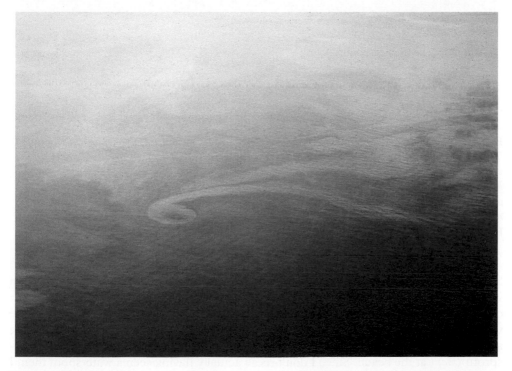

Figure 5.28 *Disturbed sea state having a comma shape, ending in a circular swath, in the Florida Keys on Aug. 24, 1993, associated with a waterspout. As viewed from a NOAA helicopter.*
Photo courtesy of author.

Figure 5.29 *Waterspout over the Florida Keys on Aug. 24, 1993, with a tail of white water trailing it. The sea state surrounding the waterspout appears to be relatively uniform, which is evidence that the waterspout was not situated along a well-defined surface wind-shift boundary. As viewed from a NOAA helicopter. It is possible to see right through the condensation funnel/spray column near the ocean surface. The dark spot on the water surface to the rear and right of this waterspout probably marks another circulation, in its beginning stage.*
Photo courtesy of author.

Subsequently, Verne Rossow at NASA, in 1970, also described the relationship between waterspouts and their parent clouds. I remember being intrigued reading about their observations when I was a beginning graduate student. My first "recorded" experience with waterspouts occurred in the 1959, when I lived on Miami Beach, and had close access to the ocean, just several blocks away. I kept a diary, and in it mentioned a "water-sprout (*sic*)" at the beach. Since I don't recall actually having seen a waterspout then, I suspect that someone at the beach must have told me about one that I had just missed. I remember being intrigued by whatever had happened. Many of the waterspouts I have seen have been in the Florida Keys when I went looking for them, under the guidance of Joe Golden.

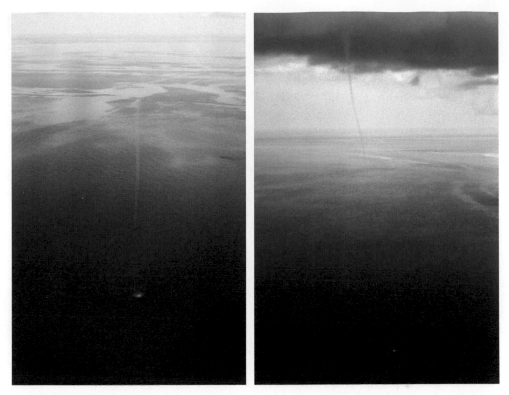

Figure 5.30 *Very narrow waterspouts. (left) A very narrow waterspout that appears to be stationary, in the Florida Keys, on Aug. 24, 1993. There is a circular ring of white water underneath it and a "bird of prey" just happened to be located in my line of sight (near the center of the image) with the condensation funnel, near the ocean surface. As viewed from a NOAA helicopter. (right) An unusually narrow waterspout in the Florida Keys, as viewed from a NOAA helicopter, on Aug. 24, 1993. It is absolutely remarkable how narrow the condensation funnel is in comparison with the height of the cloud base from which it's pendant: The ratio of the width of the vortex to its depth is extremely small. These condensation funnels may be more appropriately described as threads, rather than ropes.*
Photos courtesy of author.

I saw my first waterspout with Joe from a small airplane during the late summer of 1971 when I was a graduate student (Fig. 5.32, top panel); over twenty years later, in 1993, he took me up in a NOAA helicopter and we flew in between cloud base and the ocean surface, looking both up and down at these magnificent displays of nature (Figs. 5.29–5.31) (Golden and Bluestein 1994). I saw my first waterspout "on my own" from the roof of NOAA's National Hurricane Research Laboratory in Coral Gables, Florida (now called the Hurricane Research Division, relocated to Virginia Key) during the summer of 1972 when I was working there as a graduate student. I recall that one of the meteorologists there, Lorraine Kelly, once grabbed me, while I was visibly very excited just after I had photographed a funnel cloud/waterspout, in order to check my blood pressure (and surprisingly found that it was normal); she also kept an eye on me

Figure 5.31 *Funnel clouds from non-supercell convective clouds over the ocean. (left) Twin funnel clouds over the Florida Keys on Aug. 28, 1993, as viewed from a NOAA helicopter. Note the sharp boundary in the sea state, from glassy and highly reflective (to the left) to disturbed (to the right). It is likely that the funnel clouds formed along the edge of an outflow boundary produced by evaporating precipitation to the right. (right) Similar to the panel at the left, but with just one funnel cloud visible. Note how the condensation funnel appears to be hollow, as the center is translucent. The water droplets making up the condensation funnel, if concentrated at a fixed radius, would appear to have a maximum optical thickness where the ring is seen.*
Photos courtesy of author.

when I got too close to the ledge by the window in her office, where I often looked for waterspouts. In May 1975, when I was still a graduate student, attending my first scientific conference, I saw a number of waterspouts serendipitously from my hotel balcony in Key Biscayne, Florida, just before I was scheduled to speak (Fig. 5.27). What a good omen for my future career, studying severe convective storms and tornadoes!

During GATE (Global Atmospheric Research Program Atlantic Tropical Experiment) in the late summer and early fall of 1974, when as a graduate student, I had the exciting experience of operating a weather radar aboard a NOAA ship stationed 600 miles off the west coast of Senegal, in the tropical ocean near the Intertropical Convergence Zone (ITCZ); I spotted and photographed a waterspout there

Figure 5.32 *A few of the early waterspouts I have seen. (top) Waterspout viewed from an aircraft, under the guidance of Joe Golden, in the Florida Keys on Sept. 3, 1971. The waterspout is near a gust front created near the edge of the rain core seen to the right. It was a thrill for me to see this when I was a beginning graduate student. (bottom) Very narrow, ropelike waterspout in the western Atlantic, west of Senegal, on Sept. 18, 1974, in the intertropical convergence zone (ITCZ), viewed from the NOAA R. V. Gilliss. The funnel cloud is very narrow and does not extend all the way to the ocean surface, but there is a spray ring near the surface, evidence that water is being lofted. Note the rain core to the far left of the waterspout., as in the top panel, but where the rain core is seen on the right. It was also a thrill for me to see this when I was a graduate student, operating and collecting data with a shipborne radar from MIT.* Photos courtesy of author.

(Fig. 5.32, bottom panel). I was particularly proud when Joanne Simpson, one of the women pioneers of atmospheric science, used my photograph in one of her studies.[11]

Waterspouts have been spotted all over the world, but especially in parts of the Mediterranean, during the fall. My long-time friend and colleague, Kerry Emanuel, recently photographed some waterspouts serendipitously in the Baltic Sea during the summer while sailing, while other friends and colleagues have observed waterspouts over the Great Lakes of the US, the Great Salt Lake in Utah (when it had more water than it does at the time of this writing), and waterspouts have also been reported over lakes in Maine and New Hampshire, just to name a few other instances of waterspout sightings in seemingly unlikely places. Waterspouts are ubiquitous and may be at least as common as tornadoes are over land areas.

5.6   Tornado dynamics

"All non-tornadic storms resemble one another,
but each tornadic storm is tornadic in its own way."[12]

> *My joking, imaginary "quote," with apologies to the novelist Leo Tolstoy from his novel,*
> Anna Karenina

More information about how tornadoes behave can be gleaned from theoretical and laboratory studies about tornadoes. To the theoretical fluid dynamicist, the tornado is mostly an interesting example of what happens when a rapidly swirling column of air makes contact with the ground. To the nature observer, the tornado is mostly an incredible display of power and beauty.

Some distance above the ground, the vortex does not know about the presence of the ground, so the centrifugal outward force associated with the swirling wind is balanced by the radially inward-directed pressure-gradient force (Fig. 5.33). However, near the ground, the air is slowed down by surface friction, while the pressure field remains relatively undisturbed.[13] A consequence of the decrease in wind speed near the ground is that the centrifugal force is reduced, since it is proportional to the square of the wind speed. There is consequently an *imbalance* of forces there, such that the inward-directed pressure-gradient force overwhelms the outward-directed centrifugal force, so air is accelerated radially inward toward the center (Fig. 5.33). It is not easy to verify this observationally because measuring the wind close to the ground using a radar is difficult.

---

[11] Simpson et al., 1986.

[12] Non-tornadic storms all have regions of precipitation in common, without condensation funnels or swirling debris near the ground; tornadic storms, while having similar architectures, look different from each other much more often than they look similar to each other.

[13] Richard Rotunno, in a 2013 review paper in the *Annual Reviews of Fluid Mechanics*, attributes this finding to a section in a textbook (section 5.7) by the late, prominent, Australian fluid dynamicist George Batchelor, in which it is demonstrated that the vertical variation in the radial pressure gradient is negligible near the ground, where frictional drag is significant.

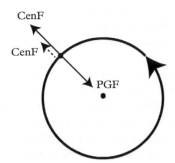

Figure 5.33 *Schematic of how surface friction disrupts the balance between the radially outward-centrifugal force (***CenF***) and the radially inward-directed pressure-gradient force (***PGF***). *CenF* is depicted by the solid arrow when it balances the* **PGF***; when friction slows the wind down, *CenF* is weaker (dashed arrow), while the* **PGF** *remains the same: The result is that there is a net radially inward directed force.*

Figure courtesy of author.

First, using safe, remote sensing by a Doppler radar, it is very difficult to "see" what is happening right near the ground owing to "ground clutter." A radar beam is not infinitely narrow and so radiation from some of its direct beam and "sidelobes"[14] are inevitably reflected from the ground and from trees, etc., overwhelming and thus contaminating the returned signal from raindrops, etc., in the path of the direct beam. From a practical standpoint, the radar cannot detect winds very well below 50–100 m above the ground owing to ground clutter. Since surface friction acts through a very shallow layer near the ground, perhaps only 10 m or even less deep, a radar is not capable in most instances of "seeing" what is happening in the surface friction layer. But, the alternative of standing in a tornado with an anemometer is probably not a wise thing to do. However, there is some evidence from photographs taken from aircraft of trees downed by tornadoes that the trees had been uprooted and blown down in the direction toward the center of the tornado, which is indicative of a stronger radial inflow than swirling flow. In fact, it might be that right near the ground in a tornado most of the airflow is actually radially inward, and that the swirling component of the wind is of secondary importance. The depth of this layer near the ground of mostly radially inward-flowing air is not known exactly, but thought to be anywhere from 10 meters of so down to just a few meters. An example of this type of "frictional inflow" may be seen in tea leaves, which gather at the center of a cup of tea when it is stirred (regardless of the direction the leaves had been

---

[14] Sidelobes are flower-petal-like patterns in the lines of a measure of constant sensitivity for a parabolic-dish radar antenna, so that while most of the sensitivity is directed in one direction, there is still some lesser sensitivity in other directions. Because the reflectivity of a tree or house is much higher than that of a bunch of raindrops, the clutter from the tree or house off beam can mask the return from the raindrops actually in the path of the direct beam. This is not the case for a lidar (similar to a radar, but for a beam of light, rather than for microwave radiation), whose beam does not spread out much with distance from the instrument (Bluestein et al. 2014).

stirred). Frictional inflow overwhelms the outward flux of cloud droplets, raindrops, and debris near the ground, so the WEH seen on radar fills in near the ground and there is a pear-shaped appearance to the weak-echo column when viewed in cross section (Fig. 5.19, right panels).

An interesting consequence of this radial inflow is that the airflow must converge at the center of the vortex. The circular geometry of the vortex and friction conspire to produce a trainwreck of air flowing radially inward from all directions.[15] In doing so, since the air can't dig into the ground, it turns upward into a rapidly moving, vertical jet of air. A vortex that makes contact with the ground such that friction is responsible for an upward jet of air is called an *endwall vortex*. The upward jet can be very intense, with wind speeds comparable to those associated with the horizontal vortex itself. A video taken from a drone of a tornado in Andover, Kansas, on April 29, 2022, from Reed Timmer, an expert storm chaser, and his group, shows in spectacular fashion the extreme rising motion in the tornado (see video on YouTube: Look for "Reed Timmer.").

When the wind speeds in a vortex increase linearly with distance away from the center, as in solid-body-like rotation, it turns out that if you were to show the vortex some affection by embracing it and squeezing it radially inward and then letting it go (love it and leave it), it would re-bound and then oscillate inward and outward, producing what are known as "centrifugal waves," because the waves owe their existence to the centrifugal force. When the ring of air is squeezed inward and let go, it finds itself in an environment in which the inward-directed pressure-gradient force is less than the outward-directed centrifugal force and it springs back to its original location but over-shoots it a bit, owing to the kinetic energy of the initial squeeze. It then finds itself in an environment in which the inward-directed pressure-gradient force is greater than the outward-directed centrifugal force and it springs back again to its original location, but also overshoots it a bit, and so on. These waves can propagate upward and downward along the edge of the vortex column, possibly explaining why we sometimes see wavelike variations in the width of the condensation funnel as a function of height (Fig. 5.34). These are sometimes seen along the edge of debris clouds, as located by their radar nature[16] or in videos.

When the speed at which air is flowing upward in the jet exceeds the speed at which centrifugal waves propagate, the flow is said to be "supercritical." [By comparison, when air flows faster than the speed of sound (what happens, for example, when air that passes by a supersonic aircraft), there are shock waves and a sonic boom is heard.] Somewhere above, there is downward motion forced by the dynamically induced low pressure at the center of the vortex (Fig. 4.89). Where the upward jet gets weaker with height aloft and even reverses direction (Fig. 5.35), there may be a sudden transition from supercritical flow to subcritical flow while the airflow becomes turbulent and the vortex widens. This phenomenon is known as "vortex breakdown" (Benjamin 1962) and is analogous to what happens when water runs down a creek, but is forced to flow over a rock, producing

---

[15] I'm pretty sure that I first heard this metaphor or one similar to it used by my friend and colleague Rich Rotunno.
[16] Houser et al., 2016.

Figure 5.34 *Wavelike perturbations in both the thickness and/or orientation of a funnel cloud (albeit a highly tilted one), on June 19, 2018, in a dissipating tornado in northeast Colorado.*
Photo courtesy of author.

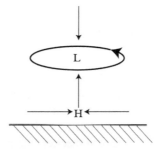

Figure 5.35 *Illustration of how radial inflow into a vortex near the ground where surface friction reduces the wind speed results in an upward-flowing jet. (Note that this mechanism for producing upward motion is not exactly the same as that responsible for inducing upward motion underneath a mesocyclone (or anticyclone) in the absence of friction, as in Figure 4.77) The converging air from all directions must come to a complete stop at the center. The only way this can happen is that if there is a relatively high dynamic pressure at the center, creating an outward-directed dynamic PGF. Air is sucked upward to the center of the vortex aloft, but it is also sucked downward above the vortex. Thus, air rises and abruptly changes direction as it "turns the corner." Near the ground, the air may actually be diverted upward slowly, not abruptly, owing to the dynamically produced high pressure area.*
Figure courtesy of author.

a hydraulic jump, a phenomenon we mentioned earlier when we looked at the airflow accompanying some orographic wave clouds and downslope windstorms. For the latter, it is water that is flowing more rapidly than shallow-water gravity waves propagate, transitioning to water that is flowing more slowly than shallow-water gravity waves propagate. In this case, the centrifugal waves play the same role as the shallow-water gravity waves. It turns out that the region where the horizontal wind speeds in a tornado should be greatest is at the level of vortex breakdown. Much of this theory is too difficult to explain clearly to the layperson. The jump in height of the water as a hydraulic jump is like the widening of a tornado vortex seen above the level of vortex breakdown.

There must be a relationship between pressure and the wind. It follows that as air flows radially inward toward the center of the vortex near the ground, it must come literally to a grinding halt as it approaches the center (à la the trainwreck metaphor). What makes it do so? We infer that there must be a dynamic pressure-gradient force that acts to slow the air down; in other words, the pressure must be *higher* at the center of the vortex, not lower (as the hydrostatic part is), so that there is a dynamic pressure-gradient force that acts in the direction *opposite* to that of the wind component in the radial direction, which is radially outward. This does not contradict what we said earlier about the pressure being lower at the center of a vortex owing to rotation because near the ground, the radially inward flowing air dominates over any swirling motion there might be. So, if the pressure at the surface is higher than that aloft, not only is air accelerated upward through an upward-directed pressure-gradient force, but also the likelihood of saturation occurring at and near the center is *less* right near the ground than it is aloft, owing to the higher pressure. In fact, when we can see the condensation funnel near the ground in some tornadoes when there is not an opaque debris cloud surrounding the center and blocking our view, and also when an intervening hill does not prevent us from seeing the ground, the funnel cloud often tapers to a point and disappears right near the ground (e.g., Fig. 5.13, top right, middle left and bottom; 5.23, bottom). It may seem amazing that the condensation funnel is truncated and may disappear right at the ground.

One of the most fascinating aspects of some tornadoes is that there can be even smaller tornadoes within them (Fig. 5.36). These smaller tornadoes are often referred to as "secondary vortices." The late, famous, and brilliant University of Chicago meteorologist Ted Fujita, who coined the term "wall cloud" and also devised the Fujita F-scale for tornado damage, suggested that a vacuum cleaner-like suction was responsible for this damage pattern and named the culprits "suction spots." It has also been thought that strong winds from tiny tornadoes rotating around a larger-scale vortex, "suction vortices," were responsible for these marks on the ground. Recently, my colleague Roger Wakimoto at UCLA examined some of our radar data in conjunction with his photographs of cycloidal ground marks and cast some doubt as to whether or not cycloidal damage patterns are in fact always due to the swirling motion in secondary vortices, based on the one case he analyzed[17]; instead, it might be that the ground markings may

---

[17] Wakimoto et al., 2022

Figure 5.36 *Multiple-vortex tornadoes and funnel clouds. (top) Multiple-vortex tornado viewed to the west, in southwestern Oklahoma, near Friendship, on May 11, 1982. (middle, left) Multiple-vortex tornado near Verden, Oklahoma, on May 3, 1999, during a major tornado outbreak. The individual, secondary vortices, as manifest as narrow, individual funnel clouds, are all leaning outward with height with respect to the center of the larger-scale, parent vortex. When we went back to look for damage associated with these small vortices, we could not find any in the rough vegetation. As many as six funnel clouds are visible at once. (middle, right) Multiple-vortex tornado east of Lake Thunderbird, Norman, Oklahoma, on May 19, 2013. From a video frame from a video taken by the author with a hand-held camera. View is to the southwest. Note how in this case the funnel clouds are not connected to*

be caused by debris that is left on the ground when air near the ground turns suddenly upward in the corner flow of the updraft (Fig. 5.37d).[18] There is no doubt, however, that secondary vortices exist in many tornadoes, which are called "multiple-vortex" tornadoes (e.g., Wurman 2002).

To understand why these multiple-vortex tornadoes form, we rely on the results of carefully controlled laboratory simulations of tornadoes using vortex chambers and of numerical simulations of these highly idealized laboratory vortices. The vortex chambers consist of enclosed cylinders in which air is sucked out from the top by an exhaust fan, while air is forced to rotate as vanes on the outer edge of the chamber channel air in from a direction a bit different from that of purely radially inward. The idealized laboratory model simulations using a computer try to mimic what happens in the actual laboratory models. The first laboratory models were made by Neil Ward at NSSL, and later by Chris Church and John Snow at Purdue University. Rich Rotunno at NCAR, Brian Fiedler at OU, and Steve and Dave Lewellen at West Virginia University pioneered the numerical simulations of tornado vortices, as did Takashi Maruyama in Japan. While the earlier simulations were made by drawing air in from the sides at the bottom and sucking it out from the chamber by an exhaust fan at the top, the more recent numerical simulations have been done by putting in a virtual buoyant bubble, which results in air being circulated within the chamber: converging underneath the buoyant bubble, rising in the bubble, diverging above the buoyant bubble, and sinking around the edges, as if the air were following the surface of a donut (as in Fig. 3.6).

We know that tornadoes, especially near the ground consist, in part, of pure, swirling motion, and in part of pure, radial motion. The relative amount of the swirling motion to the amount of inward radial motion (below some height above the ground) is called the "swirl ratio." While we don't have a full understanding of exactly what determines the swirl ratio, we do know that as the swirl ratio is increased, a point is eventually reached at which a single-cell vortex, with upward air motion at the center and sinking air motion away from the center, makes a transition to a two-cell vortex, with sinking air motion at the center, a ring of rising air motion somewhere away from the center, and sinking air motion even farther out from the center. As the amount of swirl increases near the ground, so does the drop in pressure at the center. A downward-directed pressure-gradient force then develops, which sucks air downward at the center (Fig. 4.89).

Figure 5.36 (Continued) *the cloud base above. (bottom) Multiple vortices, as evidenced by multiple funnel clouds, in a circulation that did not become a tornado, at least right away, on May 6, 2015 over Chickasha, Oklahoma. These funnel clouds also all leaned outward with height with respect to the center of rotation of the parent circulation. It did not appear as if any debris was kicked up by any of these funnel clouds, so technically there was no tornado at the time of the photograph.*

Photos courtesy of author.

---

[18] Lewellen and Zimmerman, 2008.

Vortex breakdown can occur and may eventually happen down at the ground. In addition, as the speed of the azimuthal component of the wind increases, it becomes more and more difficult to suck air radially inward, so that the vortex broadens. As the vortex widens, an annulus of strong radial shear in the winds develops. This shear becomes dynamically unstable and breaks down into secondary vortices. These vortices act like miniature, but powerful tornadoes, as they rotate around the center of the broader-scale vortex. The effect of turning up the swirl-ratio knob on the nature of the tornado is illustrated in Figure 5.37. These secondary vortices may be detected by Doppler radar, can be very intense (Figs. 5.38 (also, Wurman 2002) and 5.39), each seen as separate,

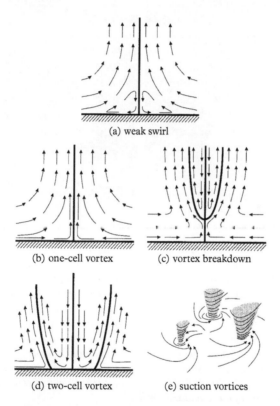

Figure 5.37 *Illustration of how the structure of tornadoes changes as the swirl ratio is increased. In (a) there is mostly radial flow. In (b) air flows up at the center of the vortex and at large distances from the center. In (c), as the swirl ration increases, downward flow develops aloft at the center, along the axis of the tornado. As the swirl ratio is increased even more, the downward flow makes it all the way to the ground and the vortex widens. In (e), multiple, secondary vortices develop and rotate around a common center. From Wakimoto and Liu, 1998, their Fig. 6. Published 1998 by the American Meteorological Society, but adapted from an earlier book chapter by R. Davies-Jones (1986) in Kessler, 1986, their Figure 10.23 on p. 229. © University of Oklahoma Press.*
Used with permission.

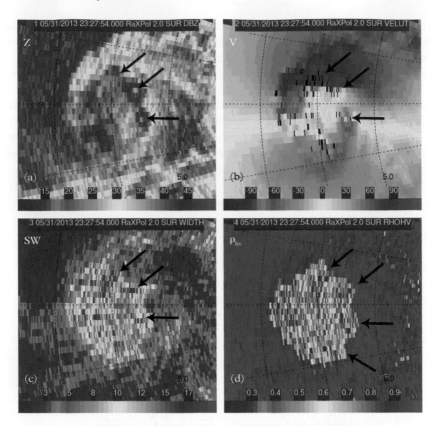

Figure 5.38 *Data from RaXPol for a multiple-vortex tornado, near El Reno, Oklahoma, on May 31, 2013. Panel (a) shows the radar reflectivity factor in dBZ. Three WEHs are pointed out. Panel (b) shows the de-aliased Doppler velocities, color-coded as red, of wind speeds around 100 m s^{-1} away from the radar, and color-coded as deep green/light blue, of wind speeds around 60 m s^{-1} toward the radar. Some tiny vortex shear-signature couplets are pointed out, each associated with a secondary vortex. Panel (c) shows the spectrum width, which depicts the amount of variation in wind speed within a radar volume. The arrows point out lobes of relatively high spectrum width associated with secondary vortices. Panel (d) shows the broad region of relatively low coplanar cross-correlation coefficient (less than 0.9), with multiple vortices pointed out, as lobes of low cross-correlation coefficient, denoting a "debris signature" (Snyder et al., 2013; Wakimoto et al., 2015; Wienhoff et al., 2020). This image is consistent with the detection of four secondary vortices.*

From Bluestein et al., 2015, their Fig. 18. © American Meteorological Society. Used with permission.

narrow condensation funnels, that often lean radially outward with height (Fig. 5.36, middle left and bottom panels).

In tornadoes, which move along with their parent storm, the increase in wind speed on the side of the vortex that has a component in the direction of the tornado, the ground-relative winds can be very intense. In a tornado that hit El Reno, Oklahoma on May

31, 2013, the one that resulted in the death of Tim Samaras, mobile radars measured wind speeds of 135 m s^{-1}, part of which was due to an amazingly rapid translation of a sub-tornado-scale, secondary vortex (~78 m s^{-1}), and part due to the tornado-vortex swirling winds themselves (~55 m s^{-1}) (Fig. 5.39). In this and similar cases, the damage swath from tornadoes is not centered symmetrically about the center of the tornado vortex, but rather shifted onto the side of the vortex that has the stronger ground-relative wind speeds. It is readily apparent from our radar data how risky it is getting near a large, multiple-vortex tornado. Even if you are well away from the center, you could get hit be a secondary vortex that is moving so quickly that there is no way you can avoid it. As a tornado moves along, the secondary vortices do not appear to make it all way around their parent vortex, as do horses in a merry-go-round. Perhaps if the tornado were stationary, then the secondary vortices might make it all the way around.

Different from multiple-vortex tornadoes, "satellite" tornadoes are actually separate tornadoes that become influenced by the wind field of a nearby tornado, which is typically stronger. The weaker, satellite vortex may then rotate around its stronger companion and then become absorbed by it (Fig. 5.40).

Finally, in some supercells, which are undergoing cyclic mesocyclogenesis (a periodic and orderly sequence of the formation and decay of mesocyclones), a series of tornadoes

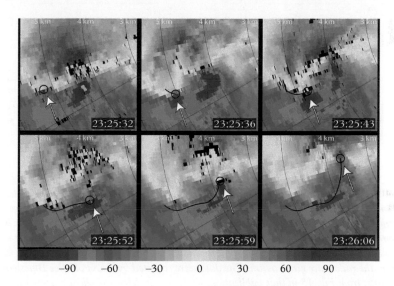

Figure 5.39 *A series of RaXPol sectors of edited (corrected for aliasing) Doppler velocities. (m s^{-1}, color-coded scale at the bottom) from the El Reno tornado of May 31, 2013 at the times indicated in UTC (6 hours later than local time). A secondary vortex within the larger-scale tornado (indicated by the broad transition from purple velocities, representing rapid flow toward, to brown velocities, representing rapid flow away) is highlighted by a circle and pointed out by a white arrow. Its path relative to the larger tornado is indicated by the solid line.*

From Bluestein et el. 2018, their Fig. 2. © American Meteorological Society. Used with permission.

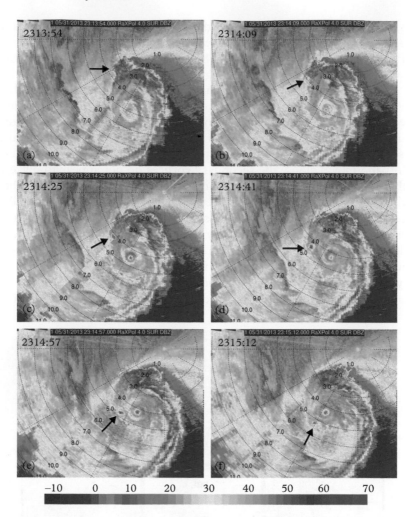

Figure 5.40 *Radar reflectivity pattern at low altitude, as a function of time (UTC on May 31, 2013) of a small satellite tornado rotating in a counterclockwise (cyclonic) manner around a large tornado near El Reno, Oklahoma. The satellite tornado is indicated by a WEH (coded green) that is identified by black arrows. The WEH associated with the larger tornado is seen as a thick green dot embedded in a comma-shaped yellow/brown radar-echo pattern having a spiral band. The range rings (at constant distance from the radar) are given in km. The data were collected by RaXPol. (See Bluestein et al. 2015 for more details; from Fig. 15 in that publication; © American Meteorological Society.)*
Used with permission.

can occur, in which case more than one tornado may be visible at one time (Figs. 5.41 and 5.42) (Dowell and Bluestein 2002b). When this happens, each tornado is likely to be in a different stage of its life. These tornadoes tend not to be satellite tornadoes

Figure 5.41 *Twin tornadoes. (top) Two tornadoes in a supercell in northern Oklahoma on April 14, 2012, viewed to the north-northwest, probably as a result of cyclic mesocyclogenesis, the production of a series of separate mesocyclones within the same parent storm. (middle) Two tornadoes in a supercell south of Dodge City, Kansas, on May 24, 2016, resulting from cyclic tornadogenesis. Details of the behavior of this storm may be found in a journal publication first-authored by one of my former graduate students, Zach Wienhoff.[a] (bottom) As in the previous panel, the tornado to the right is the older one, becoming enveloped in precipitation. The ropelike one on the left is a transient tornado that was rotating around the circulation of the mature tornado. (a frame from a video shot by the author), and probably not due to cyclic mesocyclogenesis.*

[a] Wienhoff et al., 2020.Photos courtesy of author.

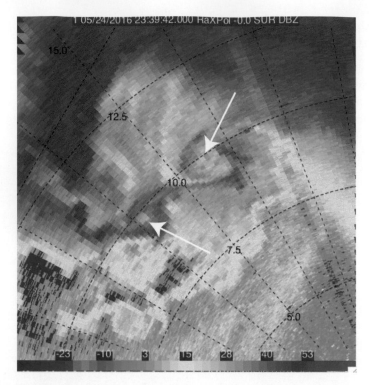

Figure 5.42 *Radar depiction of twin tornadoes in the Dodge City, Kansas tornadic supercell on May 24, 2016 (see Fig. 5.41, middle panel). Radar reflectivity from RaXPol at 0° elevation angle, The white arrows point to WEHs associated with each tornado. These WEHs are within a larger-scale hook echo.* Figure courtesy of author.

(if they are widely separated) and they are not secondary vortices of a multiple-vortex tornado. Rather, they are simply individual tornadoes that happen to be near each other. So, there is more than one way in which more than one tornado may be visible at the same time.

5.7   Types of tornadoes

In the preceding section we looked at the different appearance tornadoes can assume. In this section we note that tornadoes can form in different types of convective storms, even if they all look similar despite their *different* origins (whence my earlier "quote" similar to Tolstoy's): There is more than one way in which the same phenomenon, a tornado, can form, and no easy way to determine which way the tornado had formed, given only its appearance.

5.7.1   Tornadoes (and waterspouts) associated with mesocyclones in supercells

The most intense tornadoes form in supercells, in which the source of rotation is that of the low-level mesocyclone in the parent supercell; rotation was not present before the storm had formed, but rather was produced by the parent storm itself. It is thought that once the low-level mesocyclone forms, it may get low enough to the ground that friction reduces the wind speed, resulting in the formation of an endwall vortex with strong convergence and an upward flowing jet. The convergence acts to decrease the radius of swirling air parcels, resulting in an increase in the swirling wind speed, in accord with the conservation of angular momentum. A skater (analog to a low-level mesocyclone) pulls his/her arms inward and spins up as angular momentum is conserved (angular momentum = azimuthal wind speed × the radius at which the wind occurs; if the radius decreases, as air is brought inward, the azimuthal wind speed must speed up). Recall how in the preceding chapter we noted that the intensity of the low-level mesocyclone depends on the strength of the temperature gradient along the forward flank of the storm and the low-level vertical wind shear.

The upward motion above the low-level mesocyclone is driven by the upward force of buoyancy in the main updraft of the supercell. In the absence of surface friction, the maximum intensity of the tornado vortex is determined by the strength of the updraft. One can relate the maximum wind speed in a tornado to the intensity of the thermodynamic buoyancy of the updraft. This maximum wind speed is known as the "thermodynamic speed limit." This relationship is reminiscent of (but not identical to) that which relates the temperature of the sea surface to the maximum intensity of a tropical cyclone. With surface friction, the thermodynamic speed limit is exceeded because the ring of swirling air associated with the low-level mesocyclone can be brought even farther inward than is possible due to thermodynamic considerations alone, as a result of frictionally induced inflow (as the outward-directed centrifugal force is overpowered by the inward-directed pressure-gradient force [Fig. 5.33]).

It turns out that the tornado vortex is usually stable, so it takes work to contract a ring inward. It is similar to the problem of how much can you compress a spring: It takes work to compress it, but *infinite* work to collapse it to a point (if the coil of wires were infinitesimally narrow). When you let go of it, it will oscillate, expanding and contracting with time. The tornado could potentially be infinitely strong, if the ring could be brought inward all the way to the center. This situation might be sort of similar to that of producing a black hole, a region in space that has so much mass that light cannot escape from it. In the case of a tornado, it would take an infinite amount of work to contract a ring all the way to the center, a situation that obviously would be a waste of effort, in fact infinite effort. However, although the effects of surface friction augment the effects of buoyancy in contracting a ring of air in a vortex radially inward, it takes more and more work to contract the ring more and more, and there is some finite limit, fortunately, to how strong tornadoes can be. A conclusion from all this theory agrees with what has been found observationally. By the way, the width of a funnel cloud (or width of a tornado vortex "seen" on radar) is *not* an indicator how strong a tornado is; size does not matter. However, facing a large, wedge tornado can be frightening no matter how intense it is (Fig. 5.43).

There has never been a tornado yet (as of the time of this writing), however, with ground-relative wind speeds in excess of ~135 m s^{-1}, estimated from Doppler-radar measurements, which suggests that under typical environmental conditions, there is a physical limit to how strong a tornado can be given the typical distribution of temperature and moisture with height. As the atmosphere warms as a result of the increase in carbon dioxide and other greenhouse gases, we would expect the environmental conditions and the limit of how strong tornadoes might also change.

There is furthermore the issue of how large a tornado appears can depend on how far away it is. I recall in one instance during the first VORTEX in 1995, on April 17, in south/southwest-central Oklahoma, how we were trying to get into position to collect radar data in what appeared to be a relatively large tornado far away to our northwest,

Figure 5.43 *A broad tornado, viewed to the west, south of Dodge City, Kansas, on May 24, 2016, being admired by two unidentified storm chasers. There are some small horizontal funnel segments off to the northeast (far right) of the tornado just under cloud base.*
Photo courtesy of author.

only to suddenly realize that it was actually a narrow tornado, very *nearby*. Instead of trying to get ahead of it, we abruptly stopped and "let" it cross the road less than 100 m ahead of us. I took a video which showed that the width of the condensation funnel was only approximately the width of the two-lane road we were on and that close up it did not have sharp edges, but rather looked diffuse and ghostly (Fig. 5.44). I decided that, exciting as this experience was, I did not want to repeat it.

Supercells that are relatively shallow, particularly in cold, upper-level cyclones, in which the tropopause is very low (may be only 5–6 km above the ground) can spawn tornadoes. These shallow, low-topped supercells or "mini-supercells" demonstrate once again how in this particular case, the depth of the parent convective storms does not matter. Shallow supercells in landfalling tropical cyclones (Gentry 1983) may be similar. These tornadic supercells tend to occur in outer rainbands as distinctly discrete cells, in comparison with the more continuously looking radar echoes in bands nearer to the center of the tropical cyclone (Fig. 4.112). I have never seen a tornado in a mini-supercell, perhaps simply because I have not considered the probability high of seeing one in the environment of these mini-supercells to be worthy of a long drive. Furthermore, if I had a chance to try to photograph a tornado in an outer rainband of a tropical cyclone, it might be extremely difficult to do so owing to intervening precipitation.

Figure 5.44 *Very narrow (the condensation funnel is around the width of the road, which is probably less than 20 m), white tornado crossing the road within 100 m or so to our north on April 17, 1995, from a VHS video taken by the author from inside our van, in south-central Oklahoma, near Temple.* Photo courtesy of author.

5.7.2 Non-mesocyclone tornadoes (landspouts), quasi-linear convective system tornadoes, and waterspouts

While some tornadoes are clearly associated with low-level mesocyclones in mature supercells, other tornadoes are sometimes observed in ordinary-cell convective storms or in multicell storms. These tornadoes have been called "non-mesocyclone" tornadoes (Wakimoto and Wilson 1989); for them, the source of rotation is not produced *by* the parent storm, but rather exists *independent of* the storm, in many instances prior to the storm's birth. In other words, there may actually be a mesocyclone, but it does not owe its existence to the parent storm, although the mesocyclone may be enhanced by it subsequently. In other instances, the mesocyclone may be produced from shear along surface boundaries produced by the storm. The nomenclature "non-mesocyclone" tornado can therefore be ambiguous at times. There have been some case studies published in which there have been debates about whether or not the tornado or tornadoes spawned was or were actually from mesocyclones produced in the classic manner in supercells as described in the previous chapter, or from some hybrid processes involving both supercell dynamics and vortices produced near the ground along boundaries either produced by the parent storm or that pre-existed the parent storm.[19]

Waterspouts are frequently observed along lines of cumulus congestus in the Florida Keys. They may form even before precipitation, which is produced aloft first, has begun to reach the ground (the cumulus congestus having then been converted into cumulonimbi). Since they are so narrow and at least look weak (Figs. 5.29–5.31) (remember, though, size can be deceptive), I, upon seeing a similar phenomenon in Oklahoma back on April 22, 1981, decided to call this type of a tornado a "landspout," since it looked like a weak, waterspout, but paradoxically over land (e.g., Fig. 5.45). My memory may be faulty, and it's also possible that I heard another storm chaser first use this moniker and I unconsciously appropriated it and made it sort of official (at least into a published archive, albeit unrefereed by my peers) in a conference publication back in 1985.[20] Landspouts occur from time to time along cold fronts, and outflow boundaries, which are accompanied by converging streams of air and horizontal wind shear (which can induce rotation, i.e., vorticity) (Lee and Wilhelmson 1997). They also form along surface boundaries that owe their existence to heating along sloping terrain, particularly in eastern Colorado, near what is known as the Denver convergence-vorticity zone; this surface boundary is characterized by convergence and a pre-existing cyclonic wind shift across it. Convective clouds tend to be triggered along it during the summer and some grow into convective storms, mainly ordinary cells and multicells.[21]

It has been suggested that underneath buoyant, cumulus congestus clouds, preexisting, small-scale vortices or bands of enhanced vorticity associated with shear across the boundaries are affected by convergence near the ground underneath updrafts in

[19] Wakimoto and Atkins, 1996, Ziegler et al., 2001.
[20] Bluestein, 1985a.
[21] Brady and Szoke, 1989.

growing convective clouds, which results in strong vortices, sometimes strong enough to be tornadoes (Fig. 5.46). It is not unusual for a number of landspouts to be ongoing at the same time, but not for the same reason that several supercell tornadoes sometimes

Figure 5.45 *Non-supercell tornado in northeast Colorado, southwest of Brush, on June 6, 2014. There is an extensive debris cloud with a collar of additional debris around it, but the pressure drop is not large enough nor the amount of water vapor great enough to form a funnel cloud at cloud base. The structure near the ground, ironically, is an irrigation device.*

Photo courtesy of author.

Figure 5.46 *Idealized depiction of the formation of landspouts along a surface boundary characterized by both converging and shearing air streams. Parent vortices develop along the surface boundary (labeled "A," "B," and "C") and are stretched into stronger vortices by converging air underneath buoyant updrafts in the clouds that are formed. From Wakimoto and Wilson, 1989, their Fig. 20. © American Meteorological Society.*

Used with permission.

appear simultaneously (such as cyclic mesocyclogenesis). In the former case, the occurrence of more than one tornado along a line is due to the horizontal wind shear along the linear boundary along which the tornadoes form. Since the landspouts in Colorado tend to occur where the surface humidity is relatively low, there may not be enough moisture to condense into condensation funnels. In this case, one may see a rotating debris cloud under a rotating cloud base, without ever seeing a condensation funnel (Figs. 5.45, 5.47, and 5.48).

My favorite example of this phenomenon, a tornado without a condensation funnel, is a landspout I photographed and took videos of at Denver International Airport, while I was awaiting to board an airplane (Fig. 5.48). The debris cloud extended almost all the way up to cloud base. While we travelers in the gate areas were instructed to head for one of the in-airport shelters, I remained at the window, while periodically being told to seek shelter. I moved away from the window only when the viewing angle of the tornado became very high and when I did in fact finally become concerned about the potential for flying glass if the window shattered. It was poetic justice of sorts that I got to see this tornado at all, because I was returning to Norman to attend the Ph.D. thesis defense of my former student Jana Houser, who had studied a large tornado in her research; had I not been traveling to Norman for her defense, I would most likely *not* have been at Denver International Airport, and would not have seen this tornado.

When multiple landspouts are seen lined up under cloud base, they give the appearance of all hell breaking loose. An unusual instance of landspouts rotating around each other is seen in Figure 5.49, which also is a very dynamic event.

It is not always possible to distinguish a landspout from a mesocyclone-associated tornado just by looking at it. It should also be pointed out that the mechanism by which non-mesocyclone, landspouts are produced, may also occur in supercells; it's just that they are not associated with or near the mesocyclone in the supercell. In this case, one can be very confused, as noted earlier.

Another place non-supercell tornadoes are found is at the leading edge of quasi-linear mesoscale convective systems (QLCSs). These tornadoes are similar to landspouts in that there may be a series of them lined up. They are easily detected by Doppler radar but not easily seen owing to nearby precipitation which might envelope them. Sometimes they appear to be associated with more discrete cells within the leading line of precipitation and exhibit some supercell characteristics (Fig. 5.50). One of my former graduate students, who at the time of this writing works for the National Weather Service in Norman, Vivek Mahale, studied a typical case using data collected by the CASA (Collaborative Adaptive Sensing of the Atmosphere) network,[22] an experimental, at the time, network of low-power, X-band, polarimetric Doppler radars in Oklahoma (McLaughlin et al. 2009) (the network is now set up for operational use in the Dallas-Ft. Worth area). The storm was notable in that it occurred around sunrise, not the time of day one would expect to find tornadic storms, and was quite a surprise. I am aware of a good number of cases like this and when they occur so early in the morning when many people are still sleeping, they can be dangerous because warnings are less likely to be recognized by the public, particularly if sirens are not

[22] Mahale et al., 2012.

Figure 5.47 *Landspout/non-supercell tornado and blowing dust at the ground, east of Denver, Colorado, on Aug. 9, 1996, viewed to the east. Note the cumulus congestus, which were building upward, above.*

Photo courtesy of author.

Figure 5.48 *Non-supercell tornado at Denver International Airport on June 18, 2013. This tornado has a debris cloud underneath a truncated-cylinder-like lowered cloud base.*
Photo courtesy of author.

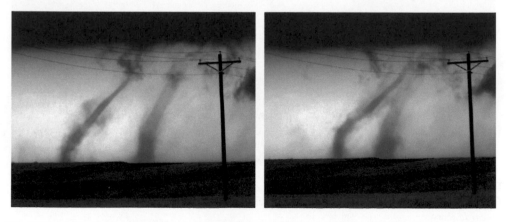

Figure 5.49 *Two landspouts that appear to be rotating around each other. (left) Two landspout-like tornadoes side-by-side, viewed to the west (from a frame from a hand-held video by the author), on May 25, 2010, near the Colorado–Kansas border, west of Tribune, Kansas, during VORTEX 2. (right) As in the left panel, but seconds later, as the southern (leftmost) tornado appears to rotate cyclonically around the northern (rightmost) tornado, perhaps in response to the flow associated with a developing, mesocyclone nearby (in which case, perhaps these were actually secondary vortices, not landspouts).*
Photos courtesy of author.

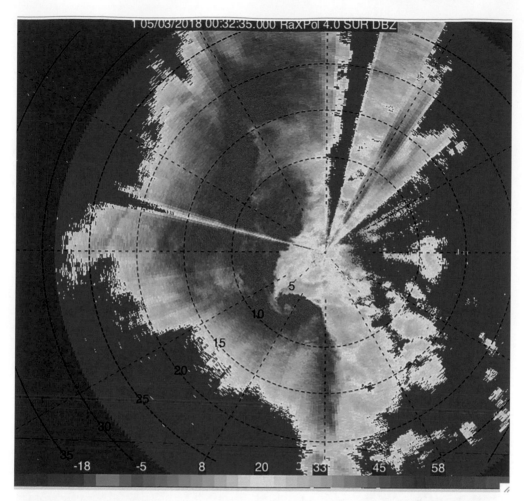

Figure 5.50 *RaXPol reflectivity at 4° elevation angle of a supercell embedded within a squall line on May 2, 2018, in south-central Oklahoma. The wall cloud is located at the center of the hook echo, to the southwest of the radar. The radar echo is heavily attenuated at the southern end as the radar beam goes through heavy precipitation for a rather long distance. This storm is just one of many embedded within a squall line on this day.*

sounded, alarms on weather radios and cell phones are not heard, and if televisions or radios are not turned on. Even when QLCS tornadoes occur during the late afternoon and early evening, they are difficult to see. Many of these types of tornadoes occur in the Mississippi Valley region and parts of the Midwest in the US and particularly from Louisiana to Pennsylvania along a curved axis (Trapp et al. 2005).

Perhaps my most interesting account of a similar QLCS tornado was in Norman, when my wife and I were having dinner at a local restaurant during the fall, several years ago. I would certainly not have been at the restaurant having a leisurely meal had I expected that there might be supercell tornadoes nearby. I recall looking at the radar display on my cell phone and seeing a hook echo and TVS along the leading edge of a squall line coming right toward us. We quickly gulped down the remains of our dinner, paid the bill, warned our server, and hightailed it back home, in the eerie, still darkness. A tornado did in fact touch down about a mile or two to our southwest, but fortunately dissipated before it ever reached us.[23]

5.7.3 Gustnadoes

The colloquial term "gustnado" was first used by storm chasers in the 1970s. It is an intense vortex along the gust front of any type of convective storm, but it is typically characterized by a rotating column of dust does that does not appear to be connected to any rotating portion of the cloud base above (Fig. 5.51). Sometimes, like landspouts, a number of them appear lined up in a row. They usually do not last long and produce only

Figure 5.51 *Two gustnadoes alongside the gust front of a convective storm in southwest Texas on June 14, 2010, during VORTEX2 (Wurman et al., 2012).*
Photo courtesy of author.

---

[23] During the evening of Feb. 26, 2023, a tornado spawned by a supercell embedded within a line of convective storms, in approximately the same location, passed through the southeastern section of Norman, Oklahoma, just missing the University of Oklahoma National Weather Center . . . and also my house, but unfortunately inflicting major damage to other houses in its path.

minimal damage. They probably owe their existence to a zone of shearing winds along the gust front and dynamic instabilities associated with the strong lateral wind shear.

5.8 Dust devils

Dust devils are intense columnar vortices like tornadoes, but instead of being driven by a saturated, buoyant updraft in a convective cloud above, they are driven by intense heating by the sun of the ground below. Dust devils rarely last more than 5–10 minutes and sometimes appear only for a few seconds. It is much too dry for any condensation funnels to appear, but dust and sand from the ground are lofted to give a similar appearance to tornadoes, especially when the dust devil is very tall (Fig. 5.52, bottom panel). Dust devils rarely inflict much damage. Cars have safely driven through them and it is indeed relatively safe to hold up an anemometer or barometer inside most of them to make wind or pressure measurements . . . and live. Dust devils are most frequently seen in desert-like or otherwise relatively arid environments such as southern Arizona, Utah, Nevada, eastern Colorado. They have even been observed on Mars.[24] Dust devils are ghostlike in that they are not connected to any cloud feature and seem to appear out of nowhere. Often more than one is visible at once (Fig. 5.53). Some are relatively squat (Fig. 5.54), while some are majestic, narrow, towering columns of dust (Fig. 5.52, bottom panel). On our planet, we find that there does not appear to be a preference for the sense of their rotation.

An important scientific question is: What is the source of the rotation in dust devils? One possibility is that horizontal shear in uniform airflow may be produced as air passes over two adjacent land surfaces which have markedly different surface roughness, for example a grove of vegetation flanked by desert sand. The air is slowed down more by the grove of vegetation than by the desert. Convergence underneath a local, dry (and invisible), buoyant updraft (thermal) can then act to increase the vorticity. It is also possible that blockage around an obstacle might create rotation. We find that dust devils typically form when the winds and consequently the vertical wind shear are relatively weak. Obstacles at the surface seems to disrupt dust devils. It has been reported that driving into a dust devil, standing in a dust devil, or otherwise putting any structure in its path will disrupt it. Not so for a tornado!

We sometimes find dust devils out over flat terrain with no obstacles or changes in surface roughness. Another possible source of rotation, particularly in this case, the tilting of horizontal vorticity due to vertical shear of the wind created by horizontal vortices associated with the individual convective cells of Rayleigh–Bénard convection when the surface is heated uniformly with respect to the air above (as described in Chapter 3). In a computer simulation study,[25] it was found that vertical shear associated with the convective cells themselves can be tilted onto the horizontal. It was also found that there must be a "symmetry-breaking event" so that vorticity of just one sign is concentrated into a

---

[24] Taylor, 2022.
[25] Fiedler and Kanak, 2001, Kanak, 2005, and Sinclair, 1969.

Figure 5.52 *Dust devils. (top) Relatively broad, but shallow dust devil north of Tucson, Arizona, on May 31, 2002. (bottom) Relatively tall and narrow dust devil north of Tucson, Arizona, on May 30, 2002.*

Photos courtesy of author.

Figure 5.53 *Invasion of the dust devils: Three simultaneously occurring dust devils in close proximity, north of Tucson, Arizona, on May 31, 2002.*
Photo courtesy of author.

strong vortex. Otherwise, counter-rotating vortices would be produced as in a supercell in an environment of unidirectional vertical shear (Fig. 4.77, top panel). It was further hypothesized that vorticity produced by tilting above the ground is diffused downward to the ground to form the dust devil.

We once collected data with a mobile Doppler radar in a large dust devil, in Texas (Fig. 5.54, right panel). A convective cloud seen nearby is consistent with the idea that Rayleigh–Bénard convection plays a role in their formation (see Fig. 5.54, left panel).

Snow devils look just like dust devils, but they are composed of snow crystals rather than dust. Snow devils, however, unlike dust devils, are not driven by buoyancy associated with the heating of the ground by the sun. Instead, they are formed when friction slows down the airflow along the slides of buildings, trees, and other obstacles to the wind. The slowing down of the airflow locally next to the obstacles creates lateral wind shear, which is associated with vorticity. This vorticity is visualized by rotating columns of snow that are shed from the sides of the obstacles and then move along with the wind. I have not had much luck photographing snow devils because they are so short lived and lack contrast with their surroundings. When I was a graduate student, my classmates and I sometimes ran outside to jump into snow devils shed along the lee side of our building, just for the fun of it. More recently, I have seen many snow devils while cross-country

Figure 5.54 *A dust devil and its radar appearance. (left) A wide dust devil in northwest Texas, near Northfield, on May 25, 1999. View is to the southeast. Note the scattered cumulus clouds over flat, homogeneous terrain, which is circumstantial evidence that Rayleigh–Bénard convection may be playing a role. (right) Radar reflectivity at 1° elevation angle of a dust devil in Texas on May 25, 1999. Data from the University of Massachusetts mobile, W-band Doppler radar. Range rings are shown every 0.10 km (100 m). This dust devil has a weak-echo hole as seen in radar data in tornadoes, and spiral bands. The weak-echo hole is probably due to centrifuging outward of dust and dirt. The dust devil ring (color-coded white) is about 100 m wide.*
Photo and figure courtesy of author.

skiing in the backcountry in clearings, but am almost always too late to get my camera out and photograph them. When there are snow devils, it tends to be windy, as it is when there is arctic seasmoke, making it feel very cold.

5.9   Steam devils

Steam devils are like dust devils (Fig. 5.55), but they form over relatively warm water surfaces when very cold air flows over them (Bluestein 1990). It is the very warm water heating the air above that drives the updrafts, not the very warm *land* surface heating the air above. The columns of rotating cloud look like flames dancing around in a fire. Arctic seasmoke (steam fog) is usually also seen at the same time, as noted in Chapter 2. Steam devils are often seen over the Great Lakes in the US during the winter when cold Arctic air flows over the much warmer water. The steam devils I have seen formed over Lake Thunderbird in Norman, Oklahoma, when near-0°F air flowed over water that was probably in the 60s °F. In this case, unlike dust devils, the surface winds were very strong, making the wind chill brutally cold and difficult to handle a camera with bare hands. I'm sure that steam devils can be seen all over the world at higher latitudes during the winter, whenever cold, continental air flows offshore from the east coast of continents, when the sea surface is not frozen, particularly in the Northern Hemisphere. Given very cold temperatures and strong winds that accompany steam devils, woe to those who willfully

Figure 5.55 *Steam devil arising out of arctic seasmoke over Lake Thunderbird on Dec. 22, 1989. (See Bluestein (1990) for more details.)*

try to go out and photograph them with bare hands and inadequate clothing. On Feb. 15, 2020, a Channel 9 television news helicopter piloted by Jim Gardner in Oklahoma City took a breathtaking video of a steam devil over Lake Hefner during an extreme cold outbreak. David Payne, the chief meteorologist at the station referred to it as an "ice-nado."[26]

5.10   Other tornado-like phenomena

Other, relatively rare, tornado-like phenomena such as fire-whirl tornadoes, high-based funnel clouds, and horizontal tubular clouds are now described.

[26] https://www.facebook.com/davidrpayne/videos/ice-nado/275864133960891/

5.10.1   Fire whirls/tornadoes

Another atmospheric vortex phenomenon similar to the steam devil and dust devil is the fire whirl (Church et al. 1980) or fire tornado, which is forced by the intense heat from a wildfire. Fire whirls are probably extreme dust devils, without the dust. Many of these were documented by video and reported in the media during the horrific, 2021 wildfire events in California. The NASA Hubble telescope has yielded images of similar structures in the Lagoon Nebula (M8, NGC 6523), which are conjectured to have been[27] created by convective instability associated with the large difference in temperature between the surface and interior of the clouds in the nebula.

5.10.2   High-based funnel clouds and horizontal tubular clouds

During my early years of storm chasing in the late 1970s, I first saw a funnel cloud suspended from a very high-based cumuliform cloud; it was so high that the cloud could not have been produced through boundary-layer processes, since it was apparent that the cloud base was far above the boundary layer. I referred to it as a "high-based" funnel cloud (Bluestein 1988, 1994) and began to record the circumstances under which similar phenomena occurred. Many come from elevated cumulus or cumulus *fractus* or from cumulus *congestus* (Figs. 5.56 and 5.57).

They are almost always unexpected (at least to me). Some are pendant from dissipating cumuliform towers and many come from cloud bases that are not sharply defined (definitely not from flat cloud bases), but rather are ragged-looking, with *fractus* elements. In some narrow cumulus congestus and cumulonimbus, particularly along the dryline, funnel clouds appear while the parent cloud is dissipating. I have never seen evidence that debris was being kicked up at the ground, so it is unlikely that they are ever associated with tornadoes, yet they do look like small tornadoes sometimes, but way above the ground. They are usually ephemeral, though some may persist for as long as several minutes. If they are associated with narrow vortices, the vortices are certainly coherent because they last much longer than it takes air to circulate around them. Their source of rotation is unknown and it is probably noteworthy that the cloud base appears to be disappearing, indicative either of a weakening updraft or the ingestion/mixing in of cooler and/or drier air from underneath or the sides of the cloud.

Having observed these funnel clouds for decades, I can now tentatively identify seven different types of high-based tubular-shaped or funnel clouds, including the ones just mentioned. The second type extends outward from a cumuliform cloud, cumulus congestus or cumulonimbus, or nearby one of them, but not from or near its base, but rather from its side or top (Fig. 5.58).

---

[27] I used the present-perfect tense because the nebula is around 5, 000 light years away. They most likely have dissipated by now.

Figure 5.56 *High-based funnel clouds. (top) Tiny, funnel cloud pendant from a small, high-based, very-shallow cumulus cloud near Pawnee Peak, in the Indian Peaks Wilderness, west of Ward, on Aug. 20, 2011. View is to the south or southeast at a very high angle. (bottom) High-based funnel cloud pendant from a cumulus* fractus, *adjacent to the edge of a cumulonimbus anvil (lower right), northeast of North Platte, Nebraska, on June 1, 2018. View is almost overhead. In both panels, the funnel cloud is approximately as long as its parent cloud is deep.*

Photos courtesy of author.

Figure 5.57 *Tiny funnel cloud (center, near top, left panel; arrow on right panel) pendant from a narrow, chimney-like, altocumulus-*castellanus *tower over Norman, Oklahoma, on Sept. 26, 2019.* Photos courtesy of author.

Sometimes they break off from the parent cloud[28] and persist for several minutes. I don't know why these clouds form and behave the way they do. They are reminiscent of tentacles sticking out from some types of underwater sea life.

The third type appears in clear air and is not *connected* to a cumuliform cloud. These tubular clouds look like the upside-down letter "U" (Fig. 5.59) and have been called "horseshoe" funnels/clouds by some observers. It is possible that they are created by buoyant, dry thermals, which tilt environmental horizontal vorticity associated with vertical shear [as moist updrafts in a supercell do (Fig. 4.77)], while the vorticity is increased as the rotating tubes of air are stretched along the leading edge of the thermals as they expand upward. The rotation may become so intense that the pressure falls enough for saturation to occur. Since these seem to be found near the base of nearby cumulus clouds, it probably does not take much of a pressure drop to produce a cloud since the relative humidity is likely relatively high there. Another possible way they may appear is when a vortex ring around a dry thermal (Fig. 3.6) is skewed about the thermal by environmental shear, so the donut ring is tilted upward on one side (upshear side) of the

---

[28] It is not always clear whether or not they exist independent of the cloud or actually are attached to the cloud and then break away.

Figure 5.58 *Tubular clouds. (top, left) Small tubular cloud that appears to be jutting out from the side of a cumulus congestus, (center, lower half of frame) on Aug. 20, 2011, near Pawnee Peak, in the Indian Peaks of Colorado, west of Ward. (top, right) Horizontal tubular cloud connected to the top of cumulus congestus clouds in Boulder, Colorado, on June 26, 2020. (middle, left) Isolated tubular cloud possibly broken off from a cumulus cloud on Oct. 17, 2020, in Boulder, Colorado, or it could be part of an upside-down U-shaped, tubular cloud (Fig. 5.66). (middle, right) Narrow, tilted, tubular cloud near a roiling cumulus* fractus *in Boulder, Colorado, June 9, 2022. (bottom, left) Quasi-horizontally oriented tubular cloud near a cumulus below and an orographic wave cloud above it, on June 9, 2022; (bottom, right) close-up view of the tubular cloud seen in the bottom, left panel. There is some evidence of helical striations, suggestive of rotation.*

Photos courtesy of author.

Figure 5.59 *"Horseshoe" funnels. (top, left) Upside-down U-shaped tubular cloud in Boulder, Colorado, on Oct. 14, 2020. View is to the west. (top, right) Upside-down, U-shaped tubular cloud over Norman, Oklahoma, on April 12, 2022. (middle, left) Upside-down U-shaped tubular cloud in Boulder, Colorado, on Feb. 9, 2003. This cloud looks like a broken section of a ring. (middle, right) Upside-down U-shaped tubular cloud in Boulder, Colorado, on July 29, 2020, amidst a crowd of cumulus* fractus. *(bottom, left) Upside-down U-shaped tubular cloud on April 9, 2021 over Boulder, Colorado, almost overhead. This cloud was unusually long compared to many others I have observed. (bottom, right) Two upside-down U-shaped tubular clouds south of Leadville, Colorado, on Sept. 26, 1986.*

Photos courtesy of author.

thermal and down (downshear side) on the other. The bottom of the ring may be broken as a result of turbulent mixing with dry air underneath the thermal. Or, it might also be that the horseshoe tubular clouds represent the top of an isolated Kelvin–Helmholtz wave/billow.

The fourth type I have seen along the Colorado Front Range in and near cumulus *fractus* (Fig. 5.60) along the edge of what appears to be a shear zone between easterly upslope caused by heating of sloping terrain during the day (as described in Chapter 2) and the prevailing westerly winds. The cumulus *fractus* appears to be roiling, like rotor clouds.

The fifth type is similar to the fourth type, but it does not appear to be connected to pre-existing cumulus *fractus* clouds, or cumulus clouds of any type to which the funnel clouds are connected. It might be, however, that it is really similar to the fourth type, just mentioned. These to me are the most amazing, since they look like ghostly tornadoes suspended high up in the air (Fig. 5.61). While ascending South Arapaho Peak west of Boulder once, I photographed and took videos of multiple, vertically oriented, funnel/tubular-shaped clouds dancing about. They reminded me of a witches' sabbath of sorts, with broomstick handles being moved around. I suspect that the experience of viewing these clouds could be enhanced by listening to the Dream of a Witches' Sabbath movement in Hector Berlioz's *Symphonie Fantastique*. These funnels suspended in clear air also remind me of the visionary image of a castle suspended in the air in the "youth" section of Thomas Cole's sequence of four mid-nineteenth century paintings called *The Voyage of Life*, on view at the National Gallery of Art in Washington, DC.[29]

The sixth type looks like the right-side-up version of the upside-down "U-shaped, horseshoe-shaped" tubular clouds. These U-shaped tubular clouds have been seen pendant from the anvils of cumulonimbus clouds (Fig. 5.62). There is some indirect evidence from Doppler-radar measurements that these also represent a horizontal vortex that is tilted by descending air. These tubular clouds look like a drooping clothesline or trapeze high wire.[30]

I surmise that the U-shaped tubular cloud is created by the tilting of horizontal vorticity by a *downdraft*, while the upside-down shaped tubular cloud is created by the tilting of horizontal vorticity by an *updraft*. I have seen a rare combination of separate upside-down U and right-side-up U tubular clouds together, but not directly adjacent to each other (not shown here).

Another *sui generis* tubular cloud is seen in Figure 5.63. This tiny tubular cloud appears to connect two stationary orographic wave clouds. This cloud looks like the fuel line connecting one aircraft to another and providing fuel during flight. I have absolutely no idea what created this cloud.

Yet another type of tubular cloud, but one that is not to my knowledge observed in pristine nature, but is rather a result of human activities, is that seen emanating from the back of the side edges of the wings or the side edges of flaps in some aircraft, as a horizontal tubular cloud that is associated with a vortex that is shed at the rear of

[29] https://www.nga.gov/collection/art-object-page.52451.html
[30] Snyder et al., 2020.

Figure 5.60 *High-based funnel cloud pendant from a roiling cumulus* fractus, *viewed to the west from the summit of one of the peaks of Twin Sisters, just to the east of Long's Peak in Rocky Mountain National Park, in northeast Colorado, on Sept. 30, 2002.*
Photo courtesy of author.

Figure 5.61 *Twin, vertically oriented tubular clouds suspended in clear air, along the southern ridge of South Arapaho Peak in the Indian Peaks of Colorado, northwest of Nederland, on Aug. 6, 2014. The* cumulus fractus *clouds underneath it are not connected to it. The cumulus clouds were forming and dissipating along what looked like a zone of air flowing upslope from the east (as part of the diurnal solenoidal mountain-valley circulation; Chapter 2) and converging with ambient westerly airflow. One of these tubular clouds tapers off at both ends, unlike funnel clouds pendant from convective clouds with flat bases.*
Photo courtesy of author.

some aircraft wings (Fig. 5.64). These clouds tend to be seen when the aircraft is in a humid boundary layer, near the ground, during takeoff or landing when only a relatively small decrease in pressure can result in the saturation of the air. These tubular clouds have been described as horizontal tornadoes and can be significant aviation hazards for aircraft flying near each other or getting near each other at ground, on a runway.

Wake vortices may be produced at the edges of aircraft wings as a result of the change with height of the horizontal gradient in the dynamic pressure around the side edges of airfoils. As described in Figure 4.50, bottom panel, relating to air that approaches a density current, the air relative to an aircraft being propelled forward comes to an abrupt halt as it hits the front of a wing. For this to happen, there must be a dynamic-pressure gradient, from high to low pressure, directed against the oncoming, relative wind to slow

Figure 5.62 *U-shaped tubular clouds underneath the anvil of a supercell. (top) Composite photograph of a U-shaped tubular cloud pendant, viewed to the north or northwest, pendant from the anvil in a tornadic supercell (the tornado is just to the left, along with the clear slot; top panel), south of Okemah, Oklahoma, on May 26, 1997. The tubular cloud several minutes later which has decreased in thickness, is seen in the bottom panel. See Bluestein (2005) and Snyder et al. (2020) for more details on this phenomenon. At the time I was photographing the tornado, I was oblivious to the tubular cloud pendant aloft to my right, and only recently pasted the two photographs together manually to form the composite image[a]; doing so was difficult because it appears as if different lenses were used for each photograph and matching the two was difficult. (bottom) The same U-shaped tubular cloud seen in the top panel, but a short time later, without the larger perspective seen in the top panel.*

[a] My attempts to paste the images together with standard Adobe Photoshop software failed, probably because the focal length of the two individual photographs was significantly different and there was too much distortion.

Photos courtesy of author.

Figure 5.63 *Tubular cloud connecting two orographic wave clouds near sunset over Boulder, Colorado, on July 6, 2009.*
Photo courtesy of author.

it down. If there is no perturbation to the dynamic pressure way out ahead of the wing, there must be a relatively high dynamic pressure right at the leading edge of the wing. Some air is therefore forced up and over the wing due to an upward-directed dynamic pressure-gradient force (as it is in a density current), and subsequently speeds up as it flows over the top of the wing. Why must it speed up? Think of how air speeds up as it is forced to flow in between buildings and underneath overpasses: The flux of air is proportional to the wind component normal to the vertical "plane" cutting vertically through the wing, multiplied by the width of the air channel over the top of the wing; the flux of air approaching and hitting the top of the wing must be the same as the flux of air flowing behind the wing (or else air would pile up unrealistically), and since the channel of air is narrower over the wing than it is ahead of the wing, *the wind must speed up* as it flows over the top of the wing. Since the air speeds up as it flows over

Figure 5.64 *Wake-vortex tubular clouds. (top) A condensation tubular cloud associated with a wing-tip vortex on July 3, 2008, when I was on the way back to Boulder, Colorado from Helsinki, Finland. The exact location of the photograph is unknown and obviously irrelevant. (bottom) A tubular cloud associated with a vortex at the edge of a flap on March 3, 2020, while landing in Atlanta, Georgia, the buildings in the downtown area of which are visible in the distance to the north.*
Photos courtesy of author.

the wing, there must be a dynamic-pressure-gradient force acting downstream, which means that there must be relatively *low* dynamic pressure above the wing. This is a special application of *Bernoulli's principle*. Also, some air is forced to flow underneath the wing, which *if it is tilted downward slightly*, must slow down and be diverted as it approaches the wing. Underneath the wing the dynamic pressure must therefore be relatively *high*, so that there is a component of the dynamic-pressure-gradient force acting to slow the air down. There is therefore an upward-directed dynamic-pressure-gradient force, the *lift*, that keeps the aircraft up in the air. The key to making the aircraft fly is that the bottom of the airfoil is relatively flat and tilted downward slightly, while the upper portion of the airfoil is curved.

Since the dynamic pressure is relatively low on top of the wing and high underneath the wing, there is a pressure-gradient force directed (toward lower pressure) at the top of the wing inward toward the center of the airplane and at the bottom of the wing out-wards from the center because away from the wing there is only a negligible pressure perturbation. Such a configuration of pressure-gradient forces leads to counter-rotating vortices on each side of the wing, which relative to the airplane are then transported backward with respect to the motion of the airplane, to the rear of the plane, thus pro-ducing wake vortices behind the wing's edges. Consistent with the strongly rotating flow, a pressure-gradient force acts inward toward the center of rotation, leading to lowered pressure and condensation and cloud droplets.

To the rear of the aircraft these counter-rotating vortices may interact with each other as described near the end of Chapter 2, in Section 2.4.4, and modulate the shape of contrails so that they have periodic protuberances as seen in Figures 2.141 and 2.142.

For *all* the small, high-based tubular-cloud fragments shown in this section, we have assumed there is a vortex associated with them and that the lowering of the pressure due to rotation is what made them visible. I must confess, however, that there could be other explanations for some of them that do not require rotation; e.g., the localized lift of humid air in a stable environment, like that associated with a pileus. It was not possible to *see* rotation in all of these tubular clouds because they were so small and far away. We must therefore be very cautious in definitely attributing all of them to rotation.

5.11 The tropical cyclone eyewall and meso-vortices

The outer edge of the eye of a tropical cyclone, the eyewall, is one of the most extraordinary sights one can see, in my "view," ranking high up in the pantheon of extremely dynamic cloud features such as tornadoes. Seen from high above by a satellite (Fig. 5.65) or from the inside by a hurricane hunter aircraft or a research aircraft (Fig. 5.66), or drone, the eyewall sometimes looks like a stadium and with striations running along its outward sloping sides (Bluestein and Marks 1987).

Intense updrafts, in large part not driven by buoyancy, push upward into the strato-sphere. I was lucky enough to fly into a number of eyes of hurricanes in the Atlantic

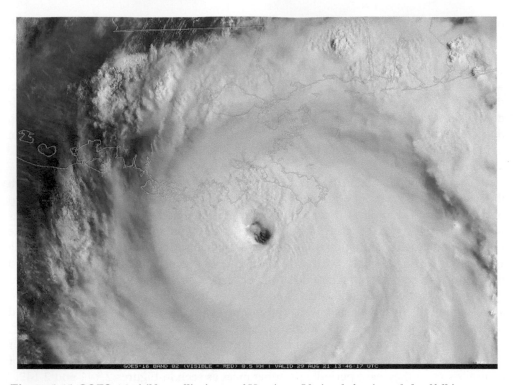

Figure 5.65 *GOES-16 visible satellite image of Hurricane Ida just before it made landfall in Louisiana on Aug. 29, 2021, in the morning (1346 UTC, 8:46 a.m. CDT) when shadows can be seen inside the eastern portion of the eyewall. The sea surface can be seen at the lower right side of the eye.* (Courtesy of College of DuPage Meteorology.)

and Eastern Pacific, with colleagues from the National Hurricane Research Laboratory (now the Hurricane Research Division of the Atlantic Oceanographic and Meteorological Laboratory of NOAA) during the fall of 1984 when I was on sabbatical leave there. At some times one can see clearly down to the ocean surface inside the eye (Fig. 5.66).

Within the eye there are sometimes mesoscale vortices that can enhance the intensity of the winds at the surface. I know of no visual cloud indications of these vortices, though I once documented a tubular cloud in the eye of Hurricane Norbert in the eastern Pacific in 1984 (Fig. 5.67) (Bluestein 1985b), however, which is not related to these mesoscale vortices. It was not, however, possible to discern definite rotation during the quick passage by this feature.

It would be fascinating to view the eyewall cloud edge from the ground, looking up. I imagine that some people have actually looked upward through a clear eye during the daytime, as some have during a tornado, but I am unaware of any of these photographs taken under these circumstances. My former graduate student Trey Greenwood got himself positioned inside the eye of Hurricane Ida in 2021 after it made landfall in

Figure 5.66 *Hurricane eyewalls. (top) The stadium effect with sloping cumulonimbi along the inner eyewall of Hurricane Diana, on Sept. 11, 1984, off the east coast of Florida, from a NOAA P-3 aircraft. (bottom) As in the top panel, but whitecaps on the ocean surface can be seen amidst the haze over the ocean at the bottom of the photograph.*

Photos courtesy of author.

Figure 5.67 *Composite of two images taken one right after another, of a tubular cloud in the eye of Hurricane Norbert in the eastern Pacific on Sept. 24, 1984, as viewed looking out and downward from a NOAA P-3 aircraft.*[a]

[a] Bluestein, 1985c.
Photo courtesy of author.

Louisiana (Fig. 5.65), but could not see eyewall clouds clearly. Having a clear eye probably requires the tropical cyclone to be at its strongest intensity, which would be most likely over the ocean, and therefore most likely seen by people on small islands (or by some tragic misfortune, stranded on a boat inside the eye). Haze near the surface would be another visual obstacle (Fig. 5.66, bottom panel) as the air is warmed by sinking inside the eye, expanding in the eye and producing a very stable atmosphere, which acts to trap "pollutants" (sea salt covered with condensate?). Just as a tropical cyclone makes landfall would be another possible time one could look up the eyewall, but being right along the coast would be very dangerous. I recall when Hurricane Carol (1954) passed by when I was very young and living in the Boston area. Many of my neighbors and I ran out into the street to assess the damage. The wind had died down, but I don't recall having seen any cloud features. When the wind shifted direction and picked up in speed, I ran home to safety, bucking the now-approaching wind. I had thought most of my life that the eye had passed directly overhead, but alas it turns out (based on the official storm track) that the center of the hurricane actually had passed by to my west. It was nevertheless a most memorable experience, probably when the wind simply shifted as the center of the hurricane had passed nearby and not necessarily the eye itself.

5.12   Summary

Small-scale atmospheric vortices are associated with centers of low pressure and may be visualized by cloud material when there is sufficient water vapor in the air and when the rotation is strong enough and the dynamic pressure drop/cooling is large enough to bring the air to saturation. The smallest are relatively narrow and much longer in width. The tubes of condensation are usually referred to as funnel clouds (or what I prefer, tubular clouds), even though they may not be tapered like a funnel except perhaps just at one end. Tornadoes and waterspouts are generally vertically oriented, but sometimes are tilted and may even have sections that are occasionally horizontally oriented. Other vortices are horizontal or deformed into both vertical and horizontal sections. While they are usually relatively smooth appearing, they sometimes are more ragged, likely indicative of turbulence and vigorous mixing.

The larger-scale vortices are visualized by cloud material being carried along in their flow and *not* because the dynamic pressure drop due to rotation is enough to bring the air to saturation. It is rather the cooling due to the lift that is responsible for the hydrostatic pressure drop sufficient to produce clouds in these large-scale vortices. Whether viewed on a large scale or small scale, they are intriguing features and herald interesting and potentially severe weather or turbulent, clear-air wind currents that can provide hazards to aviation.

References

Benjamin, T. B., 1962: Theory of the vortex breakdown phenomenon. *J. Fluid Mech.*, **14**, 593–629.

Bluestein, H., 1983: Surface meteorological observations in severe thunderstorms. Part II: Field experiments with TOTO. *J. Clim. Appl. Meteor.*, **22**, 919–930.

Bluestein, H. B., 1985a: The formation of a "landspout" in a "broken-line" squall line in Oklahoma. Preprints, *14ᵗʰ Conf. on Severe Local Storms*, Indianapolis, Amer. Meteor. Soc., Boston, 267–270.

Bluestein, H.B. 1985b: A funnel cloud in the eye of Hurricane Norbert? *Mon. Wea. Rev.*, **113**, 1238–1239.

Bluestein, H.B., 1988: Funnel clouds pendant from high-based cumulus clouds. *Weather*, **43**, 220–221.

Bluestein, H.B., 1990: Observations of "steam devils" over a lake during a cold-air outbreak. *Mon. Wea. Rev.*, **118**, 2244–2247.

Bluestein, H. B., 1994: High-based funnel clouds in the Southern Plains. *Mon. Wea. Rev.*, **122**, 2631–2638.

Bluestein, H. B., 1999a: A history of storm-intercept field programs. *Wea. Forecasting*, **14**, 558–577.

Bluestein, H. B., 1999b: *Tornado Alley: Monster Storms of the Great Plains.* Oxford Univ. Press., New York, 180 pp.

Bluestein, H. B., 2000: A tornadic supercell over elevated, complex terrain: The Divide, Colorado storm of 12 July 1996. *Mon. Wea. Rev.*, **128**, 795–809.

Bluestein, H., 2005: More observations of small funnel clouds and other tubular clouds. *Mon. Wea. Rev.*, **133**, 3714–3720.

Bluestein, H. B., 2013: *Severe Convective Storms and Tornadoes: Observations and Dynamics.* Praxis/Springer, 460 pp.

Bluestein, H. B., 2017: Tornadoes and Their Parent Convective Storms. *Oxford Handbooks Online in Natural Hazard Science.* Oxford Univ. Press (online Sept. 2017, www.oxfordhandbooks.com), 67 pp.

Bluestein, H. B., 2022: Observations of tornadoes and their parent supercells using ground-based, mobile Doppler radars. Chapter 3, *Remote Sensing of Water-Related Hazards, Geophys. Monogr. 271*, 31–67.

Bluestein, H.B. and F.D. Marks, Jr., 1987: On the structure of the eyewall of Hurricane Diana (1984): Comparison of radar and visual characteristics. *Mon. Wea. Rev.*, **115**, 2542–2554.

Bluestein, H. B. and J. H. Golden, 1993: A review of tornado observations. *The Tornado: Its Structure, Dynamics, Prediction, and Hazards, Geophys. Mono. Series*, **79**, Amer. Geophys. Union, Washington, D. C., 319–352.

Bluestein, H. B. and W. P. Unruh, 1993: On the use of a portable FM-CW Doppler radar for tornado research. *The Tornado: Its Structure, Dynamics, Prediction, and Hazards, Geophys. Mono. Series*, **79**, Amer. Geophys. Union, Washington, D. C., 367–376.

Bluestein, H. B., and A. L. Pazmany, 2000: Observations of tornadoes and other convective phenomena with a mobile, 3-mm wavelength, Doppler radar: The spring 1999 field experiment. *Bull. Amer. Meteor. Soc.*, **81**, 2939–2951.

Bluestein, H. B., C. C. Weiss, and A. L. Pazmany, 2004: Doppler-radar observations of dust devils in Texas. *Mon. Wea. Rev.*, **132**, 209–224.

Bluestein, H. B., C. C. Weiss, and A. L. Pazmany, 2004: The vertical structure of a tornado near Happy, Texas on 5 May 2002: High-resolution, mobile, W-band, Doppler-radar observations. *Mon. Wea. Rev.*, **132**, 2325–2337.

Bluestein, H. B., C. C. Weiss, M. M. French, E. Holthaus, R. L. Tanamachi, S. Frasier, and A. L. Pazmany, 2007: The structure of tornadoes near Attica, Kansas on 12 May 2004: High-resolution, mobile, Doppler-radar observations. *Mon. Wea. Rev.*, **135**, 475–506.

Bluestein, H. B., J. B. Houser, M. M. French, J. Snyder, G. D. Emmitt, I. PopStefanija, C. Baldi, and R. T. Bluth, 2014: Observations of the boundary layer near tornadoes and in supercells using a mobile, co-located, pulsed Doppler lidar and radar. *J. Atmos. Ocean. Technol.*, **31**, 302–325.

Bluestein, H. B., J. C. Snyder, and J. B. Houser, 2015: A multi-scale overview of the El Reno, Oklahoma tornadic supercell of 31 May 2013. *Wea. Forecasting*, **30**, 525–552.

Bluestein, H. B., M. M. French, J. C. Snyder, and J. B. Houser, 2016: Doppler-radar observations of anticyclonic tornadoes in cyclonically rotating, right-moving supercells. *Mon. Wea. Rev.*, **144**, 1591–1616.

Bluestein, H. B., K. J. Thiem, J. C. Snyder, and J. B. Houser, 2018: The multiple-vortex structure of the El Reno, Oklahoma tornado on 31 May 2013. *Mon. Wea. Rev.*, **146**, 2483–2502.

Bluestein, H. B., K. J. Thiem, J. C. Snyder, and J. B. Houser, 2019: Tornadogenesis and early tornado evolution in the El Reno, Oklahoma supercell on 31 May 2013. *Mon. Wea. Rev.*, **147**, 2045–2066.

Brady, R. H., and E. Szoke, 1989: A case study of nonmesocyclone tornado development in northeast Colorado: Similarities to waterspout formation. *Mon. Wea. Rev.*, **117**, 843–856.

Burgess, D. W., M. A. Magsig, J. Wurman, D. C. Dowell, and Y. Richardson, 2002: Radar observations of the 3 May 1999 Oklahoma City tornado. *Wea. Forecasting*, **17**, 456–471.

Carlson, T. N., 1980: Airflow through mid-latitude cyclones and the comma cloud pattern. *Mon. Wea. Rev.*, **108**, 1498–1509.

Church, C. R., J. T. Snow, and J. Dessens, 1980: Intense atmospheric vortices associated with a 1000 MW fire. *Bull. Amer. Meteor. Soc.*, **61**, 682–694.

Church, C., D. Burgess, C. Doswell, and R. Davies-Jones (eds.), 1993: *The Tornado: Its Structure, Dynamics, Prediction, and Hazards*. **79**, *Geophys. Monogr.*, Amer. Geophy. Union, 637 pp.

Davies-Jones, 1986, Tornado Dynamics. *Thunderstorm Morphology and Dynamics*, 2nd ed., E. Kessler, Ed., University of Oklahoma Press, 197–236.

Davies-Jones, R. P., R. J. Trapp, and H. B. Bluestein, 2001: Tornadoes. *Severe Convective Storms*, Meteor. Monogr., 28, no. 50 (C. Doswell III, ed.), Amer. Meteor. Soc., 167–221.

Dergarabedian, P., and F. Fendell, 1970: Estimation of maximum wind speeds in tornadoes. *Tellus*, **22**, 511–515.

Dowell, D. C., and H. B. Bluestein, 2002a: The 8 June 1995 McLean, Texas storm. Part I: Observations of cyclic tornadogenesis. *Mon. Wea. Rev.*, **130**, 2626–2648.

Dowell, D. C., and H. B. Bluestein, 2002b: The 8 June 1995 McLean, Texas storm. Part II: Cyclic tornado formation, maintenance, and dissipation. *Mon. Wea. Rev.*, **130**, 2649–2670.

Dowell, D. C., C. R. Alexander, J. M. Wurman, and L. J. Wicker, 2005: Centrifuging of hydrometeors and debris in tornadoes: Radar-reflectivity patterns and wind-measurement errors. *Mon. Wea. Rev.*, **133**, 1501–1524.

Fiedler, B. H., and K. M. Kanak, 2001: Rayleigh – Bénard convection as a tool for studying dust devils. *Atmos. Sci. Letters*, **2**, 104–113.

Gentry, R. C., 1983: Genesis of tornadoes associated with hurricanes. *Mon. Wea. Rev.*, **111**, 1793–1805.

Golden, J. H., 1974: The life cycle of Florida Keys' waterspouts. I. *J. Appl. Meteor.*, **13**, 676–692.

Golden, J. H., and H. B. Bluestein, 1994: The NOAA - National Geographic Society Waterspout Expedition (1993). *Bull. Amer. Meteor. Soc.*, **75**, 2281–2288.

Houser, J. B., H. B. Bluestein, and J. C. Snyder, 2015: Rapid-scan, polarimetric, Doppler-radar observations of tornadogenesis and tornado dissipation in a tornadic supercell: The "El Reno, Oklahoma" storm of 24 May 2011. *Mon. Wea. Rev.*, **143**, 2685–2710.

Houser, J. B, H. B. Bluestein, and J. C. Snyder, 2016: A fine-scale radar examination of the tornadic debris signature and a weak reflectivity band associated with a large, violent tornado. *Mon. Wea. Rev.*, **144**, 4101–4130.

Houser, J. L., H. B. Bluestein, K. Thiem, J. C. Snyder, D. Reif, and Z. Wienhoff, 2022: Additional evidence of nondescending tornadogenesis using rapid-scan mobile radar observations. *Mon. Wea. Rev.*, **150**, 1639–1666.

Kanak, K. M., 2005: Numerical simulation of dust devil-scale vortices. *Q. Roy. Meteor. Soc.*, **131**, 1271–1292.

Karstens, C. D., T. M. Samaras, B. D. Lee, W. A. Gallus, Jr., and C. A. Finley, 2010: Near-ground pressure and wind measurements in tornadoes. *Mon. Wea. Rev.*, **138**, 2570–2588.

Kessler, E. (ed.), 1986: *Thunderstorm Morphology and Dynamics, 2nd ed.* University of Oklahoma Press, Norman, Oklahoma, 411 pp.

Lee, B. D., and R. B. Wilhelmson, 1997: The numerical simulation of nonsupercell tornadogenesis. Part II: Evolution of a family of tornadoes along a weak outflow boundary. *J. Atmos. Sci.*, **54**, 2387–2415.

Lewellen, D. C., and M. I. Zimmerman, 2008: Using simulated tornado surface marks to help decipher near-ground wind fields. 24th Conf. on Severe Local Storms, Savannah, GA, Amer. Meteor. Soc., 8B1, https://ams.confex.com/ams/24SLS/webprogram/Paper141749.html.

Mahale, V. N., J. A. Brotzge, and H. B. Bluestein, 2012: Analysis of vortices embedded within a quasi-linear convective system using X-band polarimetric radar. *Wea. Forecasting*, **27**, 1520–1537.

Margraf, J., 2023: *Evolution of the Tornado and Near-tornado Wind Field of the Selden, Kansas Tornadic Supercell on 24 May 2021 Using a Rapid-scan, X-band, Mobile Doppler Radar.* M. S. thesis, School of Meteorology, University of Oklahoma, Norman, 228 pp.

McLaughlin, D., and Coauthors, 2009: Short-wavelength technology and the potential for distributed networks of small radar systems. *J. Atmos. Oceanic Technol.*, **24**, 301–321.

Rasmussen, E. N., J. M. Straka, R. Davies-Jones, C. A. Doswell III, F. H. Carr, M. D. Eilts, and D. R. MacGorman, 1994: Verification of the Origins of Rotation in Tornadoes Experiment: VORTEX. *Bull. Amer. Meteor. Soc.*, **75**, 995–1006.

Rotunno, R., 2013: The fluid dynamics of tornadoes. *Ann. Rev. Fluid Mech.*, **45**, 59–84.

Simpson, J., B. R. Morton, M. C. McCumber, and R. S. Penc, 1986: Observations and mechanisms of GATE waterspouts. *J. Atmos. Sci.*, **43**, 753–782.

Sinclair, P. C., 1969: General characteristics of dust devils. *J. Appl. Meteor.*, **8**, 32–45.

Snow, J. T., 1984: On the formation of particle sheaths in columnar vortices. *J. Atmos. Sci.*, **41**, 2477–2491.

Snyder, J. C., and H. B. Bluestein, 2014: Some considerations for the use of high-resolution mobile radar data in tornado intensity determination. *Wea. Forecasting*, **29**, 799–827.

Snyder, J., H. B. Bluestein, G. Zhang, and S. Frasier, 2010: Attenuation correction and hydrometeor classification of high-resolution, X-band, dual-polarized mobile radar measurements in severe convective storms. *J. Atmos. Ocean. Technol.*, **27**, 1979–2001.

Snyder, J. C., H. B. Bluestein, V. Venkatesh, and S. J. Frasier, 2013: Observations of polarimetric signatures in supercells by an X-band mobile Doppler radar. *Mon. Wea. Rev.*, **141**, 3–29.

Snyder, J. C., H. B. Bluestein, Z. Wienhoff, and D. Reif, 2020: An analysis of an ostensible anticyclonic tornado from 9 May 2016 using high-resolution, rapid-scan radar data. *Wea. Forecasting*, **35**, 1685–1712.

Tanamachi, R. L., H. B. Bluestein, J. B. Houser, S. J. Frasier, and K. M. Hardwick, 2012:, Mobile, X-band, polarimetric Doppler radar observations of the 4 May 2007 Greensburg, Kansas, tornadic supercell. *Mon. Wea. Rev.*, **140**, 2103–2125.

Taylor, A., 2022: The dust devils of Mars. *The Atlantic*, 29 June (available online at http://theatlantic.com)

Timmer, R., M. Simpson, S. Schofer, and C. Brooks, 2023: Design and rocket deployment of a trackable pseudo-Lagrangian drifter based meteorological probe into the Lawrence/Linwood EF4 tornado and mesocyclone on 28 May 2019. EGUsphere preprint repository. (https://doi.org/10.5194/egusphere-2023-781)

Trapp, R. J., S. A. Tessendorf, E. S. Godfrey, and H. E. Brooks, 2005: Tornadoes from squall lines and bow echoes, Part I: Climatological distribution. *Wea. Forecasting*, **20**, 23–34.

Wakimoto, R. M., and J. W. Wilson, 1989: Non-supercell tornadoes. *Mon. Wea. Rev.*, **117**, 1113–1140.

Wakimoto, R. M., and N. T. Atkins, 1996: Observations on the origin of rotation: The Newcastle tornado during VORTEX 94, *Mon. Wea. Rev.*, **124**, 384–407.

Wakimoto, R. M., N. T. Atkins, K. M. Butler, H. B. Bluestein, K. Thiem, J. Snyder, and J. Houser, 2015: Photogrammetric analysis of the 2013 El Reno tornado combined with mobile X-band polarimetric data. *Mon. Wea. Rev.*, **143**, 2657–2683.

Wakimoto, R. M., Z. Wienhoff, D. Reif, H. B. Bluestein, and D. C. Lewellen, 2022: The Dodge City tornadoes on 24 May 2016: Understanding cycloidal marks in surface damage tracks and further analysis of the debris cloud. *Mon. Wea. Rev.*, **150**, 1233–1246.

Wienhoff, Z. B., H. B. Bluestein, D. W. Reif, R. M. Wakimoto, L. J. Wicker, and J. Kurdzo, 2020: Analysis of debris signature characteristics and evolution in the 24 May 2016 Dodge City, Kansas, tornadoes. *Mon. Wea. Rev.*, **148**, 5063–5086.

Wurman, J., 2002: The multiple-vortex structure of a tornado. *Wea. Forecasting*, **17**, 473–505.

Wurman, J., C. Alexander, P. Robinson, and Y. Richardson, 2007: Low-level winds in tornadoes and potential catastrophic tornado impacts in urban areas. *Bull. Amer. Meteor. Soc.*, **88**, 31–46.

Wurman, J., D. Dowell, Y. Richardson, P. Markowski, D. Burgess, L. Wicker, and H. Bluestein, 2012: The Second Verification of the Origins of Rotation in Tornadoes Experiment: VOR-TEX2. *Bull. Amer. Meteor. Soc.*, **93**, 1147–1170.

Ziegler, C. L., E. N. Rasmussen, T. R. Shepherd, A. I. Watson, and J. M. Straka, 2001: Evolution of low-level rotation in the 29 May 1994 Newcastle – Graham, Texas, storm complex during VORTEX. *Mon. Wea. Rev.*, **129**, 1339–1368.

6

The Future of Cloud Observing

". . . clouds were there for everyone—no tax on them—free."
Alfred Stieglitz[1]

For many years cloud observing was a subjective activity, and possible without cost to anyone. Cloud types were determined by human observers, and recorded in the remarks section at the end of standardized messages from the worldwide synoptic network of surface weather observation stations (which are frequently at airports) and disseminated by teletype (as they were when I was a graduate student) and now made available over the internet. A typical message might contain remarks like "CBS ALQDS" (cumulonimbus clouds all quadrants), "ACSL N" (altocumulus standing lenticularis to the north), "WATERSPOUT E B04 E13" (waterspout to the east, began 4 minutes after the hour, ended 13 minutes after the hour), or "TCU W" (towering cumulus to the west) or "TORNADO W MOVG E, ABANDONING STATION" (Tornado to the west, which is moving to the east; the observer is abandoning the weather station!). Automatically recording surface observing sites these days, however, cannot determine cloud type, but do measure the cloud ceiling using laser ceilometers. Pilot observations are also sometimes included in the remarks section. Something has clearly been lost as most surface observing have become automated. Large airports, however, still insert remarks by human observers when considered important enough. It would be nice if all automated observing stations also recorded photographs of the sky at regular intervals during the daytime and also sent the images out over the internet in real time.

6.1 Webcams, crowdsourcing, and social media

In addition to ordinary people observing clouds (and taking photographs of them), there have in recent years been additional ways of observing clouds. There are webcams, some commercial and some set up by amateur or professional weather enthusiasts, and security cameras set up all over the world; some of the non-commercial websites are available for free, live viewing over the internet. Their web addresses may be found via your

---

[1] Alfred Stieglitz, "How I Came to Photograph Clouds," *Amateur Photographer and Photography* 56 (1923, reprinted in Richard Whelan, ed., *Stieglitz on Photography: His Selected Essays and Note* (Aperture, 2000), p. 237. https://archive.artic.edu/stieglitz/equivalents/

The Architecture of Clouds. Howard B. Bluestein, Oxford University Press. © Howard B. Bluestein (2024).
DOI: 10.1093/oso/9780198870548.003.0006

favorite search engine, but change so frequently that it would be useless to try to list them all. Some of my favorites are located in Boulder, Colorado, and have excellent views of the Rocky Mountains, where orographic wave clouds can sometimes be seen, (e.g., Fig. 6.1, boulderflatironcam.com/superiorcam) and atop Niwot Ridge (The "Tundracam"), where orographic wave clouds and blowing snow can often be seen during the winter. The latter and others may also be controlled for short times by the viewer, and rotated so as to be pointed in the desired direction. Some record images and time-lapse videos can be viewed. Some have serendipitously captured tornadoes[2] and other significant cloud types.

As digital cameras have become used almost exclusively in the past few decades, and most mobile phones, which are used by almost everybody, contain digital still and video cameras, photographs, and videos of rare clouds have become almost commonplace. It is now difficult for a spectacular cloud or tornado to go undocumented, even if the photographer is not a storm chaser or a weather enthusiast. Some of my colleagues have done studies on severe storms in which tornado photographs have been crowdsourced.[3]

Figure 6.1 *Webcam on June 20, 2020, at 1:04 p.m. MDT, looking to the west from Superior, Colorado. An isolated lenticular cloud is seen to the northwest (right), while lines of vertically trapped wave clouds are seen to the west (left and center). The partially snow-covered Continental Divide is seen on the horizon and the Foothills of Boulder are seen to the left. One can go through the archive online and find clouds on any given recent day. Eric Baer maintains this site as a "fun hobby" and we are indebted to his maintenance of this website, at the time of this writing. Some are recorded so frequently that time-lapse images are possible.*

Courtesy of E. Baer (boulderflatironcam.com/superiorcam). Used with permission.

---

[2] In particular I recall one that was captured by a webcam in southwestern Nebraska in the late spring of 2019 and by a surveillance camera in Andover, Kansas in the spring of 2022.
[3] Seimon et al., 2016.

By doing so, detailed information on the evolution of tornadoes from different vantage points has proven to be invaluable when correlating what is seen on radar with what is seen visually.

Social media has also become a great platform for viewing clouds, although not necessarily in real time, especially when the photographers are on the run. After a tornado that I did not get to see has occurred, or there has been a spectacular storm that I missed, I often comb through social media in search of a photograph or video to whet my appetite. In a few rare instances, I have corresponded with the photographer. I know that my graduate students have sometimes acquired photographs of clouds for their theses, from hitherto unknown people who have posted them for the public to see. When we collect radar data from severe convective storms, we are not always in the best viewing location, and frequently we are in transit or otherwise miss significant phenomena whose documentation is valuable for our studies. Perhaps employing artificial intelligence, machine-learning techniques could be developed that would allow us to sift through the multitudes of real time cloud images available, particularly at known fixed sites with webcams, to quickly find the type of cloud we would like to see or automatically to compile climatologies of different cloud types.

6.2   The view from drones: Spying on clouds

Tornadoes are, in my opinion, best seen from the air at low altitudes. From an aircraft at cloud base, one can look down at the debris cloud, but the visibility is often less than ideal and airplanes, like sharks, must move to stay "alive" (aloft), and one cannot hover near the tornado for very long or the aircraft engine(s) will stall. In addition, it can be very turbulent at cloud-base altitude or below, especially on a sunny day. I have told in the last chapter how Joe Golden and I viewed waterspouts from a helicopter, but this is an expensive proposition, not to be done frequently (enough). A safer and more economical method for viewing tornadoes and other clouds from the air is to use drones equipped with cameras. Drones don't have the problem we encounter on the ground of not having a suitable road to get us in position. Furthermore, trees, buildings, powerlines, and other obstructions frequently block our view of the ground. Drones can be flown high enough to see over most ground blockages. When taking photographs of clouds, I am sometimes bedeviled by trees and buildings that seem to pop up in the most inconvenient places. A drone can remove these obstacles to clear viewing. On the other hand, most inexpensive drones cannot be flown in high winds and there is sometimes danger from hail: In some environments, particularly when a tornado may form, winds not associated directly with the tornado can be very strong in the boundary layer, 50 mph or greater. In spite of this difficulty, there have been many spectacular videos of tornadoes taken from drones in recent years. My favorite, to date, is that posted on YouTube, Facebook, and Twitter (now known as X) by Reed Timmer, of the April 30, 2022, Andover, Kansas tornado. This drone video shows the incredible structural features of the tornado while it is doing damage.

6.3 Stereoscopic cloud photography: An in-depth view of clouds

Detailed studies of clouds can be done using stereoscopic cloud photography (Fraser 1968; Warner 2003). One needs to view the cloud from cameras located at two different locations, affording a view from two slightly different viewing angles. Simple geometry can then be used to discover the true three-dimensional extent (the height, depth, and width) of clouds. Using a viewer that allows the view from each camera to be cordoned off to each eye separately, one can get a good 3D sense of how a cloud takes up space. I have sometimes taken 3D photographs of clouds from an airplane window, by taking photographs at two separate times, perhaps only a second or two apart. In that short time period, the cloud may not change its appearance much, but the viewing angle may be different enough to give one the feeling of a cloud's three-dimensionality if one views each image separately in each eye. Stereo photographs of clouds are rarely taken nowadays, to the best of my knowledge, but since digital cameras are now ubiquitous, there is no excuse not to revisit stereo photography (applied to clouds), which once was in vogue, especially for landscape photography. I recall from the days of my youth, in the 1950s, that we used "View-Masters" to look at 3D images of famous tourist attractions. Virtual reality viewers are perhaps the current-day equivalent of View-Masters.

Another method of looking at clouds in depth, literally, is to produce four-dimensional holograms from time-lapse movies and videos.[4] This technique became possible a number of years ago when it first became feasible to produce holograms using actual light rather than coherent light from a laser. I don't believe that this method has been fully exploited.

6.4 Infrared cloud photography: Putting the heat on clouds

Infrared satellite imagery has been very valuable for determining the approximate altitude of the tops of clouds by comparing the satellite-observed IR temperature with that from nearby soundings, with some caveats (for example, consider the AACP discussed in Chapter 4). Looking from the *ground* at clouds from the infrared radiation they emit, rather than from the radiation at wavelengths we can actually see, information about their temperature may also be determined, if there is no intervening cloud material or precipitation: The peak wavelength of the radiation is a function of temperature and water vapor content. One of my former graduate students, Robin Tanamachi, a number of years ago, used an infrared camera (Fig. 6.2), during one of our spring field experiments (with a mobile Doppler radar) to estimate the temperature variation along

---

[4] Holle and Diamond, 1985.

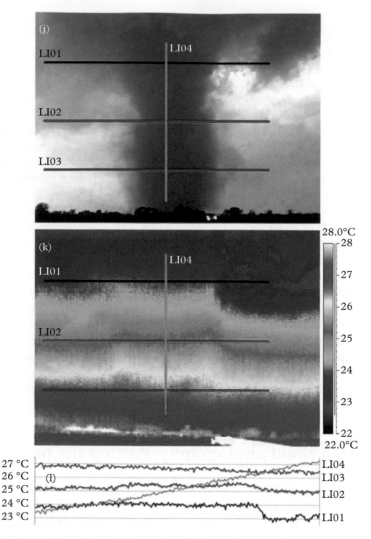

Figure 6.2 *Tornado in Kansas on May 12, 2004, viewed 3 km to the west (top panel) viewed within a few seconds of the infrared image; concurrent infrared thermal image using an FLIR S60 thermal imaging camera (middle panel); temperature (green line) along the vertical green line in the middle panel, through the condensation funnel of the tornado (bottom panel).*

From Tanamachi et al., 2006, their Fig. 9j–l. © American Meteorological Society. Used with permission.

a tornado condensation funnel.[5] We verified that the temperature varied in the vertical according to what one would expect for air that is saturated (i.e., it decreased with height at the moist-adiabatic lapse rate). It would have been most difficult, of course, to place a

---

[5] Tanamachi et al., 2006.

thermometer along the outside walls of the tornado condensation funnel and make in situ measurements. Our finding might not be of the greatest scientific importance, but there is no other way to verify easily that the vertical temperature gradient is what we would expect to find. However, it *is* an important scientific problem to know how temperature varies horizontally in storms and infrared photography offers one possibility for making appropriate measurements, though doing so is hampered when there is intervening rain, which there often is.

6.5 Emerging radar systems: Seeing the previously unseen

Radars can see what is happening inside clouds and have been used for decades to identify the nature of the cloud particles and their motion. When there are rapidly evolving processes such as tornado formation, hail formation, downdraft formation, or the buildup of electrical charge to the point at which lightning flashes can occur, rapid-scan radars are needed to resolve the behavior of what is going on inside clouds on time scales as short as seconds.

6.5.1   Electronically (rapid-scan) scanning radars

Electronically scanning (phased-array) radars can scan the complete volume of storms in around 30 seconds or less. A new mobile radar being developed at the University of Oklahoma[6] now with support from the National Science Foundation and hopefully ready for use after the second half of 2023 is called PAIR (Polarimetric Atmospheric Imaging Radar). This radar should be able to scan a good portion of the volume of a convective storm every 15 seconds or so. Operating at C-band (5 cm wavelength), it should allow us to view convective storms and tropical cyclones from both Doppler wind data and polarimetric reflectivity data, with less attenuation than at X-band.

In the future the observational surveillance network, now composed of the NEXRAD WSR-88D radars, may be replaced by electronically scanning radars. Doing so will add a new dimension to our ability to observe the innards of clouds throughout much of the US when there is precipitation, and thus allow us to add to our knowledge of cloud formation and evolution. Research and engineering efforts are ongoing to upgrade the current mechanically-scanning, S-band (10 cm wavelength) radars and their antennas and also airborne Doppler radars at C-band (5 cm wavelength) and their antennas to have electronic-scanning capabilities.[7]

---

[6] I'm part of the team that planned this radar and we will be validating it in 2023.
[7] E.g., APAR (Airborne Phased Array Radar) at NCAR.

6.5.2   Fine-scale networks of low-powered radars

The current network of surveillance radars in the US has a number of significant deficiencies. Since the radar beam is usually bent (by refraction, caused by the density variations in the atmosphere, which in turn are caused by temperature and water vapor inhomogeneities) less than the curvature of the earth, the nature of what's inside clouds near the ground cannot be determined beyond a relatively short range. By the time you get out to 200 km, at the lowest several elevation angles, most of the radar return is well above the ground, some 2–6 km above the ground, so tornadoes, strong straight-line winds (at the surface), and hail (hitting the ground), cannot be detected anywhere but high up, where only aircraft and perhaps some birds and unidentified flying objects (UFOs) venture. In addition, some radars are located just too far away from some locations, which basically have no radar coverage at all. And, in some parts of the country mountains block the radar from seeing some areas, particularly in valleys surrounded by mountains. One could, if one had an unlimited supply of money, deploy a network of high-power radars spaced so close together that there would be no gaps in coverage. A much less expensive alternative is to build networks of closely spaced, low-power radars (McLaughlin et al. 2009). Since each radar would only have to "see" out to maybe 30 km or so, low-power radars at X-band radars have been used in networks around some major metropolitan areas. The first one in the US is called CASA (Collaborative Adaptive Sensing of the Atmosphere) was deployed around central Oklahoma for several years and then moved to the Dallas, Ft. Worth area.

6.5.3   High-frequency "cloud" radars

High-frequency radars (W-band, Ka-band, Ku-band) have been used to probe clouds, because at their relatively short wavelengths they are particularly sensitive to tiny cloud droplets and ice crystals. Some have been mounted on aircraft, while others have been mounted on trucks or vans. Perhaps it is more significant that since short-wavelength radars can have antennas that yield very high spatial resolution even though they are small, they can be mounted on mobile platforms. As such, they have been put on satellites. The downside is that attenuation is severe at such high frequencies/short wavelengths. But if you're looking only locally at a cloud and there is little, if any, precipitation (which would attenuate the signal), then these radars are ideal. One can use techniques such as pulse compression to increase their sensitivity, thus rendering low-power, solid-state, small, and relatively lightweight radars possible. The smaller size, lighter weight, and less power needed also make high-frequency radars good candidates for use on satellites (Heymsfield et al. 2013). Pulse-compression (also known as frequency-modulated continuous-wave[8]/FM-CW) radars are highly sensitive because, unlike pulsed radars, they continuously send out radiation (the intensity of the backscattered signal from the cloud particles is proportional to the duration of the transmitted radar beam of radiation) and signal-processing tricks are used to separate the radar return into range bins

---

[8] A signal is sent out continuously, but the frequency of the signal is periodically ramped up and then returned to the original frequency.

(Alberts et al. 2011), which are otherwise not needed for pulsed radars. There are some drawbacks to using these radars, which include some artifacts, but most can be mitigated by clever signal-processing techniques. Another possibility is to mount such radars on high-altitude balloons, though there are some technical and cost problems that need to be solved before this can be accomplished. In addition, I think it would be useful to build some electronically scanning, high-frequency cloud radars having polarimetric capability, so that the evolution of rapidly evolving clouds can be observed.

To make the use of cloud radars truly useful, "ground-truth" verification is needed, or perhaps I should say "air-truth" verification. How do we know that the inferences we make about what's in a cloud based on polarimetric radars are correct? We need to get into the cloud and make actual measurements with in situ sensors. Doing this is difficult. First, we can't place sensors everywhere in a cloud, at any time and certainly not all the time; we can just fly into it and sample various locations at specific times. Second, highly buoyant clouds such as those associated with severe convective storms are extremely turbulent. Even if they were not very turbulent, suppose that an aircraft suddenly encounters a smooth, nonturbulent, 50 m s^{-1} updraft; the upward acceleration experienced can be enormous: A 50 m s^{-1} change in vertical velocity experienced over 2 s is more than twice the force of gravity. An aircraft flying just 50 m s^{-1}, for example, travels 100 m in 2 s, which might be representative of the horizontal scale of an updraft in a convective cloud. Finally, large hail and lightning may be encountered, further increasing the danger of flying inside the cloud. At one time, the scientific community had the use of an armored, storm-penetrating aircraft equipped with sensors (Sand 1976), but none is available at the current time. Perhaps in the future, funding will be available for a replacement aircraft.

6.6   Climate change and clouds

As global temperatures warm and the temperature gradient from the poles to the equator decreases (the polar regions are warming more rapidly than the equatorial regions), we would expect, and computer climate models indicate, that the weather patterns will change also. When this happens, we expect that there will be a change in the frequency of occurrence of different types of clouds (Rasmussen et al. 2017) in many regions of our planet. Obviously, regions that become warmer may have fewer clouds containing ice crystals and vice versa.

As the pole-to-equator temperature gradient relaxes, we expect that the intensity of the jet stream will decrease and waves in the troposphere may become more highly amplified and slow moving. Since orographic wave clouds require at least some wind to impact mountains and mountain ranges, their frequency of occurrence may decrease in some places and increase in others.

The warmer the air, the more moisture it can hold, and the warmer the oceans, the more water vapor will be evaporated into the air above. An increase in moisture in the air near the ground can affect how buoyant clouds can be, unless the air at high altitudes warms up significantly to counteract the effect of increased water vapor at low levels.

Increased moisture in the air may overall increase the number of clouds. The distribution of severe convective storms (Trapp, et al. 2007; Gensini 2021) and tropical cyclones (Emanuel 2021) may also be affected. Warmer oceans can support more intense tropical cyclones, and the susceptibility of weak disturbances to grow into tropical cyclones depends on how weak the vertical wind shear is.

It is an open question as to if or how the frequency of occurrence and geographical distribution of tornadoes will be affected (Trapp and Hoogewind 2016). Tornadoes in supercells require strong vertical wind shear, while the actual wind shear depends on the jet-stream pattern and the moisture required depends on the trajectory of air into the supercells from its source region.

Clouds also can affect weather patterns through radiative effects. How clouds will be affected by climate change is a difficult question to answer because the effects will be highly nonlinear and the feedback from clouds will also most likely be nonlinear. We look to climate models to give us some possible answers, but we are cautioned by noting that how well climate models represent clouds of all types can affect the results of the computer simulations. While the representations of the clouds are becoming increasingly accurate, better and more comprehensive observations of clouds are still needed to yield further improvements.

Changes in the jet-stream pattern can change the frequency of occurrence of droughts and floods. If droughts increase in frequency, so will wildfires, whose smoke will reduce visibility and may affect the pattern of radiation and also the nature of aerosols, needed for cloud nucleation. Smoke obviously can also ruin cloud photography. I recall that in the spring of 2003 fires in Mexico produced smoke that made the sky very hazy in the Southern Plains. I recall one day when, according to the NOAA weather radio, a tornadic supercell was heading right toward us, in Kansas, but we could not see the tornado until it got very close (within a few miles), owing to very poor visibility. In the fall of 2020 and the summer of 2021, smoke from wildfires made clouds difficult to see on many days over and near the Rocky Mountains, and elsewhere.

6.7   Summary

The future of cloud observations will depend on new technology in cloud radars and where they can be placed. Aircraft, drones, satellites, networks of low-power radars, electronically scanning (rapid-scan) radars, and new storm-penetrating aircraft will each play a role in providing more information about cloud composition and how they form and evolve. Equipping more observing sites with all-sky (or controllable platform), high-resolution digital cameras with access to the internet will allow more and more clouds to be seen without ever having to move from the comfort of your home.[9] Artificial intelligence techniques could be used to recognize cloud types and alert us to sites at which particular types of clouds may be viewed right now. Doing so would allow us to avoid

---

[9] Brotzge et al., 2022.

having to look at all the images ourselves and search for specific types of clouds. Perhaps the golden age of cloud observations is at hand . . . as long as climate change doesn't result in fewer clouds or reduced visibilities.

References

Alberts, T. A., P. B. Chilson, B. L. Cheong, and R. D. Palmer, 2011: Evaluation of weather radar with pulse compression: Performance of a fuzzy logic tornado detection algorithm. *J. Atmos. Oceanic Technol.*, **28**, 390–400.

Ashley, W. S., A. M. Haberlie, and V. A. Gensini, 2023: The future of supercells in the United States. *Bull. Amer. Meteor. Soc.*, **104**, E1–E21.

Bluestein, H. B., F. C. Carr, and S. J. Goodman, 2022: Atmospheric observations of weather and climate. *Atmos. – Oceans*, DOI: 10.1080/07055900.2022.2082369. 39 pp.

Brotzge, J., J. Wang, N. Bain, S. Miller, and C. Perno, 2022: Camera network for use in weather operations, research, and education. *Bull. Amer. Meteor. Soc.*, **103**, E2000–E2016.

Emanuel, K., 2021: Response of global tropical cyclone activity to increasing CO_2: Results from downscaling CMIP6 models. *J. Climate*, **34**, 57–70.

Fraser, A. B., 1968: Stereoscopic cloud photography. *Weather*, **23**, 505–514.

Gensini, V. A., 2021: Severe convective storms in a changing climate. *Climate Change and Extreme Events*. Fares, A. Ed., Elsevier, 254 pp.

Heymsfield, G. M., L. Tian, L. L., McLinden, M., and J. I. Cervantes, 2013: Airborne radar observations of severe hailstorms: Implications for future spaceborne radar. *J. Appl. Meteor. Climatol.*, **52**, 1851–1867.

Holle, R., and M. Diamond, 1985: Four-dimensional holograms from time-lapse cinematography. *J. Atmos. Oceanic Technolog*, **2**, 420–423

McLaughlin, D., and Coauthors, 2009: Short-wavelength technology and the potential for distributed networks of small radar systems. *J. Atmos. Oceanic Technol.*, **24**, 301–321.

Rasmussen, K. L., A. F. Prein, R. M. Rasmussen, K. Ikeda, and C. Liu, 2017: Changes in the convective population and thermodynamic environments in convection-permitting regional climate simulations over the United States. *Climate Dyn.*, **55**, 383–408.

Sand, W. R., 1976: Observations in hailstorms using the T-28 aircraft system. *J. Appl. Meteor.*, **15**, 641–650.

Seimon, A., J. T. Allen, T. A. Seimon. S. J. Talbot, and D. K. Hoadley, 2016: Crowdsourcing the El Reno 2013 tornado: A new approach for collation and display of storm chaser imagery for scientific applications. *Bull. Amer. Meteor. Soc.*, **97**, 2069–2084.

Tanamachi, R. L. H. B. Bluestein, S. S. Moore, and R. P. Madding, 2006: Infrared thermal imagery of cloud base in tornadic supercells. *J. Atmos. Oceanic Technol.*, **23**, 1445–1461.

Trapp, R. J., N. S. Diffenbaugh, H. E. Brooks, M. E. Baldwin, E. D. Robinson, and J. S. Pal, 2007: Changes in severe thunderstorm environment frequency during the 21st century caused by anthropogenically enhanced global radiative forcing. *Proc. Natl. Acad. Sci. USA*, **104**, 19,719–19,723.

Trapp, R. J., and K. A. Hoogewind, 2016: The realization of extreme tornadic storm events under future anthropogenic climate change. *J. Climate*, **29**, 5251–5265.

Warner, C. 2003: Stereo-pair photographs of clouds. *Weatherwise*, **58**, 84–89.

Appendix 1
How changes in vorticity with time can be explained without explicitly taking the pressure into account

The reason why we don't need to take into account pressure explicitly in explaining how vorticity changes with time boils down to the following mathematical property of vector calculus: the curl of a gradient is zero.

To understand this physically without resorting to equations, however, we need to dig a little more deeply. To see how the vorticity of an air parcel will change with time, we manipulate the equations of motion in the following way: The tendency for an air parcel to rotate, say in a counterclockwise manner, requires that the pressure-gradient force in the y direction increases in the x-direction as in Figure A1a. The distribution of pressure consistent with these pressure-gradient forces is shown in Figure A1b and we will ignore, for simplicity, any horizontal variations in density.

It is apparent in Figure A2b that there must also be a pressure-gradient force in the x-direction that increases in the y-direction. The latter forces rotation in a clockwise manner and the two

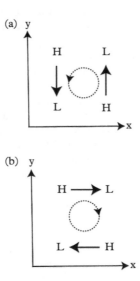

Figure A1 *Illustration of how a pressure-gradient force acting in the y-direction and increasing with x induces a counterclockwise circulation (a), but is also associated with a pressure-gradient force in the x-direction which increases with y and induces a clockwise circulation (b) of the same magnitude, thus canceling out the vorticity (tendency for rotation). For simplicity, we have neglected any horizontal variations in density.*

tendencies for rotation cancel each other out (though in this figure, we can conclude only that they do so in a qualitative sense). So, when we look at the left-hand side of the equation of motion, in which acceleration appears, when we consider the tendency for rotation to occur based on the pressure-gradient forces (on the right-hand side of the equation), we find that they contribute nothing.

Appendix 2
Camera equipment, film, and the evolution of cloud photography techniques

a. Camera equipment and film

I have used many cameras over the years to take the photographs that appear in this book. Although the following list is complete, the periods during which I used the cameras are sometimes only approximate, especially for the older cameras, and some of the film information may contain errors.

Film/analog cameras

Old box camera of unknown brand (film—Kodak black and white and Kodacolor 120, all negative film and Kodacolor 120 negative film) (~1956–1959/1960s)

Miranda Sensomat (film—Tri X ASA 400 B&W negative, Plus X 125 ASA B&W negative, Panatomic X ASA 32 B&W negative, Kodachrome II/ASA 25, 100 color "reversal"[1] slides and occasionally Ektachrome ASA 64, 160 color slides) 1969–1981

Nikon FM (film—Kodachrome 25, Fuji Velvia 50, 100 color slides) (1982–?)

Nikon FM-II (film—Kodachrome 25, Fuji Velvia 50, 100 color slides) (?–?)

Pentax 67 medium format (film—Fuji Velvia 50 color slides) ~2000 (?)–2013 or slightly later

Sony VHS analog video camera (analog tape cassettes) (1986/1987–2002)

Digital cameras

Fuji Fine Pix A303 digital point-and-shoot camera (2002–2006)

Sony Handycam DCR-TRV50 digital video camera (digital tape cassettes) (May 2002–2008/2010)

iPhone digital camera (~spring 2010–April 2013?)

Fuji Fine Pix E900 digital point-and-shoot camera (March 2006–?)

Sony Handycam HDR-SR12 digital movie camera (hard drive) (~2008–2010)

iPhone 5 digital camera (April 2013–May 2018)

Canon Power Shot digital camera 3500 (~2010–2013)

---

[1] "Reversal" film (I dropped the qualifier for all subsequent slide films because they were all reversal film) produced positive images, rather than negative images.

Sony Cybershot DSC-HX10V digital camera (2013–2019)

Nikon D800E digital camera (2013–now[2])

Sony a6000 digital mirrorless camera (~2016–now)

iPhone X digital camera (May 2018–now)

Sony FDR-AX53 4K digital movie camera (2017–now)

Go Pro HERO8 Black digital video camera (spring 2020–now)

Sony RX100 vii digital camera (2020–now)

b. Some thoughts on taking cloud photographs: A brief personal history

The use of film, in retrospect, was very difficult compared to our current use of digital techniques. Most rolls of 35-mm film could allow only about 36–38[3] frames of images, and some, especially of the medium-format type ("120," or ~120 mm), could handle only around ten.[4] One therefore had to be very sparing when taking photographs and have an extra camera handy, especially when clouds were rapidly changing appearance (like tornadoes), because taking the time to remove exposed film and re-load the camera with unexposed film could take much too long and if one did so, one risked having the cloud disappear (especially tornadoes) in the interim.

It sometimes took days to receive back processed film and on a few occasions the film was lost in transit or lost or damaged by the film processor. Since the image was made directly on the film, dust and hair would often contaminate the photograph negative or slide if one had to change the film, especially outside when it was windy, which it often is when out in the field looking at storms.

Before the availability of digital, image-processing techniques, one had difficulty correcting for dust, etc. and correcting for bad exposure, etc. To accomplish the latter, one had to take successive frames rapidly while changing the exposure or shutter speed or both slightly, a process called "bracketing" (typically by changing the shutter speed or changing the aperture on the lens over a range of three or so stops) to correct for possible under- or overexposure,[5] but doing so meant fitting fewer images of the same cloud on a roll of film and therefore increasing the chance you would run out of film. I found that underexposing the film by a half-an-f-stop (or more sometimes) was usually necessary when taking photographs of clouds so that they would not be overexposed, especially when parts of the ground, which are typically darker than the sky, appeared in the frame and the light meter sensor might consequently overweight the dark areas. This procedure is not necessarily a hard and fast activity; for example, when viewing the silhouette of a tornado backlit by bright light: In this case, it is usually acceptable to overexpose the background, but let enough light in so that the tornado and cloud-base features are not washed out. Some photographers used filters that darkened only part, for example the top part of the frame, so that one could expose, the top of an image, where there are bright clouds, less than the ground, which is darker, thus yielding an image in which features *both* at the ground (dark) and in the sky (bright) are properly

---

[2] By "now" I mean the time of this writing, spring 2022.

[3] I often squeezed 37–38 images onto each roll of "36," but at the risk of having some early or late frames truncated or contaminated by light.

[4] The number of frames that could fit on a roll depended on the size of the camera; my 6×7 Pentax camera allowed only around ten images. I reserved this camera only for photographing clouds that demanded extreme spatial resolution and fine grain and were rare and spectacular.

[5] Bracketing is easily done with digital cameras also.

exposed and discerned in the film. I have tried this technique and it was relatively difficult and time consuming.

In order to make some clouds stand out, especially when the clouds were bright white and the sky was blue, I used a polarizing filter to increase the contrast between the clouds and sky. Doing so was possible in certain parts of the sky only where the light was most polarized (in the plane normal to that of the line of sight) and the air relatively clean. Use of a wide-angle lens and a polarizer is not advisable, since only a small portion of the frame is highly polarized. One had to remember to compensate for the loss of light in the filter and not underexpose the image; an image that without the filter could be taken without the use of a tripod might require a tripod, owing to the less light let in and the longer exposures needed. On the other hand, there are times when allowing light of the *same* polarization that the filter excludes can increase the definition of a cloud.

In addition, film was sensitive to heat and it was therefore not advisable to leave cameras and film inside a car on the beach on a sunny, hot day. The spatial resolution of the film was limited by the graininess, and in general, the faster the film speed, the grainier the photograph. Finally, the films with the fastest speeds could be contaminated if placed through an X-ray machine at an airport security area. But, on the other hand, one had a physical representation of an image that one could hold and instantly examine in light, and didn't have to have a computer handy and worry about memory limitations or batteries. Some of my Kodachrome 25 slides have lasted for 50 years or more without much image degradation, especially if the film had been processed properly and stored in a cool, dry, place and not exposed to much light when examined.

Cloud photography was vastly improved and become much easier when digital cameras became available, at a relatively inexpensive cost, early in the 2000s. Just prior to this time, when personal computers became widespread in the 1990s, digital scanners allowed older film negatives and slides to be digitized and stored on digital memory. Thus, while a physical negative or slide could deteriorate, the digital representation did not, though it has been advisable to make copies in a number of different places and periodically store files on newer and new storage media, just to be safe in case the storage media themselves deteriorated. Scanning slides was/is a painstaking, slow process, and dust, etc., on the film made obtaining clean images difficult. I used Adobe Photoshop[6] to get rid of as many imperfections as I could find and correct for under- and overexposure.

Working with digital files from digital cameras, I have found that even grossly over- or underexposed images can be corrected using Photoshop when the dynamic range of the camera sensor is very large. I once took photographs of a comet at night, which looked completely black and I was very disappointed and was about to give up. However, on a lark, I increased the brightness of the image many times and then was actually able to see the comet and stars clearly. This experience reminds me of the 1966 movie *Blow-Up* by Michelangelo Antonioni, in which a murder was captured inadvertently by a photographer, who subsequently spends a lot of effort blowing up and enhancing a photograph to look for evidence, which is not readily apparent at first. I am also reminded of an episode of the crime television program in the US, *Columbo*, in which a frame from a surveillance camera photograph is enhanced and used to catch a murderer who is caught driving a car with lighting inconsistent with the suspect's story on where he was at that particular time.

In addition to adjusting for light, color can also be "corrected," and contrast adjusted. With good digital cameras, the use of polarizing filters is no longer needed as much, except perhaps in

---

[6] Adobe Photoshop is not the only software available to manipulate digital images, but it was the first one I learned how to use and is very powerful and easy to use.

a few instances when it is very hazy. The contrast can usually be increased to sharpen the difference between white clouds and a blue-sky background without having to use a polarizing filter. However, at times the contrast must be increased so much that the images lose their sharpness and become grainy looking, so an optical method of increasing contrast is preferred. Finally, photographs can be taken sometimes in relatively low light without the use of a tripod because the sensitivity can be increased substantially, though at the cost of some increased noisiness introduced into the image. Some digital cameras have the ability to record images in "raw" format, which allows for even greater control over correcting and enhancing images, at the expense of increasing the size of image files. Haze and dynamic range limitations can be overcome by using special digital filters.

Yet another advantage of digital photographs is that their times and GPS locations may be recorded, which aids significantly in identifying the photographs for scientific use. One of my old analog cameras allowed me to time stamp each photograph, but with the cost of removing part of the picture in each frame (e.g., Fig. 5.44).

Another technique often used by storm chasers is to take digital videos and select frames from the video or take a screen capture of the display of a video that is paused. Using high-resolution technology such as 4K digital video, one can recover spectacular images of tornadoes that may be missed using still photography, or recover images of lightning at night and images of tornadoes at night intermittently made visible by lightning flashes.

Despite the many advantages of digital photography, there has recently been an increase in a general interest in old-fashioned analog photography using film. I learned about this recently in a documentary movie called *Grain: Analog Renaissance* (2022). The renewed interest might simply be a knee-jerk reaction to the prevalence of digital photography or a wish to remain true to what we see in the face of digital manipulation. Would you rather see a photograph of a tornado as it is, including contrails crossing the sky in the background, or utility poles or wires crossing the condensation funnel, or one in which unwanted sections are deleted and contrast enhanced to make the tornado artificially seem even more fearsome?

Also relevant to cloud photography is framing. Using Adobe Photoshop, one can crop images to obtain the desired framing. On the other hand, it is always desirable to make sure that some ground-based object is also in the frame to provide perspective to the cloud photograph. Sometimes, doing this is impossible, and one must settle for a photograph of a cloud, particularly when high above the horizon or almost overhead, without a horizon or any ground-based objects.

Getting a large cloud into a frame might be difficult using regular lenses, so a wide-angle or ultra-wide-angle (fish-eye) lens must be used. The downside of using an ultra-wide-angle lens is that there can be considerable distortion, such that utility poles, etc. seem to lean over at the edges of the image.

Another option is to obtain a series of overlapping images around a considerable (even as much as a 360° field of view) field of view and digitally stitch them together. In this book, there are some wide-angle photographs and some panoramic images composed of stitched-together, side-by-side photographs using special software, while others are obtained automatically using special software already built into the camera. These stitched-together panoramas, however, are not without some distortion, as can be seen in some of the photographs of long orographic wave clouds in Chapter 2.

More on techniques of cloud photography can be found in the following reference:

Galvin, J. F. P., 2003: Picturing the sky. *Weatherwise*, 58, 80–83.

Glossary of key terms

The definitions of selected terms listed below are mostly those of the author, but in many other cases they are modified versions based on those in the American Meteorological Society's *Glossary of Meteorology* (http://glossary.ametsoc.org).[1]

Anvil Upper portion of mature convective clouds that typically contains non-buoyant blowoff, usually cirrostratus, from an earlier buoyant updraft or updrafts. Typically refers to the anvil shape of the cirrostratus emanating from the top of a cumuliform cloud, though it is not always accompanied by precipitation reaching the ground anywhere in the parent cloud. Anvils may be composed of ice crystals or, to a lesser extent, water droplets, usually at high-enough altitudes that they are supercooled.

Altocumulus Patches or parallel bands of clouds at midlevels typically composed of liquid-water droplets. Their patchy elements are smaller than those of stratocumulus, but tend to be larger than those of cirrocumulus. Their patchiness is thought to be due to radiation effects and vertical wind shear.

Altocumulus *castellanus* Altocumulus whose upper parts appear to be cumuliform and buoyant and which have a common flat base, often arranged in lines.

Altocumulus *floccus* Altocumulus composed of tufted elements looking like pieces of cotton with a somewhat cumuliform or rounded appearance. The bottom part of the cloud is ragged and not flat.

Altocumulus *lenticularis* Altocumulus that are shaped like lenses or kidney beans and have smooth surfaces. They are often associated with stationary, orographically induced gravity waves in the lee of mountains, but may also be seen elsewhere. When they are stationary, they are referred to as *ACSL (altocumulus standing lenticularis)*.

Anticyclone A gyre of wind rotating in the clockwise direction in the Northern Hemisphere and in the counterclockwise direction in the Southern Hemisphere. The gyre rotates in the opposite sense as the rotation of the Earth about its axis.

Arcus Wedge-shaped band of cloud at the leading edge of a gust front. It is usually smooth-looking and may have striations parallel to the band. Also called a *shelf cloud*.

Asperitas Wavy cloud elements that often appear as wavy striations underneath a solid stratus cloud base at low levels. The waves usually appear to be chaotic.

Backscattering See *scattering*.

Banner cloud A plume of cloud that extends downstream from an isolated, sharp, mountain peak, especially when it is shaped like a pyramid. The airflow over the top of the mountain reverses direction in the lee side and curls back up the mountain.

Bar cloud A colloquial expression for a stationary, orographic wave cloud that is relatively narrow and has well-defined ends. It has been used to describe clouds in the lee of the Colorado Front Range. It is most common when there is a substantial along-the-mountain-range wind component; in Northeast Colorado, there is a substantial northerly wind component at mountain-top level.

---

[1] I was the author of a number of the original definitions in the *Glossary* for terms related to severe convective storms.

Bear's cage   Storm chaser colloquial expression describing the low-level mesocyclone near the ground when precipitation envelops much of it.

Bore   Gravity-wave phenomenon that occurs when a stable low-level air mass is forced upward and over another, but more-stable air mass.

Boundary layer   The layer of air affected by turbulent friction at the Earth's surface and the turbulent exchange of heat and water vapor with the Earth's surface. It may be very shallow or as deep as several kilometers or more.

Bow echo   A line of convective storms that bulges outward in the direction it advances and is associated with swaths of damaging, strong, straight-line winds.

Bubble, thermal   A discrete, buoyant mass of air. It may be unsaturated or saturated, but the former is in clear air and the latter in cloudy air.

Buoyancy   A vertical force that occurs when a parcel of air has a different density than that of the surrounding air. The excess density (negative buoyancy) or density deficit (positive buoyancy) may be due to differences in temperature, water vapor, or liquid water, or ice content.

Centrifugal force   The radially outward force experienced by an air parcel, in its reference frame, when it is rotating or following a curved trajectory. It is equal and opposite to the *centripetal force*.

Centrifugal waves   In a tornado, oscillations in the width of the vortex that may occur when there is an imbalance between the radially inward-directed *pressure-gradient force* and the outward-directed *centrifugal force*.

Centripetal force   The radially inward force experienced by an air parcel, in a fixed reference frame, when it is rotating or following a curved trajectory. It is equal and opposite to the *centrifugal force*.

Cirrus   High-altitude clouds mostly composed of ice crystals, which often have a wispy appearance.

Cirrostratus   A broad sheet of cirrus clouds having a relatively uniform texture.

Cirrocumulus   Thin cirrus clouds composed of small elements or bands.

Cloud microphysics   The dynamics of individual cloud particles and their nuclei, as opposed to the dynamics of parcels of air, which are much larger.

Cold pool   A relatively shallow (usually <2–3 km) layer of air underneath a convective storm that has been cooled when rain falls into unsaturated air and evaporates or when frozen precipitation melts on the way down.

Condensation   The conversion of water vapor into liquid water.

Conditional instability   The state of being stable with respect to vertical displacements if the atmosphere is unsaturated (with respect to water vapor), but unstable if it is saturated. This condition occurs when the lapse rate is less than the dry-adiabatic, but greater than the moist-adiabatic.

Contrail   "Condensation trail" in the form of a line of cloud emanating from the exhaust of a jet engine in an aircraft flying at high altitude.

Convection   In its broadest sense, a transport of heat and momentum by air motion. In this book it refers specifically to the transport of heat and momentum, mainly by turbulent eddies that have buoyancy (either positive or negative), particularly in clouds but also in the clear-air boundary layer.

Convective condensation level (CCL)   The altitude at which surface air (represented by the average water vapor mixing ratio near the surface) first becomes saturated when heated.

Convective precipitation   Precipitation formed in buoyant updrafts and which falls near the updrafts. It is usually mentioned in contrast with *stratiform precipitation*.

Coriolis force   A force that is felt on a ground-based reference frame that is a result of the rotation of the Earth about its axis.

Corner flow   The region in the boundary layer of a strong vortex such as a one-cell tornado where the air turns abruptly from radially inward to upward near the center of the vortex.

Crow instability   A dynamic instability in which the counterrotating vortices shed in the rear of flying aircraft interact with each other.

Cumulus   Clouds that have bubbly-looking tops as a result of turbulent motions due to upward buoyancy and often are arranged in a checkerboard pattern. They are usually composed of water droplets, which may turn into ice crystals under certain conditions.

Cumulus *congestus*   Cumulus clouds that extend vertically into cloud towers. Also called *towering cumulus (TCU)*.

Cumulus *fractus*   Cumulus clouds that appear ragged and irregular looking.

Cumulus *humilis*   "Fair-weather" (humble) cumulus; cumulus that do not extend vertically very much.

Cumulus *mediocris*   Cumulus that have some vertical development: more than that of cumulus *humilis*, but less than that of cumulus *congestus*.

Cumulonimbus   A relatively deep cumulus cloud that produces precipitation. It frequently extends up to the tropopause and may be composed of just water droplets or both water droplets and ice crystals. It may or may not produce lightning and thunder.

Cyclic mesocyclogenesis/tornadogenesis   The periodic generation and dissipation of mesocyclones and tornadoes in a supercell through a well-defined life cycle.

Cyclone   A gyre of wind rotating in the counterclockwise direction in the Northern Hemisphere and in the clockwise direction in the Southern Hemisphere. The gyre rotates in the same sense as the rotation of the Earth about its axis.

Density current   A moving fluid (such as air) that is driven by the pressure gradient created from an area of relatively high hydrostatic pressure at the surface, especially of a *cold pool* that intrudes into a warmer, environmental region of lower hydrostatic pressure. *Gust fronts* in convective storms are examples of density currents; also known as gravity currents.

Derecho   A mesoscale convective system that generates damaging, straight-line winds at the surface over a path at least 650 km long and 100 km wide.

Deposition   The process of converting water vapor to ice, bypassing the liquid state: the opposite of *sublimation*.

Detrainment   The transfer of air from inside a cloud to its surroundings: the opposite of *entrainment*.

Dewpoint   The temperature of an air parcel, initially unsaturated, if it is cooled at constant pressure and constant water-vapor content to saturation.

Doppler radar   A radar that measures the component of the wind along the line-of-sight of its beam using the principles of the Doppler effect, by which the backscattered radiation from receding targets is shifted down in frequency and the from approaching targets is shifted upward in frequency.

Doppler velocity   The component of the wind along the line-of-sight of a radar or lidar.

Downdraft   Downward component of the wind in a convective cloud or storm. Downdrafts may be forced through negatively buoyant air parcels associated with temperature colder than that of their environment, especially due to evaporation or sublimation of precipitation as it encounters unsaturated air, melting, precipitation loading, or downward-directed (non-hydrostatic) pressure-gradient forces.

Dry-adiabatic   A process in which there is no exchange of heat and moisture is not considered. Changes in temperature are due only to the gain or loss of heat energy through compression or expansion of the air experienced in vertical motions.

Dryline Surface boundary separating relatively cool, moist, marine air from hot, dry continental air over gently sloping terrain, at which the top of the marine air intersects the ground. Convective storms are sometimes triggered along and just east of the dryline.

Dust devil Small-scale, short-lived vortex column driven by heat from the sun at the ground, when it is dry and contains dust or sand that can be picked up by the wind.

Entrainment The transfer of environmental air into a cloud. The opposite of *detrainment*.

Equilibrium level The altitude of an air parcel ascending in a convective storm that reaches the point at which its buoyancy is reduced to zero.

Eulerian framework Measurements made in a reference frame fixed to the ground. Compare with a *Lagrangian framework*.

Evaporation The process of converting liquid water into water vapor, especially when precipitation falls into an unsaturated layer of air or when liquid-water cloud droplets abut against unsaturated air.

Extratropical Midlatitude, typically near or within the 30–55° latitude belt.

Foehn cloud/foehn wall A stationary cloud edge formed when relatively humid air ascends a mountain range and cools to saturation, but then dissipates as it descends and warms on the lee side so the humidity decreases and the cloud droplets evaporate. It is visible from the lee side as a sharp cloud boundary.

Fog A cloud that is present at or near the Earth's surface.

Fog, advection Fog that is formed when relatively moist air flows over a colder surface and is cooled to saturation (the dew point).

Fog, ground Fog whose cloud base is at the ground and obscures visibility.

Fog, radiation Fog that is formed when the air is cooled due to radiation loss, to or below its dewpoint. It is typically relatively shallow and occurs over land areas when the surface winds are weak and there is little or no cloud cover above.

Fog, steam Fog that occurs when very cold, unsaturated air flows over a relatively warm water surface, as the two air masses, each of which is unsaturated, mix with each other at constant pressure. It is also called *arctic seasmoke* or just *seasmoke*. It forms because the vapor pressure of an air parcel when it is mixed with water vapor from the water surface varies linearly with temperature, while the variation of the saturation vapor pressure with temperature varies nonlinearly.

Fog, steaming Fog that forms on wet surfaces as the sun heats a wet ground, evaporating water into the air until it becomes saturated. It is not to be confused with *steam fog*.

Forward-flank downdraft Negatively buoyant air that descends underneath the leading edge of a supercell, typically on its downshear side and usually underneath the anvil.

Friction force A force that occurs when air flowing over a non-moving surface, such as the ground, is slowed down by the non-moving surface. Air having zero momentum at the surface is mixed turbulently with air having higher momentum aloft to produce a mixture of air that flows less rapidly than it would have otherwise.

Front/frontal zone An elongated interface or transition zone between air masses of different density; the density gradient is typically due to differences in temperature. The magnitude of the temperature gradient across the front or frontal zone is typically an order of magnitude or more than that in the atmosphere at large.

Funnel cloud A cloud that is typically shaped like a funnel, pendant from the base of a cumuliform cloud. It is associated with a rotating column of air and pressure lower than that of the surrounding air. If the air column rotates rapidly enough to cause damage at the surface, even if the condensation funnel does not extend all the way to the ground, it is called a *tornado*.

Funnel cloud, high-based   A funnel cloud that is pendant from a cumuliform cloud whose base is well above the condensation level associated with the lift of air from near the surface or from heating of surface air. This type of funnel cloud is not associated with tornadoes.

Geostrophic   The wind speed and direction associated with the turning of the Earth about its axis of rotation (Coriolis force) and the synoptic-scale pressure-gradient force, when there is a balance between the Coriolis force and the pressure-gradient force.

Geostrophic wind   The wind speed and direction associated with *geostrophic* balance, a balance between the Coriolis force and the pressure-gradient force. In the Northern Hemisphere the geostrophic wind blows such that higher pressure lies to the right and lower pressure lies to the left; the opposite happens in the Southern Hemisphere.

Glaciated   What a cloud becomes when supercooled liquid-water cloud droplets are converted to ice crystals.

Glory   Small rings of light at the surface of a cloud composed of water droplets, particularly when an airplane blocks the sun and seen at the point at the end of a line extending from the sun through the aircraft to the cloud surface.

Gravity wave   A wave in the atmosphere that bobs up and down, occurring when there is an imbalance between the vertical component of the pressure-gradient force and buoyancy in a stable atmosphere.

Gust front   The leading edge of an advancing *cold pool* in a convective storm. It is an example of a *density current* or *gravity current*.

Halo   A circular ring of light around the sun of different colors, caused by the reflection or refraction of sunlight by ice crystals in cirrostratus clouds. The halo is typically at an angle of 22° from the sun, but may be visible at other angles and also as arcs of light rather than circular, rings.

Horseshoe vortex   Vortex whose axis of rotation is bent into a U shape or the shape of a horseshoe. Horseshoe vortices may be sometimes seen as upside-down U-shaped, tubular clouds.

Hydraulic jump   The sudden jump in height of the interface between a dense fluid at low altitudes and a less-dense fluid aloft, when air moving faster than surface gravity waves slows suddenly to speeds less than that of surface gravity waves. Air accelerating down a mountain at speeds greater that of surface gravity waves makes a transition from laminar flow to turbulent flow, suddenly turns upward and may form a cloud.

Hydrostatic   The state of the atmosphere when there is a balance between the downward force of gravity and the upward force due to the decrease in pressure with height.

Instability   A state in which an air parcel, if forcibly moved in a certain direction, continues to accelerate in that direction even if the force is removed.

Kelvin–Helmholtz wave/billow   A periodic, wavelike disturbance in the shape of a cloud, often in the form of a curlicue, which occurs when buoyant air parcels are acted on by vertical shear.

Lagrangian framework   Measurements made in a reference frame moving along with an *air parcel*. Compare with an *Eulerian framework*.

Laminar   Smoothly varying in space and time; not turbulent.

Lapse rate   Rate of decrease of temperature with height.

Landspout   A colloquial term for a tornado, which is typically relatively weak, and is formed in a convective cloud or storm that does not produces a mesocyclone. The source of vorticity, instead of being produced by the storm, is pre-existing in the boundary layer, typically as horizontal shear along a surface boundary. Landspouts typically occur during the growth stage of a parent cloud or storm and are called landspouts because they look like weak Florida Keys waterspouts, but form over land.

Lenticular cloud   Wave cloud shaped like a lens or kidney bean. See also *altocumulus lenticularis*.

Level of free convection (LFC)   Altitude at which an unsaturated air parcel, when lifted dryadiabatically to its lifting condensation level (LCL) and then moistadiabatically, first becomes warmer than its surroundings.

Lifting condensation level (LCL)   Altitude at which an unsaturated, moist air parcel, when lifted dryadiabatically, first becomes saturated.

Lowering   Colloquial expression used by storm chasers to describe the lowering of the cloud base, usually flat, under a convective cloud, often as scud clouds form under cloud base and then get attached to the cloud base above. When the lowering rotates, it is called a *wall cloud*.

Mamma (also called mammatus)   Pouch-like, downward-hanging protuberances underneath cirrus, cirrostratus, and anvils of cumulonimbus clouds in particular. They look like upside-down convective clouds, but with a much smoother appearance.

Mesocyclone   A cyclone typically around 2–5 km in diameter, which is found especially in supercells.

Mesocyclone signature   Couplet of approaching and receding Doppler velocities, in the reference frame of a convective storm along or nearly at the same range, indicating azimuthal shear in the Doppler velocities, which is indicative of a mesocyclone. Azimuthal shear is typically around 5×10^{-3} to 10^{-2} s^{-1} and exhibits continuity in both the vertical and in time.

Mesoscale   Pertaining to atmospheric features ranging from several to several hundred kilometers in diameter and lasting for tens of minutes up to periods of time less than a day or so long. Typical mesoscale features are convective storms and complexes of storms, precipitation bands, orographic gravity waves, and circulations induced by differential heating, such as sea and land breezes and diurnal upslope and downslope circulations around mountains and sloping terrain in general. The width of fronts and the dryline are also on the mesoscale.

Mesoscale convective system (MCS)   A complex of convective storms that produces a region of precipitation that is around or greater than 100 km long and may be as wide as 100 km but usually less. MCSs may harbor areas of both stratiform precipitation and convective precipitation.

Microburst   A strong, negatively buoyant downdraft in a convective storm that is less than around 4 km across and lasts for about 2–5 minutes. *Dry* microbursts are driven mainly by evaporative cooling as precipitation falls into very dry air, while *wet* microbursts are driven mainly by the loading of liquid water in a very moist environment or by melting.

Mie scattering   The process by which electromagnetic radiation is backscattered off spherical targets, such as small raindrops, whose diameter is comparable to the wavelength of the radiation.

Mixing   As applied to cloud dynamics and thermodynamics, it refers to the process by which the properties of an air parcel (e.g., temperature, water vapor content, and momentum) are determined by the interaction of smaller, turbulent air parcels having different properties.

Moist-adiabatic   A process that is *adiabatic*, but when the atmosphere is saturated and may contain water or ice. If the total water substance is conserved the process is reversible; if it is not conserved, for example, if precipitation falls out, then the process is irreversible and called *pseudo-moistadiabatic*.

Multicell   A type of convective storm that is typically composed of a series of ordinary convective cells in various stages of their life cycle. It is favored in an environment of relatively weak vertical wind shear over the depth of the troposphere, but there must be some vertical shear at low levels. The production of new cells is associated with lift along the leading edges of the *cold pools* produced by previous cells. Multicell convective storms typically last for much longer than ordinary cells, on the order of several hours.

Multiple-vortex tornado   A tornado that contains smaller, secondary vortices moving and propagating around its central axis.

Ordinary cell   The basic building block of a convective storm in an environment of weak vertical wind shear, in which a buoyant updraft produces precipitation that falls into unsaturated air and may also melt, leading to a *cold pool* and the eventual destruction of the buoyant updraft and demise of the convective storm, over a 30–50-minute period.

Orographic wave cloud   A cloud, typically at middle or high altitudes, that is produced by gravity-wave motion triggered when air is forced over mountains in a stable atmosphere. The clouds are typically stationary, but some may propagate or move. Many are laminar and tend to be long bands or, when isolated, be shaped like lenses or kidney beans.

Overshooting top (penetrating top)   A transient protrusion of a buoyant updraft in a cumulonimbus above the *equilibrium level*, which is usually at the top of the anvil, near the tropopause.

Parcel, of air   A small (hypothetical) volume containing a very large number of air molecules, each having approximately the same properties (e.g., temperature, number of water vapor molecules, number of dry air molecules, and momentum). One may think of the size of the volume as being somewhere on the order of a meter on a side or less and that as it moves along it carries along its properties also. The concept of a parcel is applied to analyses of motions in fluids in general and requires a different type of analysis we employ when considering the motions of rigid bodies, since air parcels may be deformed.

Penetrating top (overshooting top)   See *overshooting top*.

Pileus   A smooth-appearing cloud that looks like a cap forming above a rising, buoyant, cumuliform, convective cloud tower when stable air above it is cooled to saturation, especially when there is a moist layer of air above the convective cloud. Sometimes there are multiple layers of pileus, especially if there are vertical variations in moisture.

Plume   A buoyant jet of air forced by a point source of heating.

Polarimetric radar   See *Radar, polarimetric*.

Pressure, atmospheric   The force per unit area acting on a side of a volume of an air parcel. Part of it is due to the random motion of air molecules and part is due to the macroscopic motion of air parcels (the dynamic part).

Pressure, dynamic   The pressure exerted on an air parcel that is a result of the motion of the air.

Pressure, hydrostatic (see also hydrostatic)   The pressure exerted on an air parcel when it is in *hydrostatic* balance. This pressure is the force per unit area exerted by the mass of air in a column above.

Pressure–gradient force   A force acting from high to low pressure, owing to the difference in pressure across an air parcel.

Propagation   The apparent motion of a line of a constant quantity (such as pressure or temperature) as result of its increasing with time in one direction and decreasing with time in the opposite direction, rather than a result of air parcels having that constant quantity actually moving and carrying that quantity along. For example, a convective storm may propagate because a new storm forms on one side of the storm while the storm itself dissipates, giving the illusion that the storm is moving.

Pseudo-moistadiabatic See *moist-adiabatic*.

Quasi-linear convective system (QLCS) A mesoscale convective system that is aligned approximately along a line, also known as a *squall line*.

Radar In meteorology, a device that detects distant regions of precipitation (and cloud particles, dust, debris, insects, and birds) by sending out electromagnetic radiation and determining the range of the precipitation (or other scatterers) from the radar; in the case of pulsed radars, it is based on the time delay of the returned signal after each pulse has been sent out. It is an acronym for Radio Detection And Ranging; it is also a palindrome.

Radar, Doppler (see Doppler radar) A *radar* that measures the speed of the scatterers it senses in the direction of the radar beam, from the change in frequency of the backscattered signal, according to the *Doppler effect*.

Radar, polarimetric A radar that is capable of transmitting radiation that is polarized and measuring the relative amount of radiation backscattered having orthogonal polarizations (usually horizontal and vertical). The relative amount of radiation received from each channel and the phase difference between both channels of the transmitted and received signal contain information about the nature of the scatterers, such as whether they are large raindrops, small raindrops, hail, insects, etc.

Radar reflectivity (factor) A normalized measure of the intensity of a signal received back at a radar that is due to *Rayleigh scattering*.

Rain, ice process The formation of rain when there is a mixture of supercooled liquid-water droplets and ice crystals: Ice crystals grow at the expense of the water droplets as the water droplets evaporate and the additional water vapor in the air is deposited onto the ice crystals. The ice crystals grow and fall out as snow, but may melt on the way down to become raindrops. The ice process of rain formation is also called the *Bergeron-Findeisen-Wegener* process.

Range The distance from a radar to a collection of scatterers.

Rayleigh scattering The process by which microwave electromagnetic radiation sent out by a radar is backscattered off spherical targets, such as small raindrops, whose diameter is much less than the wavelength of the radiation.

Rear-flank downdraft (RFD) A downdraft due in part to negative buoyancy associated with precipitation loading and cooling from evaporation of precipitation or melting, and in part to downward-directed dynamic pressure-gradient forces, in the rear flank of supercells.

Ridge, of high pressure A line along which there is a relative maximum in pressure with respect to the direction normal to the line. The opposite of a *trough*. May be used also to describe the temperature and moisture fields, among others.

RKW theory A theory explaining how low-level vertical wind shear can act to make updrafts along the leading edge of a cold pool more erect (less at an angle to the vertical) and more likely to trigger new buoyant updrafts, thereby sustaining a convective system for a long period of time. The theory was proposed by Richard Rotunno, Joe Klemp, and Morris Weisman, from which the RKW designation comes.

Roll cloud An *arcus* or *shelf cloud* that is completely detached from the parent convective storm's cloud base.

Rotor An elongated, horizontal vortex sometimes found underneath vertically trapped, orographic gravity waves or hydraulic jumps in the lee of mountain ranges. Rotors are hazardous to aircraft because the lift experienced may change significantly when flying across them.

Rotor cloud A narrow, continuous, or broken, narrow cloud band composed of ragged-looking elements associated with *rotors*. They are sometimes seen underneath smooth-looking orographic wave clouds.

Scattering The process by which electromagnetic radiation is absorbed by precipitation, dust, and insects, etc. sent out from a radar and then is absorbed and re-radiated back to space. That portion of this radiation that is scattered in the direction back to the radar is said to be *backscattered*.

Scud Low clouds that are ragged-looking and change shape rapidly, often below a more well-defined cloud base. They may take the form of cumulus *fractus* (broken cumulus). They are often found in the vicinity of precipitation and might be a sign of relatively humid air originating in the precipitation area in which rain has evaporated, or from water on the ground, or from raindrops that have broken up into smaller water droplets, but is then sucked up into an updraft while mixing with drier air, causing irregular-shaped clouds to form, as in a *lowering* or a *wall cloud*.

Shelf cloud See *arcus*.

Squall line A *quasi-linear mesoscale convective system* that is composed of a solid or broken line of convective storms. It frequently also has a trailing area of stratiform precipitation.

Steam devil A vertical column of rotating air driven by the heating of very cold air by a relatively warm surface body of water. Similar to a *dust devil*, but for much cooler air masses and surfaces.

Stratus A rather uniform-looking cloud mass, often gray and gloomy appearing, with an approximately uniform cloud base.

Stable The state in which an air parcel, if forcibly moved in a certain direction, accelerates in the direction *opposite to that which it was moved* if the force were suddenly removed. If the air parcel overshoots its equilibrium point, it may oscillate about its equilibrium point, producing a wavelike phenomenon.

Starting plume A time-dependent buoyant *plume* with a well-defined advancing leading edge, unlike an ordinary *plume*, which is steady state and has no leading edge.

Steady state When there are no changes with time of quantities at locations fixed with respect to the ground.

Storm splitting A process in which convective storms in an environment of strong vertical wind shear split into *cyclonically* rotating and *anticyclonically* rotating members, which separate with time. Developing *supercells* tend to split into symmetric members when the direction of the vertical shear vector does not change with height. When it changes with height, one member is favored over the other. Some storms may undergo successive periods of splitting.

Straight-line winds Winds that blow in approximately the same direction, with little if any curvature in the airflow. *Straight-line* winds are distinctly different from winds associated with vortices such as *tornadoes* and *mesocyclones*.

Stratiform precipitation A relatively uniform region of moderate precipitation often with mesoscale dimensions and found at the rear of *mesoscale convective systems*. Vertical velocities are much weaker than those in the leading line of *convective precipitation*.

Stratosphere The layer of Earth's atmosphere extending from roughly 10–17 km, which extends from the *tropopause* to the bottom of the mesosphere at around 50 km. The stratosphere is characterized by high stability with respect to vertical displacements of air since the temperature does not change much with height or increases with height. Its structure is controlled to a large extent by ozone and the radiative transfer processes associated with it.

Streamwise vorticity The tendency for air to rotate about an axis along which air is moving, yielding helical trajectories.

Streamwise vorticity current (SVC) In a supercell, near the ground, a current of air that is moving along the edge of or within the cold pool in the forward flank of the storm, along which air acquires rotation.

Sublimation   The process of converting ice into water vapor without any intermediate melting into liquid water. The opposite of *deposition.*

Sun dog (also called a *parhelion* or *mock sun*)   An optical phenomenon characterized by a bright, colored area at the elevation angle of the sun. There are sometimes two, one on either side of the sun, especially along the 22° halo. Sun dogs are often found in relatively thin cirrostratus.

Supercell   A convective storm characterized by a persistent, rotating updraft. Supercells typically form in an environment of strong, deep, vertical wind shear. They also form in the outer bands of landfalling tropical cyclones in an environment of strong low-level shear. Supercells tend to move to the left or right of the mean wind averaged over the depth of the troposphere. They typically spawn the strongest tornadoes and largest hail.

Swirl ratio   In a tornado, a measure of the relative amount of airflow swirling around the tornado (the *azimuthal component*) to the airflow going toward the center of the tornado (the *radial component*). The character of the airflow (the horizontal and vertical variations in the wind field) depends on the *swirl ratio.*

Synoptic scale   A term applied to weather systems on the order of thousands of kilometers across and persisting for time scales on the order of days. Most extratropical cyclones and anticyclones at the surface and upper-level troughs and ridges tend to be synoptic scale.

Tail cloud   A colloquial term for a horizontal, solid or broken line of *scud* clouds or a smoother-appearing cloud band that feeds into a wall cloud, usually from the precipitation region in the forward flank of a supercell. The *tail cloud* is usually long, but may be so short that it appears just as a short appendage jutting out from the wall cloud.

Thermal (see also *bubble, thermal*)   A discrete, small volume of buoyant air, typically clear and unsaturated (cloud free). Thermals also exist inside cumulus clouds, where the air is saturated.

Tornado   A rapidly rotating column of air pendant from a cumuliform cloud overhead that is capable of inflicting wind damage at the surface. It may or may not be associated with a funnel cloud. Wind speeds may be as low as around 30 m s^{-1} or as high as 135 m s^{-1}. Tornadoes may be as narrow as tens of meters across or as wide as several kilometers across.

Tornadic vortex signature (TVS)   Couplet of approaching and receding Doppler velocities, in the reference frame of a convective storm, indicating very strong shear in the Doppler velocities along or nearly along the same range ring, which is indicative of a tornado. Azimuthal shear is typically around 0.2 to 0.9 s^{-1}; the vorticity associated with a TVS for a circularly symmetric vortex is twice the azimuthal shear. The intensity of the TVS depends not only on the intensity of the tornado but also on the distance of the tornado from the radar, the resolution of the radar in range (in terms of the duration of the pulse, if the radar is pulsed), and the width of the radar beam.

Tornadogenesis   The formation (birth) of a tornado.

Towering cumulus: See Cumulus *congestus*

Tropical cyclone   A cyclone typically hundreds of kilometers across whose winds overall decrease in intensity with height above the boundary layer; its center aloft is warmer than the air surrounding the cyclone. Intense tropical cyclones are called *hurricanes, typhoons,* or *cyclones,* depending on their geographical location.

Tropopause   The boundary between the troposphere and stratosphere, which is usually characterized by an abrupt change in the lapse rate, such that there is an increase in static stability in the stratosphere. The tropopause is typically around 10–20 km altitude and is highest in the tropics and lowest in polar regions. When the tropopause appears at more than one level it is said to be "folded."

Troposphere   The bottom layer of the Earth's atmosphere, which extends from the surface to 10–20 km altitude. The temperature overall decreases with height in the *troposphere*.

Trough, of low pressure   A line along which there is a relative minimum in pressure with respect to the direction normal to the line. The opposite of a *ridge*. May also be used to describe the temperature and moisture fields, among others.

Tubular cloud   An elongated cloud that is narrow and smooth-looking. It may be accompanied by a vortex whose axis of rotation is aligned with the cloud. It is similar to a *funnel cloud*, but its ends are not tapered as in a funnel.

Turbulence   Irregular, chaotic variations in time and space, of the wind field.

Updraft   An upward-moving current of air, typically in a convective cloud, particularly in a cumulus *congestus* or cumulonimbus. Updrafts can be as strong as 50 m s^{-1} or more and as wide as several kilometers.

Virga   Precipitation falling from a cloud, but evaporating or sublimating completely before reaching the ground.

Vortex   A rotating, cylindrical volume of air about some axis, which typically is vertically oriented, but it could also be horizontally oriented or any other orientation in between.

Vortex breakdown   In some tornadoes, the transition region where air flows upward faster than the speed of *centrifugal waves* to one where the air flows vertically more slowly than the speed of *centrifugal waves*. In the lower region, upstream-propagating waves are not possible, while in the upper region, downstream-propagating waves are possible. The transition region is marked by a widening of the vortex and the flow becoming turbulent.

Vortex signature   Couplet of approaching and receding *Doppler velocities*, in the reference frame of a storm along or nearly along the same range ring, indicating azimuthal shear in the *Doppler velocities*, which is consistent with a vortex, either a *cyclone* or an *anticyclone*. See also *mesocyclone signature* and *tornadic vortex signature*.

Vorticity   A measure of the wind field for its ability to induce rotation. In vector calculus, it is the curl of the wind field. *Vorticity* may be vertical or horizontal, or somewhere in between.

Wall cloud   A lowered cloud base under the updraft of a convective storm, which is usually rotating. On the downshear side it often has a vertical face, from which its name is derived.

Warm rain   The formation of rain through collisions of liquid-water droplets and the growth of some by accretion. The formation of rain through this process may be countered by the breakup of raindrops into smaller drops and droplets.

Waterspout   Any tornado over water. Most often, a tornado over water in an ordinary-cell or multicell convective storm or growing cumulus congestus, but it may also be in a supercell.

Water-vapor mixing ratio   The ratio of the mass of water vapor to the mass of dry air.

Weak-echo column (WEC)   A *weak-echo hole* that extends vertically in the radar reflectivity field associated with the center of a tornado.

Weak-echo hole (WEH)   A local region of relatively weak radar reflectivity at the center of a strong vortex such as a tornado, as a result of the centrifuging radially outward of the most radar-reflective scatterers such as raindrops and debris. The WEH tends to appear above the boundary layer of a tornado but not near the ground, where radial inflow tends to overwhelm the radially outward motion of the larger scatterers.

Suggested further reading[1]

American Meteorological Society, 2013 to current ("living document" online): *Glossary of Meteorology*. http://glossary.ametsoc.org.

Bader, M. J., G. S. Forbes, J. R. Grant, R. B. E. Lilley, and A. J. Waters, 1995: *Images in Weather Forecasting and Radar Imagery*. Press Syndicate, Univ. of Cambridge, 499 pp.

Battello, M., A. Brunner, L. Mercalli, R. Niccoli, C. Solito, and J. Trifoni, 2006: *The Sky*. Cubebook, White Star Publishers, Vercelli, Italy, 736 pp.

Burt, C. C., 2004: *Extreme Weather: A Guide & Record Book*. W. W. Norton & Co., New York, 304 pp.

Durran, D., YouTube videos on wave clouds: (1) "Rocky Mountain Wave Clouds" and (2) "Trapped Lee Waves Over the Western US."

Dunlop, S., 2006: *Weather*. Thunder Bay Press, San Diego, California, 288 pp.

Ferrand, J., Jr., 1990: *Weather*. Stewart, Tabori & Chang, New York, 240 pp.

Gilliam, H., 2002: *Weather of the San Francisco Bay Region (2nd ed.)*. University of California Press, 106 pp.

Hamblyn, R., 2001: *The Invention of Clouds: How an Amateur Meteorologist Forged the Language of the Skies*. Farrar, Straus & Giroux, New York, 403 pp.

Henson, R., 2007: *The Rough Guide to Weather*. Rough Guides, New York, 422 pp.

Higgins, G., 2007: *Weather World: Photographing the Global Spectacle*. David and Charles, Cincinnati, Ohio, 256 pp.

Hobbs, P. V., and A. Deepak, 1981 (eds.): *Clouds: Their Formation, Optical Properties, and Effects*. Academic Press, 497 pp.

Houze, R. A., Jr., 2014 (2nd ed.): *Cloud Dynamics*. Elsevier/Academic Press, 432 pp.

Keen, R. A., 1992: *Skywatch East: A Weather Guide*. Fulcrum Publishing, Golden, Colorado, 204 pp.

Keen, R. A., 2004: *Skywatch West: The Complete Weather Guide (Rev.)*. Fulcrum Publishing, Golden, Colorado, 263 pp.

LeMone, M. A., 1993: *The Stories Clouds Tell*. Project Atmosphere, Amer. Meteor. Soc., 25 pp.

Ludlum, D, M., 1991: *The Audubon Society Field Guide to North American Weather*. Alfred A. Knopf, New York, 656 pp.

Ludlum, D. M., 2001: *Weather*. Harper Collins Publishers, London 664 pp.

Mayhew, D., 2017: *Storm Chaser: A Visual Tour of Severe Weather*. Amherst Media, Inc., Buffalo, New York, 127 pp.

Mogil, H. M., 2001: *Tornadoes*. Voyageur Press, Stillwater, Minnesota, 72 pp.

---

[1] This list of books, articles, and websites about clouds is meant to be a starting point for readers interested in learning, in more depth, about clouds, their characteristics, and dynamics, and is by no means intended to be, nor is it, comprehensive. Some of these books are mainly collections of photographs, while others are more technical; they make up part of my personal-library collection. They have, in part, inspired me to author this book. Disclosure: My cloud photographs appear in a few of these references.

Mogil, H. M., 2007: *Extreme Weather: Understanding the Science of Hurricanes, Tornadoes, Floods, Heat Waves, Snow Storms, Global Warming and other Atmospheric Disturbances*. Black Dog & Leventhal Publishers, New York, 304 pp.

Parviainen, P., 2004: *Clouds*. Barnes & Noble Books, New York, 319 pp.

Pretor – Pinney, G., 2019: *A Cloud a Day*. Batsford, London, 368 pp.

Pretor – Pinney, G.,2011: *The Cloud Collector's Handbook*. Chronicle Books, San Francisco, 144 pp.

Pretor – Pinney, G., 2006: *The Cloud Spotter's Guide: The Science, History, and Culture of Clouds*. Pedigree, Toronto, 319 pp.

Pruppacher, H. R., and Klett, J. D., 1978: *Microphysics of Clouds and Precipitation*. Reidel, Dordrecht, Holland, 714 pp.

Reed, J., 2007: *Storm Chaser: A Photographer's Journey*. Abrams, New York, 192 pp.

Rogers, R. R., and M. K. Yau, 1989 (3rd ed.): *A Short Course in Cloud Physics*. Pergamon Press, 290 pp.

Sachweh, M., 2016: *Storm Chasing: On the Hunt for Thunderstorms*. Delias Klasing Verlag, Bielefeld, Germany, 162 pp. (preface by H. Bluestein, pp. 6–7).

Schulz, K., 2015: Writers in the storm. *The New Yorker*, 23 Nov., 105–110.

Scorer, R., 1972: *Clouds of the World: A Complete Colour Encyclopedia*. David & Charles, London, 176 pp.

Scorer, R. and A. Verkaik, 1989: *Spacious Skies*. David & Charles, London, 192 pp.

Seaman, C., 2018: *The Big Cloud*, Princeton Architectural Press, Hudson, New York, 176 pp.

World Meteorological Organization, current online: *International Cloud Atlas: Manual on the Observation of Clouds and Other Meteors*, WM)-NO, 407. http://cloudatlas.wmo.int

Index